Dmitry Kaminskiy

Biomarkers of Human Longevity

Data Science for Accelerating Aging Research and R&D
The Critical Catalyst for Practical Human Longevity
Tangible Investment Decision-making and De-Risking

DKG
2023

Biomarkers of Human Longevity.

Deep Knowledge Group
www.dkv.global

ISBN 978-1-9163917-7-2

Table of Contents

Table of Contents

Foreword by Dmitry Kaminskiy

In my first book, ***Longevity Industry 1.0: Defining the Biggest and Most Complex Industry in Human History***[1], we distilled the complex assembly of deep market intelligence and industry knowledge that ***Deep Knowledge Group***[2] and its Longevity-focused subsidiaries (including ***Longevity.Capital***[3] and ***Aging Analytics Agency***[4]) have developed over the past eight years, into a comprehensive exposition of the Global Longevity Industry. We showed the public exactly how our international consortium of commercial and non-profit entities managed to define the overwhelmingly complex and multidimensional Longevity Industry for the first time. We also showed how we created a tangible framework for the systematization and forecasting of the industry.

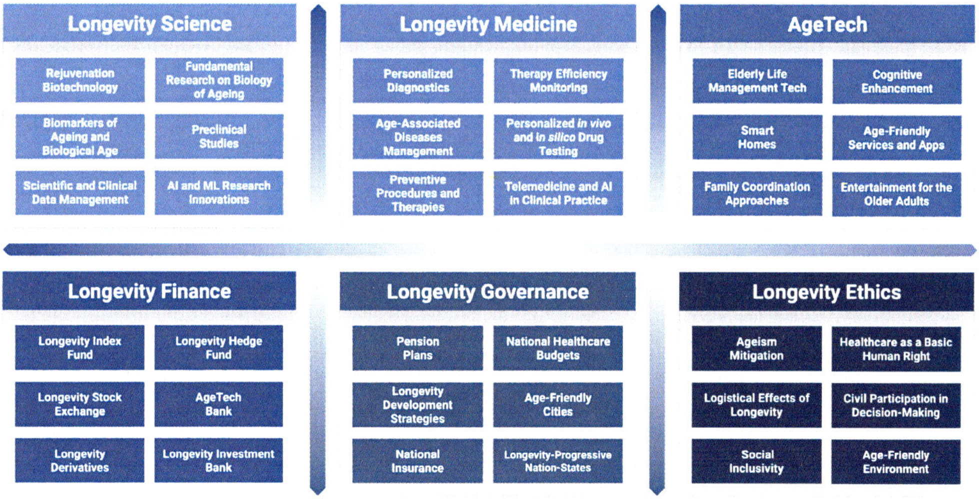

Source: www.longevity-book.com

That book emphasized the potential of Biomarkers of Human Longevity to serve as a significant catalyst and accelerator of the translation of

[1] www.longevity-book.com
[2] www.dkv.global
[3] www.longevity.capital
[4] www.aginganalytics.com

theory into practice across almost every significant aspect, domain, and driving force of the Global Longevity MegaTrend. These aspects and domains include scientific R&D; personal health optimization and life extension; the Longevity business and industrialization; investment decision-making and the financial industry; and even Longevity politics.

Except for AI, there is no single technology or industry that has greater potential to impact the Longevity Industry across all its sectors and domains than Biomarkers of Human Longevity. They represent the major key to translating aging science theory into practical Longevity applications over the next several years. They are of equal value to all types of industry participants: scientists, entrepreneurs, investors, and even governments.

Landscape of Companies Developing Longevity Biomarkers

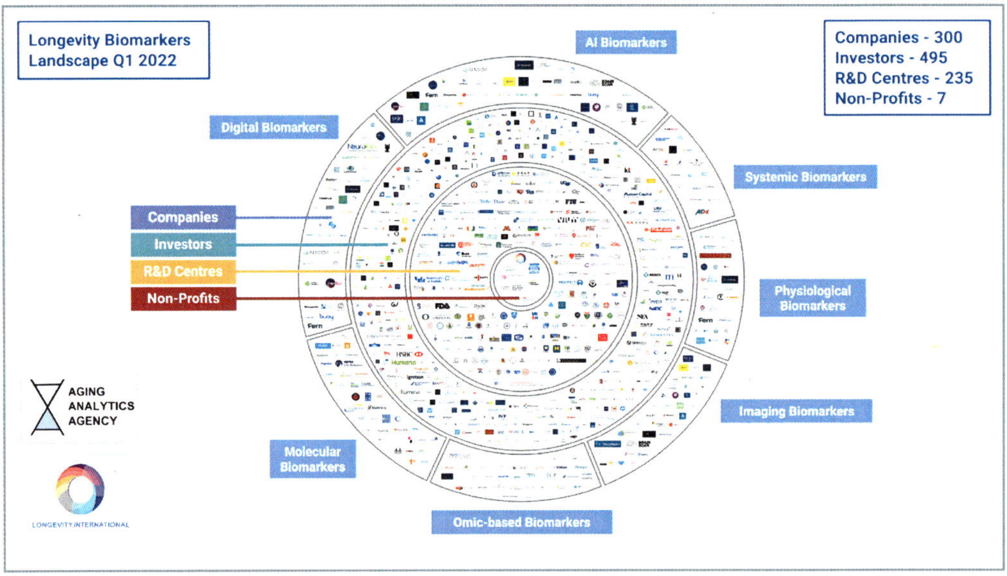

Source: Longevity Biomarkers Ecosystem[1]

Yet, despite the significant accelerative potential of biomarkers, the scientific and business community currently has no clear consensus regarding standardized sets of Longevity Biomarkers and how they could and should be used.

Therefore, the time has come for a dedicated book that identifies the true potential of this technology to increase individual and national

[1] www.longevity.international/biomarkers

Health-Adjusted Life Expectancy and Quality-Adjusted Life Years. It could optimize strategic decision-making both for start-ups and corporations, and also it could optimize investment decision-making for investors. Such an analysis would pave the way for a much more tangible, scalable, and sustainable Global Longevity Industry in which the socially inclusive, humanitarian impact of Longevity is maximized and potential ethical and socioeconomic concerns are neutralized.

Deep Knowledge Group[1] and its Longevity-focused subsidiaries and affiliates – including its analytical subsidiary ***Aging Analytics Agency***[2]; its specialized investment arm ***Longevity.Capital***[3]; its portfolio companies ***Longevity Banking Card***[4] and ***Longevity Financial Advisory firm***[5]; and the international non-profit association ***Longevity.International***[6], which Deep Knowledge Group is supporting and sponsoring – have prioritized the need for Biomarkers of Human Longevity and their revolutionary potential. They have integrated them in various ways into their overall scope of activities and strategic agenda for several years now. They are expertly positioned to provide a fundamental understanding of the significant challenges and opportunities to be faced within this domain and how they can be applied by individuals, institutions, and even entire governments to the maximum level of benefit, neutralizing potential pitfalls.

I have seen the power and impact of a biomarker-based, data-driven, algorithmically-enhanced, and AI-mediated approach to validating, optimizing, and adaptively re-tuning my own personalized diagnostic, monitoring, lifestyle, and therapeutic regime. I can, therefore, understand the broader ways in which this domain can be leveraged by institutions, investors, and governments.

Longevity is characterized by unprecedented degrees of complexity, dealing as it does with the very cutting edge of biomedicine (involving not single disease targets but complex hierarchical networks of the biological processes underlying aging), and being poised at the intersection of a variety of domains of human activity (science, technology,

[1] www.dkv.global
[2] www.aginganalytics.com
[3] www.longevity.capital
[4] www.longevity.cards
[5] www.longevity.finance
[6] www.longevity.international

finance, and policy). In any field, the best way to deal with the challenge of complexity is to develop industry-standard frameworks for reliable assessment, benchmarking, analysis, and predictive forecasting. We have seen several modern and historical examples of fields facing significant levels of complexity benefit tremendously from the development and onboarding of specialized metrics in general and, in the case of the BioTech Industry, the onboarding of biomarker panels in particular. The oncology industry serves as a good example of this trend, where the onboarding of consensus in standards for biomarkers helped define and establish the best practices to identify pitfalls, risks, and areas of potential of the industry in a way that is completely understandable to doctors, entrepreneurs, investors, government regulators and policy-makers. In a very short time this has resulted in rapid industry maturation, investment de-risking, acceleration of R&D and clinical translation efforts, and the implementation of an effective real-world impact on actual patients.

The Global Longevity Industry is in a very similar state to fields like the oncology sector before consensus for framework of cancer biomarkers was on-boarded. We can clearly see low-impact sectors being hyped up at the expense of more disruptive and accelerative sectors. We also see a lack of common understanding as to the definition of Longevity and of which sectors, domains, processes, technologies, and therapies it actually encompasses; we also see little understanding of the sophisticated frameworks required for its analysis.

It should be clear that having these frameworks and metrics for distinguishing hype from reality and for measuring success, progress, relevance, and reliability is the minimum requirement for any industry or field of human endeavor – especially one that is characterized by high levels of complexity and rates of innovation. We have many examples of industries that possess these frameworks – IT-, tech-, and engineering-focused sectors – being able to make tangible progress in a self-sufficient manner; we also have examples of an industry like SpaceTech, which is thriving now, but success in space exploration would be unthinkable without universally accepted standards on metrics and general parameters.

What this book seeks to make clear is that, in the Longevity Industry, Biomarkers of Human Longevity are the most low-hanging, market-ready, and tangible metrics to use as systemic framework for assessment, due diligence, validation and forecasting.

The points I have made thus far may seem to form an argument for adopting methods of accelerating the industry's existing growth and diversification, complementing the current efforts of industry players. However, I wish to be very clear that this is actually fundamentally necessary for maintaining the industry's forward momentum. While these approaches do have the potential to accelerate industry development, they are first and foremost about industry de-risking, stabilization, and sustainable growth.

We are seeing many large and reputable investors with extensive experience in BioTech seeking exposure to the Longevity Industry, expecting the standard frameworks for due diligence and company valuations that enabled their previous success to still hold true. Meanwhile, due to the global megatrend of Longevity popularization, we are now even seeing new investors without previous experience in biomedical sectors enter the industry. It is clear that these factors, in combination with the overwhelming complexity of the industry, have the potential to lead to many negative consequences. Together, they pose the risk of a Longevity bubble, in which investor sentiment declines before the industry's true potential can be actualized in practice.

Besides the broader importance of Biomarkers of Human Longevity as a crucial tool for effective strategic decision-making within my role as General Partner of Deep Knowledge Group, the topic of Longevity biomarkers is also important to me personally. Over the course of my work since 2013 to develop a sophisticated, robust, therapeutic, and methodological regime to optimize my personal Healthy Longevity, biomarkers stand out as the most critical component for orchestrating, validating, adjusting, and re-tuning technologies to my gender, age, ethnicity, health, and performance profile.

Having direct access to the research produced by Aging Analytics Agency over these years, I decided to write a dedicated book that takes Biomarkers of Longevity as its central concern. In this book, I will distill all major conclusions, analytical frameworks, and forecasts developed and applied by Deep Knowledge Group on this matter. By doing this, I aim to share some of the key principles of our group on how this technology can be effectively used to optimize strategic decision-making, showing exactly how Biomarkers of Human Longevity are one of the most prospective market-ready tools for the following:

- Achieving tangible short-term results in personal Practical Human Longevity
- Facilitating and de-risking Longevity investment decision-making
- Re-tuning the business models of financial corporations (insurance companies, pension funds, and investment banks) at the intersection of the Opposed MegaTrends of Population Aging and Healthy Longevity

I urge all relevant potential players – from investors to leading Longevity experts, scientists, entrepreneurs, practitioners, AI specialists, and more – to join the rapidly-growing Longevity biomarkers sector. Biomarkers should now be recognized as a fundamental precondition for the further sustainable growth of the entire Longevity Industry and the attainment of increased Healthy Human Longevity.

Chapter I: Biomarkers of Human Longevity as the Industry's Critical Catalyst for Accelerating the Practical Results in Aging Research Needed for Industry Growth

In this first chapter, we will begin by discussing how Biomarkers of Human Longevity will serve as the primary catalyst for the short-term change, development, and harmonization within the Longevity Industry, able to optimize the strategies and practices of Longevity companies, investors, and policymakers. They will enable the first tangible, real-world results in the translation of Longevity science and theory into practical Healthy Human Longevity.

Biomarkers of Human Longevity: The Main Catalyst for Accelerating Longevity Industry Development and Neutralizing Its Biggest Systemic Risks

As the Precision Health Industry grows, we will see an increased emphasis on creating and validating a wide range of Biomarkers of Longevity. This will enable the extension of Healthspan and the maintenance of optimal health for the population via continuous AI-empowered monitoring of fluctuations in personalized Biomarkers of Aging. P4 Medicine (Preventive, Predictive, Personalized and Participatory), being the cornerstone of lifespan and healthspan extension, will be the central platform for the use of a wide array of Longevity biomarkers by general public and progressive governments seeking to optimize their population-level National Healthy Longevity.

The development of human Longevity biomarkers has been a priority of Deep Knowledge Group since 2019, when it was quickly becoming apparent that this was the critical roadblock in the growth of the Longevity Industry.

As such, we decided to provide Aging Analytics Agency, Deep Knowledge Group's Longevity-focused analytical subsidiary, with the budget necessary to conduct a unique case study on the pivotal role of biomarkers in the road ahead. The first iteration, ***Biomarkers of Longevity: Current state, Challenges and Opportunities Landscape Overview 2019***[1], marks the first public disclosure of Aging Analytics Agency's ongoing study into the current state and future trajectory of this domain. It was completed in October 2019 and presented at Longevity.International's ***First International Policy and Governance Summit***[2], at which I was the keynote speaker.

Biomarkers of Longevity: Distribution by Amplitude Level

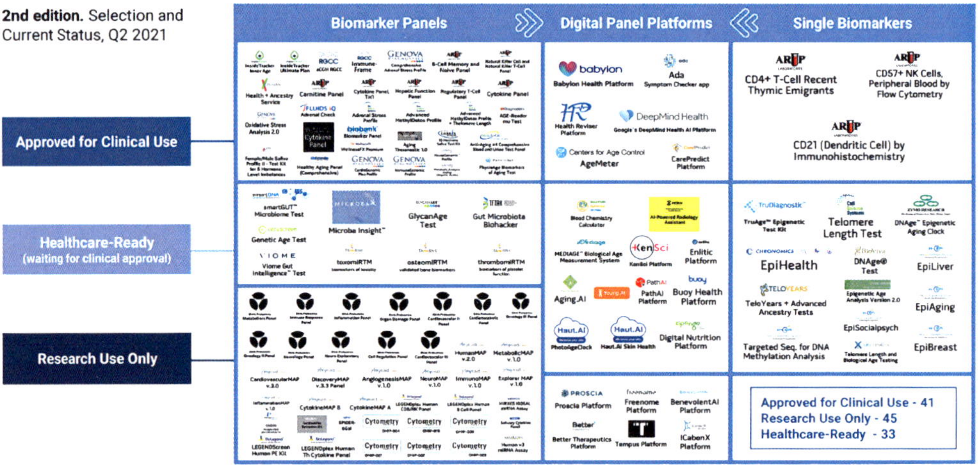

Source: Longevity Biomarkers Landscape Overview Q4 2021[3]

The special case study performed a benchmarking of human Biomarkers of Aging and Longevity according to their degree of actionability (a weighted metric that analyzed their level of precision and accuracy with those factors that determine how easily they can be implemented on a massive scale, such as cost, invasiveness, and broad ease of adoption).

[1] www.aginganalytics.com/biomarkers-of-longevity
[2] www.longevity.international/longevity-governance-summit-2019
[3] www.longevity.international/biomarkers

The study garnered such favorable reception that in Q4 2021, Aging Analytics Agency produced an extended and enhanced edition, titled *"Longevity Biomarkers Landscape Overview Q4 2021"*[1].

The project uses a comprehensive analytical framework to rank and benchmark existing Panels of Biomarkers of Aging, Health, and Longevity according to their ratios between accuracy and actionability. It identifies the panels of biomarkers that could have the most significant impact on increasing both individual and national Healthy Longevity in the next few years.

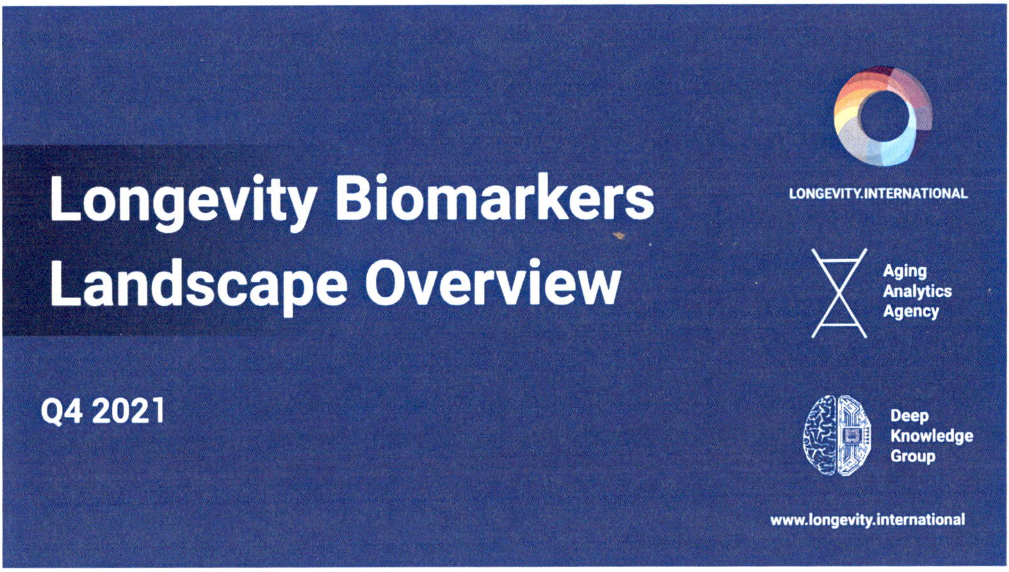

Source: Longevity Biomarkers Landscape Overview Q4 2021[1]

Relevant stakeholders, scientists, and researchers viewed the topic as having such wide applicability that we agreed it extends beyond mere commerce. So we decided to have Aging Analytics Agency transfer this body of knowledge to the non-profit organization Longevity.International[2], which has since become the owner of the resulting database. Longevity.International is an independent distributed knowledge IT-Platform for enhancing industry interconnectivity. It is a new opportunity for different expert specialists, companies, investors and policy-makers to see who is doing what, enabling the synergetic cooperation needed to support industry growth. Moreover, I believe it to be a perfectly appropriate method

[1] www.longevity.international/biomarkers
[2] longevity.international

of information exchange since we believe there should be no competition in Longevity, only collaboration and cooperation.

The importance of benchmarking Longevity Biomarkers and Biomarker Panels by their ratio of accuracy vs. actionability, rather than just their accuracy, cannot be overstated. For this domain of technologies to accelerate the translation of Longevity theory into practice and enable short-term progress in the extension of Healthy Human Longevity, they need to consist of biomarkers that are market-ready and accessible to the majority of individuals.

Biomarkers and biomarker panels are now market-ready and actionable enough (i.e., with sufficiently low cost and invasiveness) to be developed, applied, and used at scale. They are also accurate enough to prove useful in validating the safety and effectiveness of lifestyle, behavioral, and Longevity medicine-focused interventions as well as tracking their changes on individual and population-level Healthy Longevity. We can expect to get fully implemented, sophisticated, and robust panels within the next 2–3 years.

With these panels in place, Longevity Industry participants, including national governments, have no excuse not to use them for therapeutic validation and population health optimization.

This is the reason why Aging Analytics Agency transferred this project to ***Longevity.International***[1] to foster international collaboration and transparency in the field of aging research.

We hope that releasing the report and IT-Platform in an open-access manner via ***Longevity.International***[1] will encourage scientists, companies, and other industry stakeholders to contribute to the platform to arrive at a robust consensus framework eventually.

This data provides advice to industry leaders for the conception, development, and maturation of their action plans; provides insurance companies with the tools to help improve their customer services and risk pricing principles; and provides policymakers with the information necessary to combat population aging and realize the opportunity of National Healthy Longevity.

[1] www.longevity.international/

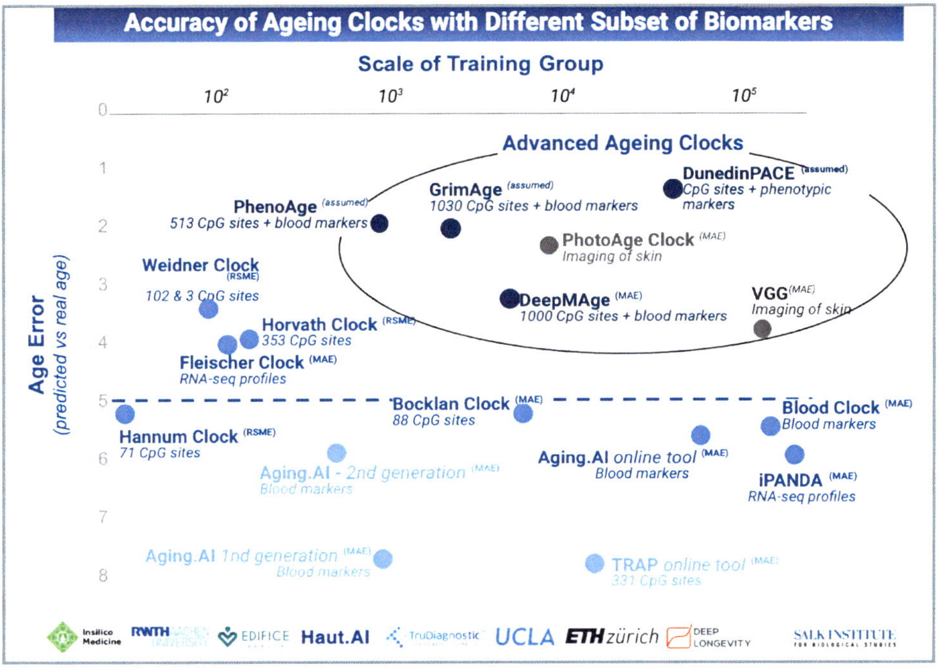

Source: Longevity Biomarkers Landscape Overview Q4 2021[1]

The project offers key strategic recommendations regarding technologies and biomarker implementations within reach of companies, entities, and nations to help them optimize their developmental action plans and strategies, providing specialized guidance for business and investment core decisions. It delivers the following:

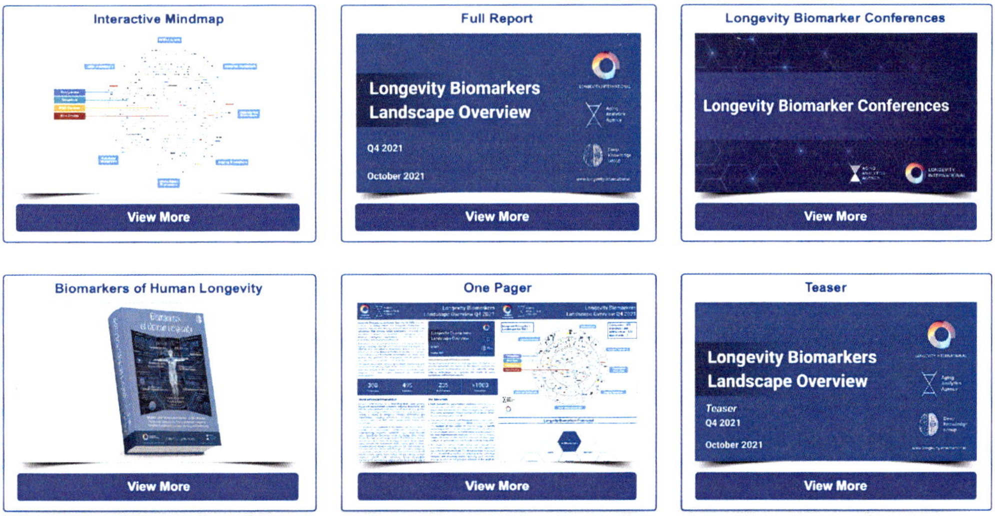

Source: www.longevity.international/biomarkers

[1] www.longevity.international/biomarkers

- A "most comprehensive" list of single Biomarkers of Aging and panels; their advantages, strengths, disadvantages, and weaknesses; and future perspectives, challenges, and opportunities, with a focus on technologies currently used for assessment.
- Concrete Analysis of recent novel Biomarkers of Aging that are just now entering R&D processes; emerging tools; and novel assay platforms awaiting approval or standardization for clinical implementation, one step away from being market-ready within the next several years.
- Highlights regarding why AI platforms will come to be a necessary component of Longevity biomarker discovery, research, and development.
- An overview of different categories of panels, whether for Research Use Only or Approved for Clinical Use.
- Introduction to Technology Readiness Levels (TRL), a type of measurement system used to assess the maturity level of a particular technology. Each technology project is evaluated against the parameters for each technology level and is then assigned a TRL rating based on the project's progress. There are nine technology readiness levels. TRL 1 is the lowest, and TRL 9 is the highest.
- Classification of most advanced aging biomarkers (aging clocks); their combinations with AI systems; and SWOT analysis and Technology Readiness Level (TRLs) assessment of each category, with reviews of example use cases and action points. Overview of the aging biomarker mobile applications currently existing on the market.
- A brief overview of aging clock application in the insurance industry, concerning technological and ethical aspects.
- Conclusions and practical recommendations regarding the application of aging clocks.

The special analytical case study and IT-Platform has been constructed to provide key strategic decision-makers across science, R&D, industry, investing, and public policy with up-to-date expertise and insights into the current state of the aging biomarkers' technology and market; the relative technological and economic traits of each biomarker group and their applicability in various healthcare industries; and the insurance industry (which is discussed more thoroughly in *Chapter IX: Longevity Economy 2.0: Biomarkers of Human Longevity as Major Catalyst for Increase Investments and Ensure Sustainable Growth of Longevity Financial Sector*).

TRL Count per Category

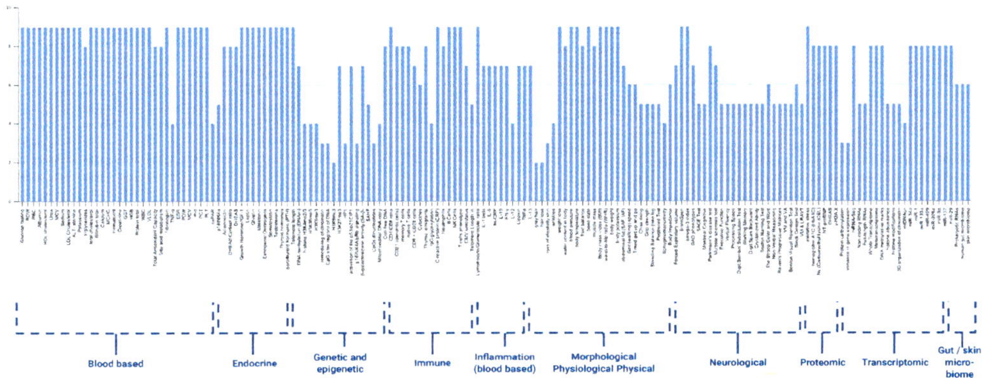

Source: Longevity Biomarkers Landscape Overview Q4 2021[1]

The project provides all stakeholders, including companies, investors, governmental and policy-focused organizations, and the general public with:

- Concrete deep analysis of which biomarkers and biomarker panels are available today; the strengths and weaknesses of each; their accuracy, availability, and current actionability; their technology readiness levels (TRLs); and the peculiarities of each type of Longevity biomarker regarding its uses for real-time and precision monitoring of health status and biological age.
- Tangible estimates of which biological age biomarkers and implementations are consolidated, or their current conditioning stage for precision assessment of health status and endpoints of clinical trials and therapies and their use in insurance risk assessments.

[1] www.longevity.international/biomarkers

- Highlights regarding the role of digital biomarkers and AI platforms and how they will become necessary components of aging and Longevity biomarker discovery, research, development, and daily use – also an overview of mobile apps containing actionable biomarkers or aging clocks.

Artificial Intelligence as the Major Catalyst for Biomarkers of Human Longevity

Of all the component technologies and toolsets driving progress in the Global Longevity Industry, the one with the greatest potential to create a real-world impact on human Longevity in a short timeframe, and the one with the highest cost-effectiveness ratio, is the application of AI and data science to Longevity. Unfortunately, despite being the component with the greatest promise, it happens to be underrepresented and underfinanced within the Global Longevity Industry.

There are many reasons for the enormous potential of AI in Longevity generally and of AI for Human Biomarkers of Longevity in particular:

- Longevity is unprecedentedly complex, both as a science (dealing with the deepest levels of biology, health, and disease) and as an industry (being composed of the intersection of many distinct, individually complex domains of frontier science and technology). AI, data science, and mathematics are being applied in R&D to process data that is too voluminous and complex for humans to address manually – it is the engine not only for neutralizing complexity but also for yielding its power to create new positive results.
- As aging research advances, so too does the number of specific Biomarkers of Aging and Longevity. AI will become the only tool for managing this enormous volume of data as it applies to P4 Medicine, Precision Health, and Longevity R&D.
- AI is an industry that, as a whole, is very well-financed, with leading nations currently competing to win the global AI race, to develop and secure the most advanced AI technologies and IP, and to capture the highest densities of AI specialists. Ongoing developments in core AI innovation are themselves rapidly implementable (with AI being a virtual, digital technology that can be

replicated, transmitted near-instantaneously, and utilized at zero material cost once developed), and, therefore, capable of having immediate accelerative impacts on Longevity.

- AI is an evolving and self-accelerating technology in the sense that the latest advances in AI make it easier to develop the next paradigm shift in AI, prompting an exponential effect.
- Many technologies and methods for extending Healthy Human Longevity, practicing preventive medicine, and maintaining an optimal state of precision health are already invented, validated, and ready for use; however, they lack the infrastructure for scaling them to the masses. This is why we predict that the vast majority of practical, real-world effects in terms of extending Healthspans in the next several years will come from existing, validated technologies, making it a data aggregation, analysis and management challenge rather than a biomedical or BioTech R&D problem.

In our opinion, AI for Longevity is the "smart money" sector of the industry that can achieve tremendous results and accelerate timelines in the progress of tangible, real-world Healthy Human Longevity, even with modest levels of funding compared to other sectors. We predict that this is the exact central role AI will play in the Longevity space within the next several years, and very likely by the year 2025, when the level of industry development has reached what we could consider " Longevity Industry 2.0." This evolved form of the Longevity Industry will include the aggregation, development, and deployment of Biomarkers of Aging, Health, and Longevity; Preventive Medicine diagnostics and prognostics; Precision Health technologies and therapeutics; and integrated Wealthspan-extending AgeTech and WealthTech solutions for financial wellness across extended periods of Healthy Longevity.

What the Longevity Industry will look like at that point, as well as the ways in which industry development can be optimized and accelerated to reach that point sooner than expected, is the subject of our next book: Longevity Industry 2.0: DeepTech Engineering the Accelerated Trajectory of Human Longevity: The Blueprint and Pathway from Longevity Industry 1.0 to 2.0[1].

Our previous book, Longevity Industry 1.0 – Defining the Biggest and Most Complex Industry in Human History[2], distilled the complex assembly of

[1] www.longevity-book.com/two
[2] www.longevity-book.com/one

deep market intelligence and industry knowledge that Deep Knowledge Group has developed over the past eight years into a full-scope understanding of the Global Longevity Industry. It publicly disclosed how we managed to define the overwhelmingly complex and multidimensional Longevity Industry for the first time. Longevity Industry 2.0 presents the framework for safeguarding the sector's current upward trajectory and ensuring its optimized, sustainable growth toward its next stage and the realization of its practical benefits for humanity by the year 2030.

The apex of AI for Human Biomarkers of Longevity, and its most robust and advanced embodiment, will be the enabling force for creating a so-called digital avatar of the full human body, using thousands if not tens of thousands of personalized biomarkers, with at least several hundred precise Biomarkers of Aging and Longevity. These will include not only biological but also psychological and behavioral biomarkers; the full range of sectors of biomarkers sectors is illustrated on our front cover.

AI and Digital Platforms: Toward a "Digital Avatar"

Source: Longevity Biomarkers Landscape Overview Q4 2021[1]

It is clear that AI will soon become not just a complementary but a fundamental tool for developing, refining, and applying Biomarkers of Human Longevity. It will act as the trigger in a chain reaction that will lead to rapid progress in the translation of theoretical research on aging into Practical Human Longevity. Indeed, the metaphor of the nuclear chain reaction is not out of place here: in our previous analytical reports and Longevity policy proposals, we have argued that what the Longevity Industry needs most from national governments is a

[1] www.longevity.international/biomarkers

fully-fledged commitment to transform the challenge of population aging into the opportunity of optimized National Healthy Longevity, on the same scale and with the same ambition as the Manhattan Project and the creation of the atomic bomb. This remains true today, and such a commitment would create change just as fundamental and widespread as the Manhattan Project, although in a positive rather than negative direction this time.

Summary

This chapter introduced actionable, market-ready panels of Biomarkers of Human Longevity as a scientific and technological domain with the strongest prospects of accelerating the translation of Longevity theory into practice – and accelerating the short-term trajectory of real-world applications in Practical Healthy Human Longevity. The key points of this chapter are:

- We now have biomarkers and biomarker panels that are market-ready; actionable enough to be developed, applied, and used at scale; and accurate enough to prove useful in validating the safety and effectiveness of lifestyle, behavioral, and precision medicine-focused interventions and tracking their changes in individual and population-level Healthy Longevity.
- P4 Medicine is the central platform for the utilization of a wide array of Biomarkers of Longevity in healthcare.
- AI for Longevity is the "smart money" sector of the industry that can achieve tremendous results and accelerated timelines in terms of progress in tangible, real-world Healthy Human Longevity.
- The apex of AI for Human Biomarkers of Longevity will be the enabling force for creating a so-called digital avatar of the full human body, using thousands if not tens of thousands of personalized biomarkers.
- All future growth of the Longevity Industry, including all future decision-making and all future business development, must involve reference to human biomarkers. This is unavoidable.

Chapter II: The Longevity Industry's Fundamental and Systemic Risk: Over-Reliance on Research Based on (Mouse) Animal Model Results

The present chapter will discuss what Deep Knowledge Group considers the largest systemic risk and a potential source of future destabilization in the Longevity Industry: the widespread use of outdated and unreliable approaches to clinical validation.This, in combination with Chapters 3 and 4, will provide the necessary context for the sixth chapter, which delivers an overview of how AI-driven Biomarkers of Human Longevity – combined with other modern tools, methods, and technologies – can neutralize these risks and pave the way for more stable and balanced developmental prospects for the Global Longevity Industry, de-risking investments, stabilizing economic and industry outcomes, and enabling the influx of trillions of dollars worth of investment capital.

The Longevity Industry's Most Flawed Assumption Is Also Its Biggest Fundamental and Systemic Source of Risk

The Longevity Industry faces a fundamental, widespread, and systemic problem that threatens to destabilize its current upward trajectory of growth, diversification, and increasing investment volumes: namely, the well-established but highly outdated practice of experimentation on model organisms with the expectation that their outcomes will translate to humans.

This has been the consensus paradigm followed in the traditional BioTech Industry for the past 50 years. However, the last decades have shown consistently decreasing ratio of successful translation of drugs to humans safely and effectively enough to bring them to market in the majority of cases.

Given the current 90%–92% failure rate of clinical translation into humans in the comparatively simple BioTech Industry – which deal with interventions with far fewer complex components (e.g., single molecules) targeting far simpler biological systems (e.g., individual disease targets) – it is reasonable to expect higher failure rates for multidimensional therapies with many "moving parts" that target significantly more complex biological processes of aging.

The source of this highly anticipated failure rate is twofold. The first fundamental factor is the vast biological (genomic, epigenomic, physiological, etc.) differences between humans and model organisms, both in general and with regard to aging in particular. That is a scientific problem. The second is not so much a scientific problem, but a problem rooted in human error – by current estimates, as many as 70% of scientific experiments, studies, and articles are not repeatable[1] due to either well-intentioned mistakes or intentional fraud. This phenomenon is often referred to as science's modern "replication crisis"[2] and will pose even more challenges for the expected success rate of human-focused Longevity therapeutics.

If this trend continues, it could lead to general pessimism about the Longevity Industry among investors and harm by association the prospects of the entire field, including non-biomedical sectors. This will weaken investor confidence and drive down the market cap of the increasing number of Longevity companies holding IPOs before demonstrating human safety and efficacy. It also creates the risk of the rise of a Longevity boom and bust cycle.

There is, however, still time to neutralize these risks by embracing new, modern, and more sophisticated approaches to investment analytics, due diligence, company valuation, and – most importantly – the technological and scientific validation of Longevity therapies.

[1] www.bbc.com/news/science-environment-39054778
[2] www.en.wikipedia.org/wiki/Replication_crisis

Mice vs. Men: The Rise and Risk of Longevity Hype

Back in 2013[1], Silicon Valley tech giant Google promised the world it would "solve the problem of death." We have entered a new decade now, and there is little evidence that Google is significantly nearer to its goal than in 2013.

There is a lot of positive hype on the subject: many people, ranging from the general public to scientists, entrepreneurs, and investors, are confident that we are at the brink of creating genuine human life extension techniques that will soon translate into real-world, accessible applications.

Here, we will systematically examine the evidence for whether or not this is actually the case and offer a set of guidelines to help separate the hype from the reality around this proposition in a concrete, logical, and tangible way.

Source: Time Magazine[2]

In fact, many Longevity-focused scientists have achieved rather significant progress over the last decade with respect to stalling the aging process and, in some cases, even achieving rejuvenation (restoration of a young phenotype) in certain model organisms such as yeast, worms, flies, and mice.

For instance, ***in January 2020***[3], researchers at MDI Biological Laboratory, in collaboration with scientists from the Buck Institute for Research on Aging and Nanjing University in China, identified synergistic cellular pathways for Longevity that amplify lifespan fivefold in C. elegans, a nematode worm used as a model in aging research.

---

[1] www.content.time.com/time/magazine/article/0%2C9171%2C2152422%2C00.html
[2] www.content.time.com/time/subscriber/article/0%2C33009%2C2152422%2C00.html
[3] www.mdibl.org/in-the-news/pathways-that-extend-lifespan-by-500-percent-identified-2

Source: Phys.org

In another example[1] (October 2019), Maria Blasco and colleagues managed to extend the lifes pan of mice by 24% by breeding a set of chimeric mice using embryonic stem cells with telomeres twice as long as usual.

The number of headlines highlighting substantial increases in both the lifespan and Healthspan of model organisms has grown dramatically in the past decade.

Source: Kaiser Health News[2]

[1] www.nature.com/articles/s41467-019-12664-x

[2] www.khn.org/news/a-fountain-of-youth-pill-sure-if-youre-a-mouse

However, while we see significant progress in model organisms, this is not the case with respect to humans. Silicon Valley-based Longevity companies' confidence that positive results in model organisms will translate into comparable outcomes in humans is a major destabilizing factor in the industry and the most dangerous underlying assumption in the entire industry ecosystem.

Such an assumption is something that ***Deep Knowledge Group's***[1] Longevity-focused investment arm, ***Longevity.Capital***[2], takes specific measures to avoid by mandating preliminary positive results in humans as a core factor of its scientific and technological due diligence for biomedical companies. It also strategically prioritizes other sectors with high degrees of market-readiness that are not exclusively focused on biomedicine, such as AI for Longevity, AgeTech, and Longevity FinTech.

While the transferability of positive outcomes in model organisms to humans has long been a concern for some of the more forward-thinking specialists in the field, it is only recently that hard data on the matter has come to light. For instance, on February 14, 2020, Science published a study under the title ***Translating preclinical studies to humans***[3].

The article summarizes many of the issues at the heart of the difficulties in translating drug successes in model organisms to humans. It provides some suggestions on how AI could be used to shift toward more human-centered approaches: "Systems biology and machine learning (ML) can be used to translate relationships across species. Instead of attempting to 'humanize' experimental animal models, which is possible only to a limited extent, greater success may be obtained by humanizing computational models derived from animal experiments." Suffice to say that we at Aging Analytics Agency have always been skeptical regarding the results in model organisms and the overconfidence in their ability to translate to humans. Because of this, we welcome the latest science, which is validating those concerns more and more with each passing day.

---

[1] www.linkedin.com/pulse/longevity-hype-silicon-valley-traditional-approach-biotech-kate-batz

[2] www.longevity.capital

[3] www.science.sciencemag.org/content/367/6479/742

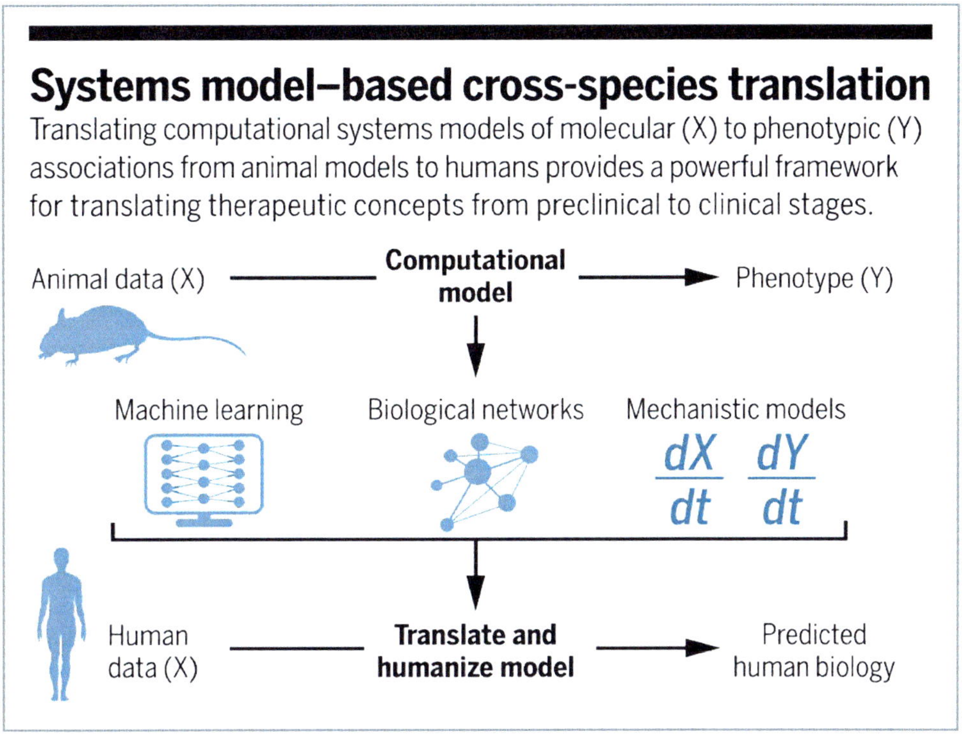

Source: Translating preclinical models to humans[1]

Meanwhile, despite the growing volume of positive results in model organisms, advances in practical human life extension and understanding of human Longevity are at a relative standstill.

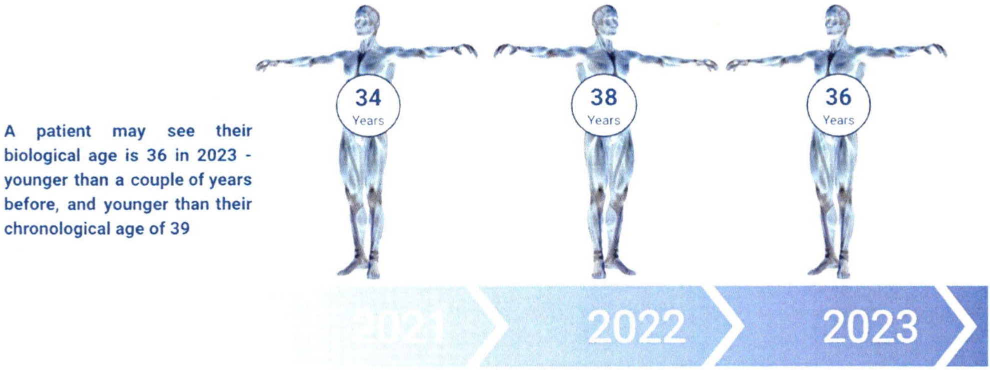

Source: Longevity Biomarkers Landscape Overview Q4 2021[2]

In our view, one of the factors behind this comparatively slow rate of progress is that resources are often wasted on purely theoretical and

[1] www.science.sciencemag.org/content/367/6479/742
[2] www.longevity.international/biomarkers

curiosity-driven research. With the enormous volume of resources behind big tech and research giants such as Google and Amazon, we should be able to see useful results by now. But instead, all we see is meager theoretical advances.

The above image illustrates the nature of this waste. The earlier investment in tech-driven approaches and methodologies might have achieved a much greater decrease in human biological age and human Healthspan over the past decade.

Furthermore, the fact is that retail investors are already beginning to feel the tangible results of these risks and dangers on their own bank accounts, as a result of the public share price of public Longevity companies (who made their IPOs based solely on the results of model organism studies) plummeting due to post-IPO failures in clinical (human) translation and validation.

More cautious investors may wait to see if similar BioTech and Longevity companies can succeed with further clinical studies.

Geroscience aims to prevent and repair the actual damage that causes aging. This damage can include somatic DNA damage, telomere attrition, transposon-related genomic instability, epigenetic drift, and more. Therefore, we will see an increasing emphasis on creating and validating a wide range of Biomarkers of Aging, which will enable the extension of Healthspan and the maintenance of optimal health for the population via continuous AI-empowered monitoring of fluctuations in personalized Biomarkers of Aging.

The use of precise, actionable, and available aging biomarkers with a high TRL score is essential to be able to support the transition from the "classical" medical paradigm (diagnosing and treating diseases when they occur) to the prevention medicine paradigm; to P4 Medicine; and, more generally to Precision Health.

By now, it should be clear that the Longevity Industry faces a deep and widespread problem: the majority of leading Longevity investors are still operating under the paradigm of the kind of therapeutic validation (and, by extension, company valuation) that has worked in the broader BioTech and BioPharma industries previously but does not work today in the Longevity Industry and Longevity medicine in particular.

This fundamentally flawed and risky industry-standard paradigm has led to unjustified company valuations following successful pre-clinical trials in model organisms such as mice.

Due to the complexities of the biology of aging and the underlying science and technology – as well as the high levels of scientific and technological intersectionality in the Longevity sphere – many companies are likely to fail. This will lead to much higher clinical trial failure rates and, ultimately, a failure of many players to provide ROIs proportional to their valuations.

The perception of what the term Longevity encompasses varies significantly in different geographical regions, and this is even truer for what can be viewed as Longevity technology and the Longevity Industry. The case of California, for example, illustrates this point very well. In Silicon Valley, there is a lot of talk from companies, entrepreneurs and scientists about human life extension, but in reality the majority of these activities (and, indeed, most Longevity-related media attention) is focused on positive results in mice and other model organisms. And, unfortunately, the perception among industry participants and the general public does not have any doubt that these positive results will inevitably translate to impacts on human life extension, and typically these two outcomes are somehow, quite unreasonably, considered almost equal. This reflects a significant disproportion between perception and reality within the industry, and is much more prominent in Silicon Valley compared to other regions (e.g., Switzerland), where the size of this disproportion is much more balanced and reasonable.

Scientific research, if performed on model organisms and not, in any capacity, with in silico experiments on humans, can be considered a science for science's sake as it does not bring any results that can be reliably predicted to work, safely and effectively, in humans. Companies use the results of such research as a basis for their valuation and as a core element of arguments for positive due diligence decisions. Unfortunately, they are doing what is suitable for their current business model, and not for business models that have Longevity in humans as their foremost concern. They use the same paradigm that existed in the BioTech Industry, and this approach can be good for their business but is not enough in the current situation. If companies aim to deliver genuine, actionable results to humans in terms of Longevity, they should be seeking to do more.

A paradigm shift from the BioTech mindset is needed to fit the current needs of human biomedical progress. The science of aging is at the frontier of the most advanced biomedical and BioTech science, and scientists working in this field should pursue more modern, up-to-date frameworks for forecasting the outcomes of their experiments.

The same can be applied to companies and investors making strategically important decisions based on model organism results only. For us, this means they are not much interested in genuine, actionable results, and the ecosystem and industry business models of prominent pharma players may be lagging behind the actual progress in science. Unfortunately for many investors, the actual results of the research performed on humans do not fit into their business model, which is more a BioTech business model than a Longevity business model.

Surprisingly, in BioTech, initial positive intermediate results are suitable enough to attract investors. This needs to be changed, and it can only be changed with a paradigm shift that takes into account the translational gap between model organisms and humans, and the practical realities of human Longevity.

Summary

In this chapter, I have outlined what Deep Knowledge Group has always believed to be the largest fundamental and systemic risk and source of potential market destabilization for the Longevity Industry: namely, the overwhelming reliance on model organism studies (e.g., mice rather than humans) and the assumption that positive preclinical results in model organisms will translate into positive results in humans. The key points to take away from this chapter are:

- The Longevity Industry faces a fundamental and widespread problem: the majority of leading Longevity investors are still operating under the paradigm of therapeutic validation (and, by extension, company valuation) that worked in the BioTech and BioPharma industries. The well-established but highly outdated practice of experimentation on model organisms results in a 90% failure rate of clinical translation into humans.

- This paradigm is a fundamentally flawed and risky industry standard that has led to unjustified company valuations following successful preclinical trials on animal species such as mice.
- Many companies should consider the need to address in a much more advanced and modern way the complexities of the biology and underlying science and technology around aging and the high levels of scientific and technological intersectionality in the Longevity sphere. Otherwise, they are likely to fail, leading to much higher clinical trial failure rates and ultimately a failure of many players to provide ROIs.

CHAPTER III: Reality Check – The Current State of Longevity Clinical Trials

Modern-day approaches to clinical trials are, unfortunately, not modern at all; in fact, they are quite outdated. During the 1920s and 1930s, medical researchers began to conduct "cooperative investigations" designed to overcome errors attributed to individual observers working in isolation, and replace them with standardized evaluations of therapeutic results in hundreds of patients.

In 1938, a widespread disaster resulting from the distribution of untested, unsafe sulfanilamide products used to fight streptococcal infections, prompted the US Congress to pass the Food, Drug, and Cosmetic Act. Although the new law did not specify any particular testing methods, it did require that drugs be studied by "adequate tests by all methods reasonably applicable to show whether or not the drug is safe."

This resulted in robust, statistically-based clinical trials that became a critically important part of evidence-based medicine in the U.S. following WWII. The US Food and Drug Administration (FDA) had authority under the 1938 Act to establish rules governing the use of investigational drugs, although it did not employ this authority to regulate clinical trials and clinical trial methodology until 1961.

Since then, the same basic approach has been used. This approach worked comparatively well in the early years of BioTech, but it has also become increasingly outdated and inefficient as the complexity of drugs and interventions has increased during the last few decades.

Interestingly, the FDA itself has acknowledged many of these shortcomings, including the need for more personalized approaches that move away from the "blockbuster drug" model, as well as the increasing need for more nuanced approaches to clinical trials that take into account the aged and elderly, and the impact of the aging population. In their white paper, ***FDA and Clinical Drug Trials: a Short History***[1], accessible on www.fda.gov, they write:

[1] www.fda.gov/media/110437/download

“Health care costs are rising at rates far higher than the rate of inflation and the nation faces the challenge of promoting the health of the ‘boomer’ generation during its retirement years... In the past, the drug industry has concentrated on developing so-called ‘blockbuster drugs. The large-scale, randomized clinical trial has been critical in demonstrating the safety and efficacy of these drugs. Many, however, are predicting that the future of medicine points toward developing drugs and diagnostics to treat subsets of patients who may respond to one treatment but not another because of genetic and other factors... The randomized clinical trial is an excellent methodology if you want to understand, on average, whether one treatment is better than another treatment ... but if we think about the distribution of outcomes, no single person in the healthcare system is the average.”

Despite these kinds of official recognition by the FDA, the actual frameworks in place to evaluate drugs and conduct clinical trials under FDA regulations have not yet been changed to account for the increasing need for a more nuanced and personalized approach to trial design.

As it stands in 2021, clinical trials are still using the same basic framework (i.e., randomized clinical trials that fail to account for the personalized needs of specific demographic groups) that has been in place since the 1960s.

In this and the following chapters, it will be formally shown that there are vanishingly few clinically validated drugs that reliably extend Healthy Human Longevity. It will also be shown that there exists an alternative road ahead, including (among other things):

- The use of a larger number of biomarkers to validate clinical trial results.
- Changes to clinical trial regulatory frameworks that allow for a higher degree of clinical trial diversity, and the testing and documentation of safety and efficacy in a more precise, personalized and diverse manner for different demographics (age-groups, ethnicities, cohorts with specific pre-existing conditions and co-morbidities, cohorts with specific multi-omic profiles, etc.).
- The more thorough and widespread use of AI to predict clinical trial outcomes[1].

[1] www.researchgate.net/publication/342354346_Multimodal_AI_Engine_for_Clinical_Trials_Outcome_Prediction_Prospective_Case_Study_Summer_2020

We estimate that the success ratio of translation of relevant drugs from mice to humans is 10% in BioTech in general. **However, we forecast that it will amount to a mere 1% in case of anti-aging drugs, meaning that 99% of Longevity relevant clinical trials will fail in humans irrespective of their occasional success in mice.** As such, the overall approach to clinical trials is itself a great limiting factor in BioTech, and, in the case of Longevity, it is so completely obsolete as to require a total overhaul. It, therefore, needs to be rebuilt from the ground up to maintain any degree of modern relevance or reliability.

The BioPharma value chain for small molecule drug discovery

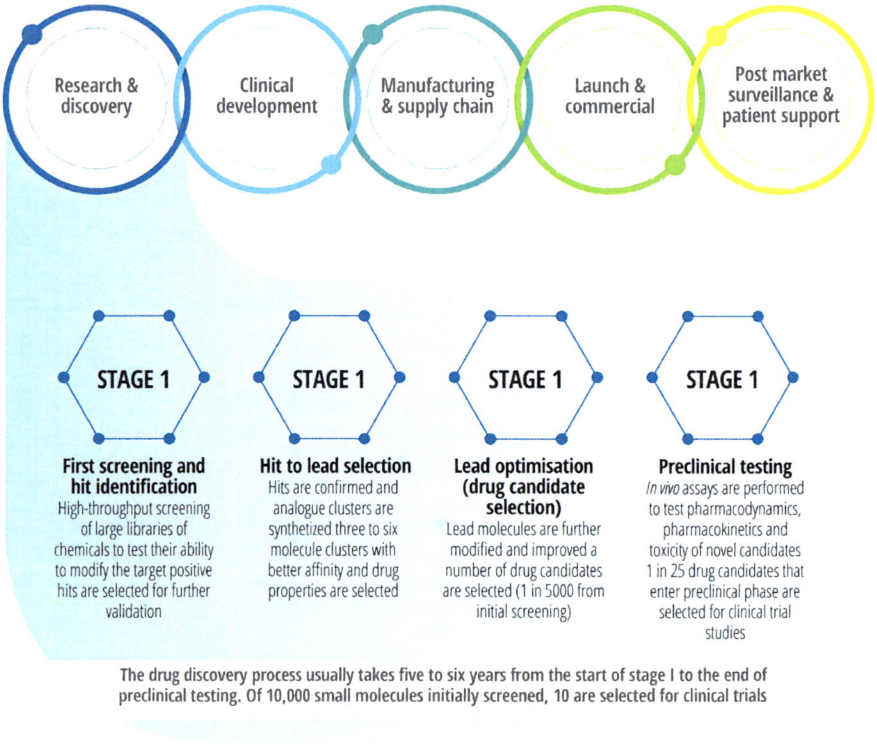

Source: Deloitte Insights[1]

The success ratio of traditional clinical trials has begun to decline gradually, no longer exceeding 8-10%, while the failure rate is up to 92%. While the overall cost of Pharma R&D is rising, the total cost of bringing any new drug to market is increasing, and the overall num-

[1] www2.deloitte.com/content/dam/Deloitte/ch/Documents/life-sciences-health-care/deloitte-ch-en-intelligent-drug-discovery.pdf

ber of new drugs being brought to market is declining. If the recent trend in terms of clinical trial success is preserved, in combination with the above-mentioned trends, their financial profits will risk plateauing. This will mean that the cumulative cost of bringing a new drug to market, combined with total R&D spending (including the cost of failed drugs within the same category), may result in Pharma R&D becoming a mostly-unprofitable industry – again, unless some central factor leading to this plateau is effectively neutralized.

Pharma Efficiency is Declining Steadily

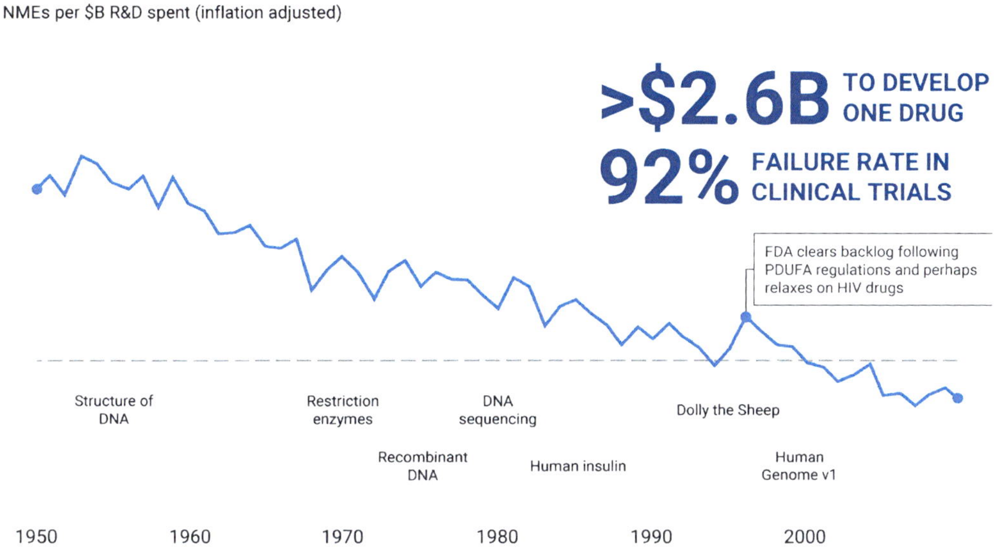

Source: Deep Knowledge Analytics[1]

This efficiency deficit isn't helped by outdated approaches to clinical trials. Under current practices, randomized controlled trials – once held up as the "gold standard" of clinical research – last 9 years and cost $1.3 Billion on average. And despite the time and capital invested in trials, only 1 in 10 drugs that enter Phase I of a clinical trial in the USA will be approved by the FDA. Furthermore, there are circumstances where randomized controlled trials are not ethically applicable – experimentation on newborns, for example. One of the key challenges in delaying clinical trials is recruiting and retaining patients. Approximately 80% of clinical trials fail to meet enrolment deadlines, while 30% of Phase III studies last longer than expected due to challenges in recruitment.

[1] www.analytics.dkv.global/data/pdf/Declining-Efficiency-of-R-and-D-in-Pharma-Corporations.pdf

Other inefficiencies in clinical development emerge because of the complex regulations that govern trials, as well as outdated data input processes, which waste the time of busy trial investigators.

Meanwhile, some experts argue that we are also seeing rising clinical trial inefficiencies in part because the majority of simple small-molecule drugs (i.e., low-hanging fruit) have been found, necessitating the development of more complex molecular structures to modulate drug targets that are still inaccessible for small-molecule drugs. This general trend is exacerbated by the increasing need to go beyond molecules and traditional biopharmaceuticals (of whatever complexity) to more complex therapeutic modalities (e.g., stem cell and regenerative medicine therapies). It is also compounded by the increasing levels of personalization and precision required to have a substantive therapeutic effect, as well as the ongoing decline of the "blockbuster drug." Clinical trial processes and regulators will need to modernize and catch up with scientific progress in order to ensure medical breakthroughs can move beyond ideas and actually save patients' lives.

It has been clear for a number of years now that the only reasonable solution to stalling Pharma R&D efficiency and cost-efficiency is the industry-wide onboarding of AI for Drug Discovery[1], both generally across all phases of drug development, and in preclinical target identification in particular. AI has the power to increase efficiency and cut R&D costs dramatically while also accelerating drug development timelines exponentially.

By using AI in drug development, it is possible to accurately predict which drugs will be safe and effective for specific patient subgroups. AI accelerates the drug development cycle by generating drug candidates for which we already have some evidence of effectiveness. Traditional pharmaceutical companies screen through many possible leads, testing all of them in the hope that one will work. By contrast, AI for Drug Discovery companies leapfrog this trial-and-error process by using *in silico* modeling of targets to hone in on drug candidates with some level of likelihood that they will work. Then, they proceed to preclinical trials in model organisms and, in some select cases, use AI to design drugs with specifiable pharmacokinetic and pharmacodynamic properties that increase their probability of being effective for specific disease targets.

[1] www.deep-pharma.tech/ai-for-dd-q2-2021-subscribe

Deep Knowledge Group has been actively monitoring this space and the transformative role of AI in the pharma and healthcare industries since 2014, when its venture arm made its first investment in the AI for Drug Discovery and Biomarker Development sector. The group's internal analytics and market monitoring on this topic expanded over the ensuing years to the point of necessitating the launch of a dedicated analytical subsidiary, Pharma Division[1] of Deep Knowledge Analytics[2], in 2017, which quickly established itself as the dominant analytics provider in this domain.

Artificial Intelligence for Drug Discovery, Biomarkers Development and Advanced R&D in the BioPharma Industry

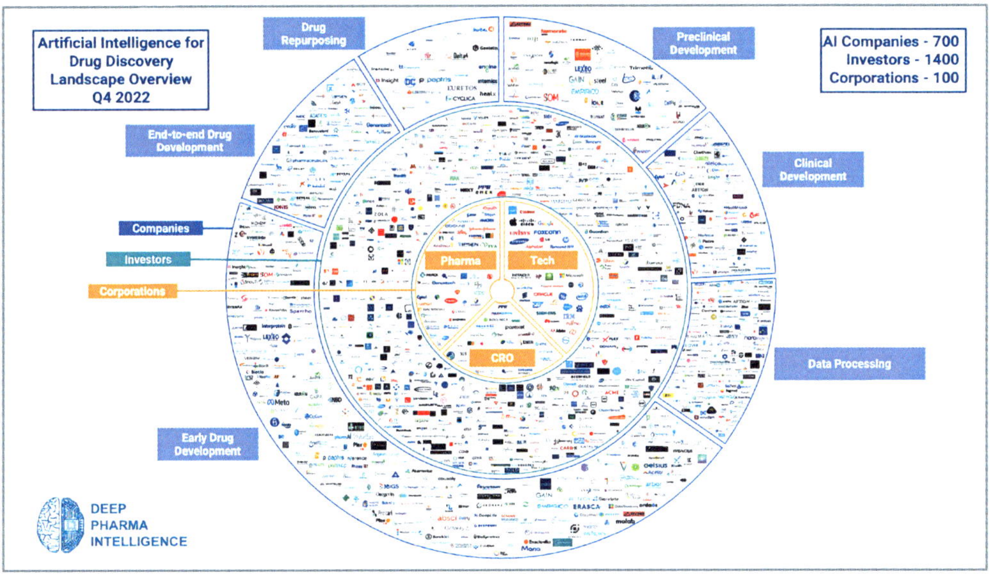

Source: Deep Pharma Intelligence[3]

From 2017–2020, Deep Knowledge Analytics Pharma Division coordinated the systematic production of comprehensive industry landscape reports, becoming a go-to source of deep analytics and market intelligence for top Pharma Industry executives and decision makers, and receiving regular media coverage and acknowledgement by Forbes, The Financial Times, MIT and others. In 2020, the analytical assets developed by the company were inherited by a new

[1] www.ai-pharma.dka.global
[2] www.dka.global
[3] www.deep-pharma.tech/landscape-overview-q2-2021

spin-out, ***Deep Pharma Intelligence***[1] (a joint venture between Deep Knowledge Analytics and BiopharmaTrend), which since its formation has produced a high volume of ***open-source industry reports***[1], ***proprietary analytics***[2], ***investment digests***[3], and ***Big Data Analytical Dashboards***[4].

Deep Pharma Intelligence now serves as the analytical and market intelligence leader of this domain, producing regular analytical reports on major areas of high potential in the Pharmaceutical and Healthcare Industries, maintaining ratings of companies and governments based on their innovation potential and business activity in the BioTech space, and providing strategic consulting and investment intelligence services to top-tier clients, including major investment funds and banks, family offices, insurance companies, government organizations, and big pharma companies, among others.

Throughout this time, Deep Knowledge Group and its analytical subsidiaries witnessed a very clear evolution in the way that pharma executives and decision makers view the impact that AI will have on their R&D approaches, drug development practices and business models. In 2014, BioTech scientists and executives alike were cautious of the real-world usefulness of even rudimentary forms of AI (e.g., big data analytics). Between 2015–2016, however, while the majority of scientists had recognized its disruptive potentials, pharma executives were still skeptical.

Things changed in 2017–2018, when enough success stories had accumulated, and the number of AI for Drug Discovery start-ups and financing rounds had grown to the point of catching the attention of pharma executives, who at this point considered it a potentially interesting R&D domain to experiment with, but certainly not the major driver of future industry growth.

By 2018/2019, however, the majority of large pharma corporations began onboarding AI for Drug Discovery in major ways (via either internal AI departments and divisions, collaborations with AI and Tech corporations, or via M&As, partnerships and contracts with leading start-ups), and now it has become indisputable industry consensus. Today, in 2021, it is literally unthinkable to be a pharma corporation

[1] www.deep-pharma.tech
[2] www.deep-pharma.tech/proprietary-analytics
[3] www.deep-pharma.tech/investment-digest
[4] www.deep-pharma.tech/public-companies-dashboard

without dedicated resources in this arena, and a business model that accounts for its fundamental (as opposed to just accelerative or complimentary) role in the future of the industry.

AI for Drug Discovery Boom – 2017 vs. 2023

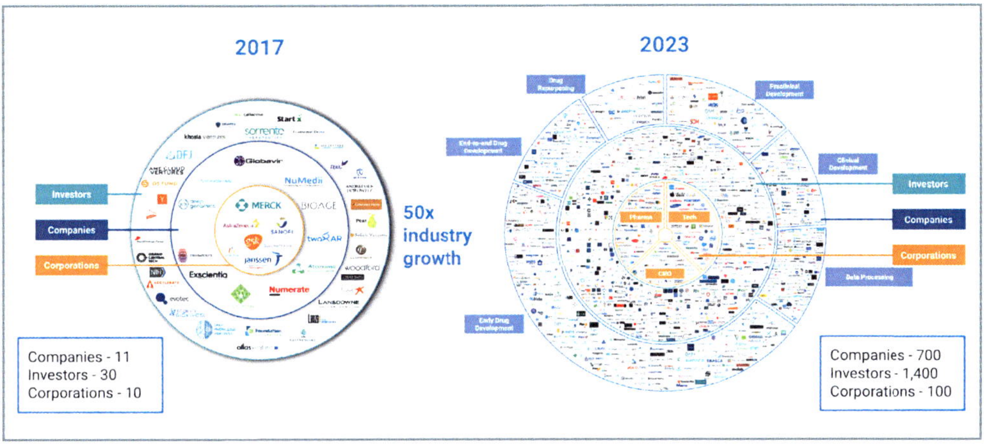

Source: Deep Pharma Intelligence[1]

We will discuss this matter at greater length in Chapter 5 ("Biomarkers as the Major De-Risking Tool in Biomedicine: Examples from Oncology and Other Industries"), specifically within the context of AI for oncology research and drug development, given that oncology is one of the largest, fastest-growing and most market-ready sectors within the larger AI in Pharma arena, and one of the best examples of the power of AI to accelerate industry development cycles.

Note: our group's AI in Pharma industry monitoring, market intelligence and analytical activities (from 2014-2017 in an internal and informal capacity, and from 2017-2021 in a formal and public capacity) have given us a substantial volume of tangible data-driven insights and forecasting capabilities on the applications and impact of AI not just for pharma and healthcare, but for Longevity research, Longevity biomarkers and anti-aging drugs as well, which we will be exploring in greater depth in a number of other chapters in this book.

AI for Drug Discovery and Biomarker Development is currently entering its golden age of practical applicability, and has seen increasing

[1] www.deep-pharma.tech/landscape-overview-q2-2021

levels of real-world validation over the past several years, culminating in the 2019 publication ***Nature Biotechnology*** of a landmark study in which Insilico Medicine generated and validated a novel small molecule drug candidate in just 46 days[1]. It designed the drug from the scratch based on specified molecular properties using AI in just 21 days, compared to the more traditional timeline of 2–3 years using standard Pharma approaches.

Note: Deep Knowledge Ventures was the first investor in ***Insilico Medicine***[2] in 2014-2015.

This was the first example of such an achievement and marked a historical moment for the AI in Pharma industry that was widely dubbed "Pharma's AlphaGo moment" in the media. This was in reference to the AI program developed by DeepMind (acquired by Google in 2014 for $500 million) that beat the top human player of the game Go, which still serves for many as the strongest real-world example of AI's disruptive and accelerative power.

During 2020 several other companies, including Deep Genomics, Deep Mind, Healx, and Exscientia, announced that they successfully been able to design new drugs using their AI systems.

The power of AI is not just limited to the development of new drugs – in fact, it cuts across almost every aspect of Pharma R&D. Its power, for example, to find better drug targets (as opposed to drug candidates for existing, established disease targets) is also apparent, with companies like Deep Genomics having managed to use AI to find drug targets previously deemed "undruggable" by the Pharma Industry. This achievement has, on average, resulted in an unprecedented 70% of the company's research projects resulting in therapeutic leads, while drug development programs have progressed from target discovery to drug candidates in less than a year.

The recognition of AI as the key solution to declining Pharma R&D framework and business models has taken place not just in the minds of smart investors in Drug Discovery start-ups but also as an option of large pharmaceutical corporations, which have been increasingly adopting AI for Drug Discovery over the past three years.

However, beyond traditional Pharma, and on the subject of clinical

---

[1] www.nature.com/articles/s41587-019-0224-x

[2] www.insilico.com

trials generally and Longevity clinical trials in particular, it is clear that the legacy frameworks for clinical testing – developed decades ago and still in use today – need to be modernized and updated.

But first, a reality check is necessary to evaluate exactly how radical a change in approach is needed. We need to ascertain what is actually happening now in clinical trials under the standard (legacy) approach, why this approach is still managing to attract and satisfy investors, and what its outcomes would be for the sustainable future development of the Longevity Industry.

The object of modern biogerontology research is to slow down the fundamental biological processes that accompany aging, such as cell senescence, impaired cellular energy regulation, mitochondrial dysfunction, age-related stress resistance, impaired proteostasis, and many others. To deal with the above-mentioned causes of aging, modern research on drug development focuses on the following areas:

Genomic Instability – Aging can be the consequence of increased DNA damage accumulation due to physical, chemical, and biological agents and DNA replication errors, spontaneous hydrolytic reactions, and reactive oxygen species (ROS).

Telomere Attrition – Telomeres are the regions located on the ends of chromosomes. They tend to become shorter after each DNA replication. When this sequence ends, the cell dies. Telomerase deficiency in humans is associated with age-related diseases.

Epigenetic Alteration – Epigenetic changes involve alterations in DNA methylation, post-translational modification of histones, and chromatin remodeling.

Loss of Proteostasis – Proteostasis involves mechanisms for the stabilization of correctly folded proteins and mechanisms for the degradation of proteins. These processes tend to change during aging.

Deregulated Nutrient Sensing – Nutrient sensing includes trophic and bioenergetic pathways, such as insulin and IGF-1, signaling pathways, and other systems (mTOR, AMPK, and sirtuins).

Mitochondrial Dysfunction – During the process of aging the cell faces a noticeable reduction in ATP generation and increased electron leakage in the respiratory chain.

Cellular Senescence – Cellular senescence can be defined as a stable arrest of the cell cycle. The accumulation of senescent cells in aged tissues can lead to age-related disease progression.

Stem Cell Exhaustion – Stem cells are cells from which all other cells with specialized functions are generated. There is a substantial decrease in the number of stem cells during life. Recent studies suggest that stem cell rejuvenation may reverse the aging phenotype.

Altered Intercellular Communication – Neurohormonal signaling tends to be deregulated in the elderly. This happens because the level of inflammatory reactions increases and immunosurveillance against pathogens and premalignant cells declines.

Regenerative Longevity Clinical Trials

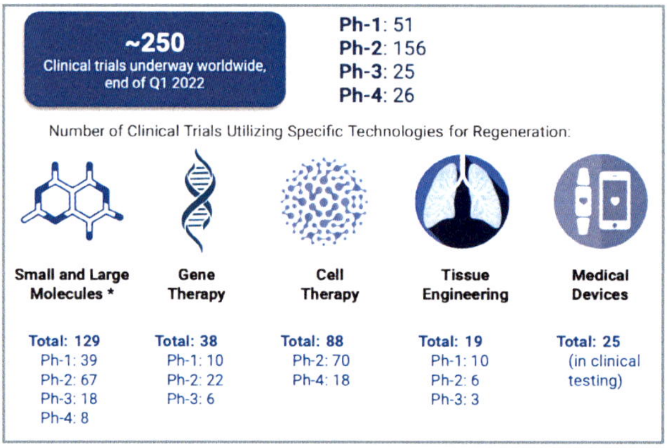

Source: Regenerative Medicine Landscape Overview Q1 2022[1]

As can be seen above, there is a definite asymmetry between the Hallmarks and Clinical Phases. The reasons are somewhat related to the fact that some of the approaches are relatively new to the clinical trials market and, therefore, it is hard for them to make it through the whole process quickly. At the same time, if analyzed statistically, such asymmetry might give the scientist additional insights into the efficacy of existing approaches.

Preclinical Development

The preclinical study is a stage of drug development that precedes clinical trials. It is necessary because there is a need to study the potential for toxicity before clinical trials take place.

[1] www.aginganalytics.com/regeneative-medicine

The main goal of preclinical development is to test different characteristics of a potential drug, such as pharmacodynamics and pharmacokinetics, ADME (absorption, distribution, metabolism, excretion), and safe dosage for human testing.

There are two types of preclinical studies: in vitro and in vivo. ***In vitro*** means that biological agents (proteins, enzymes, cells) are outside their everyday biological context. It is also called a "test-tube experiment." ***In vivo*** means that an experiment is conducted on the animals.

The most significant preclinical Longevity research is listed here:

Company	Drug Candidate	Company	Drug Candidate
Revel	**Enzymes** that break glucosepane to restore tissue plasticity	JumpStart Fertility	**Medicines** intended to improve egg cell quality and restore fertility in women of advancing age
Continuum Biosciences	**Mitochondrial/ Metabolic therapies** for aging, to restore normal function of different cells	sens research foundation	**GlycoSENS**: destroys glucosepane crosslinks to restore movement to the structural proteins
Antoxerene	**Senolytic drugs** to target the pathways of aging in senescent cells, thus reducing senescence potential	Underdog Pharmaceuticals	**Cyclodextrins** to fight 7-ketocholesterol and remove artery plaques, which causes heart diseases
Atropos Therapeutics	**Senosuppressors** – drugs to decrease the effect of senescent cells on an organism	attralus	**Peptope and CAR-M immunotherapies** against transthyretin amyloidosis
Covalent Bioscience, Inc	**Cardizyme** destroys transthyretin amyloid (misTTR) which causes age-related heart diseases	Yumanity Therapeutics	**USP14 inhibitors** to treat Alzheimer's and Parkinson's diseases
Covalent Bioscience, Inc	**Alzyme** destroys amyloid beta, a misfolded prot ein that forms Alzheimer's plaques	AgeX Therapeutics	**Telomerase therapy** and proprietary induced tissue regeneration (iTR) technology

Source: Longevity Clinical Trials[1]

[1] www.aginganalytics.com/longevity-clinical-trials

Company	Drug Candidate	Company	Drug Candidate
LEUCADIA Therapeutics	**Arethusta** promotes the flow of cerebrospinal fluid through the brain to treat Alzheimer's disease	salk. Where cures begin	**OSKM factors**: epigenetic cells' reprogramming into induced pluripotent stem cells (iPSCs) by the expression of Yamanaka genes
selphagy Therapeutics	**Medication** to improve the removal of damaged or misfolded proteins that accumulate with aging	turn.bio	**OSKMLN** factors: epigenetic cells' reprogramming into induced pluripotent stem cells (iPSCs) by the expression of Yamanaka genes, which cause heart diseases
UCSF	**Klotho protein**, supressor of insulin and insulin-like growth factor 1 (IGF1)	ICHOR THERAPEUTICS	**LYSOCLEAR**: an enzyme that disrupts lipofuscin accumulation.
cnio stop cancer	**Telomerase gene therapy** – to reduce cancerogenesis, degenerate tumors	CLEARA	**FOXO4-DRI** causes apoptosis in the senescent cells
Spotlight Biosciences	**Using machine learning** to help it map new genes that were not previously recorded	iduna THERAPEUTICS	**Epigenetic reprogramming** for tissue rejuvenation
senolytx	**Senolytics** – therapy to remove senescent cells, lowering the impact of senescence on an organism	Berkeley UNIVERSITY OF CALIFORNIA	**Gene apheresis** to restore stem cell reserves
sens research foundation	**MitoSENS** to prevent mitochondrial aging and thus stabilize all cells	SIWA	**An immunotherapy** for removing cancer and senescent cells
animal BIOSCIENCES	**Mitochondrial/ Metabolic therapies** for restoring normal functions of individual cells	SENISCA	**Modulating RNA splicing** to rejuvenate aged cells

Source: Longevity Clinical Trials[1]

[1] www.aginganalytics.com/longevity-clinical-trials

Company	Drug Candidate	Company	Drug Candidate
OISÍN BIOTECHNOLOGIES	The plasmid-based senolytics (**SENSO-lytics™**) to remove senescent cells	Elevian	**Medication** to restore regenerative capacity, with the potential to prevent age-related diseases
jumpstart	**Developing a drug** that restores egg cell quality and fertility	TELOCYTE	**Gene therapy** for short telomere restoration

Source: Longevity Clinical Trials[1]

As can be seen above, most drugs in the Preclinical Phase now focus on cell-related processes. Besides the standard approaches, like senescent cell removal, apoptosis promotion, and gene therapy, some drugs are based on metabolic processes and cerebrospinal fluid (CSF) unique features. Telomere and proteostasis restoration, as well as epigenetic alteration prevention, are among the most common techniques being used

The preclinical study is an essential phase of the drug development process. Its main goal is to filter and then choose the most promising candidates that may enter the Clinical Phase. The Clinical Phase itself is very challenging and requires enormous investment, hard work, and time. That is why drugs in the Clinical Phase are of great interest to both the industry and the scientific community.

Clinical Trials

Clinical trials are experiments that are performed as a part of clinical research to **determine the safety and efficacy of new medications**, healthcare devices, diagnostic products, or behavioral interventions.

The first stage of drug trials is called preclinical development. This stage cannot provide 100% reliable information about pharmacokinetics, pharmacodynamics, and toxicity. Scientists can receive only approximate conclusions from their preclinical research. That is why they **need to do human experiments**.

[1] www.aginganalytics.com/longevity-clinical-trials

Phases of Clinical Trials

Number of Phase	Purpose	Group Sizes
Phase 0	Pharmacokinetics and pharmacodynamics	10-15 patients
Phase I	Toxicity and safety	20-100 patients
Phase IIa	Optimal dosage with minimum side effects	50-300 patients
Phase IIb	Preliminary efficacy of the drug in comparison with placebo	50-300 patients
Phase III	Efficacy of the drug in comparison with standard treatment	300-3000+ patients
Phase IV	Pharmacovigilance and technical support of a drug, addition purposes	-

Source: Longevity Clinical Trials[1]

Clinical trials are separated into phases. Three phases are obligatory – these are called the main phases: phases I, II, and III. Two phases are optional: phases 0 and IV.

There are two types of clinical trials: observational and interventional studies. In an observational study, the investigators only observe the subjects and measure their outcomes; they do not actively manage the study. In an interventional study, the investigators give the research subjects an experimental drug or therapy, thus potentially influencing possible outcomes. Then the researchers assess how the subjects' state of health changes.

There are three main phases of clinical trials, of which Phase I is the earliest. Phase I Clinical trials are necessary to determine a safe dosage of medication for healthy people (a group consists of 10 people). Phase I trials are not randomized and, thus, are vulnerable to selection biases. These trials are often conducted in a clinical trial clinic since the level of safety of the drug is still unclear, and full-time medical observation is mandatory both for patient safety and investigation needs. Contract research organizations

[1] www.aginganalytics.com/longevity-clinical-trials

(CROs) are also essential players in the process of conducting clinical trials for a given drug or therapeutic technique. These companies allow their clients to run any phase of the clinical trials quicker and more efficiently, since they tend to specialise in different services, including clinical trials. The subject who receives the drug is usually observed until several half-lives of the drug have passed. This phase is designed to assess the safety (pharmacovigilance), tolerability, pharmacokinetics, and pharmacodynamics of a drug. Phase II is the testing of a drug on volunteers to assess its efficacy and side effects (100-300 participants). Sometimes Phase II trials are designed as case series, demonstrating a drug's safety and activity in a selected group of participants. Also, unlike Phase I, some Phase II trials are designed as randomized controlled trials, where some patients receive the drug, device, or other therapeutic solution and others receive placebo or standard treatment for a given condition. But randomized Phase II trials have far fewer patients than randomized Phase III trials. The development process for a new drug failure usually occurs during Phase II trials when the drug is discovered not to work as planned or to have toxic side-effects. Phase III, the largest clinical trial, is held to assess the efficacy, effectiveness, and safety of a drug (300–3,000 participants with a specific disease). Phase III studies are randomized controlled multicenter trials on large patient groups (300–3,000 or more depending upon the disease/medical condition studied) and are aimed at being the definitive assessment of how effective the drug is, in comparison with the current "gold standard" treatment. Because of their sample size and comparatively long duration, Phase III trials are the most expensive, time-consuming, and difficult trials to design and run, especially in therapies for chronic medical conditions. Phase III trials of chronic conditions or diseases often have a short follow-up period for evaluation relative to the period of time the intervention might be used in practice.

The Clinical Trials Database contains 45,000 articles on Longevity research. Almost all preclinical research is concentrated in Europe (45.2%) and the USA (26%). Asia now accounts for 16.4% of all preclinical research, a percentage that continues to grow. As can be seen in the infographics below, Europe and the USA are leaders in clinical trials. The reasons for this can be divided into two categories: companies' origin and regulatory norms. Since most of the pharmaceutical companies are located in Europe or the USA, it is somewhat inefficient to relocate clinical studies, since it is harder for clients to control them. At the same time, every country has its own unique regulatory norms. Therefore it's also inconvenient to relocate such studies. On the other hand, we should mention that clinical trials themselves could be executed in a variety of countries, since it's important to gather data from people of different ethnicities due to the genome polymorphism observed in different populations.

Clinical Trials Distribution by Phases

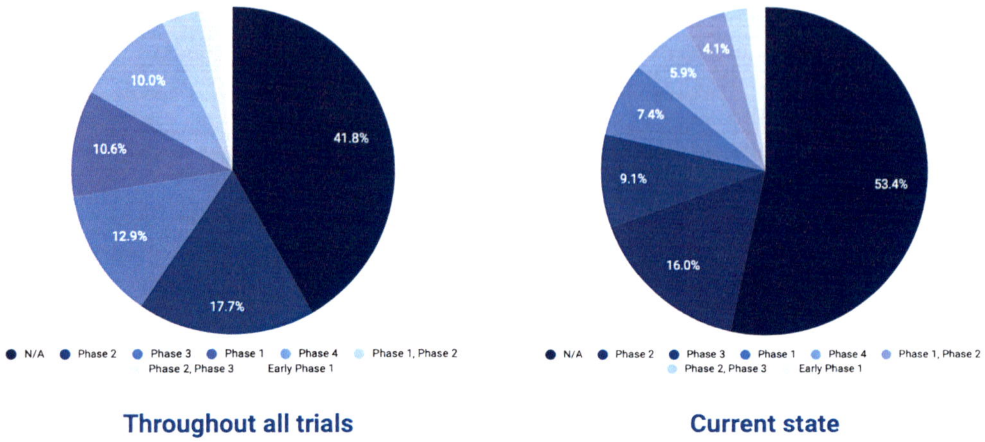

Source: Longevity Dashboard Clinical Trials[1]

Longevity Clinical Trials distribution during 2021

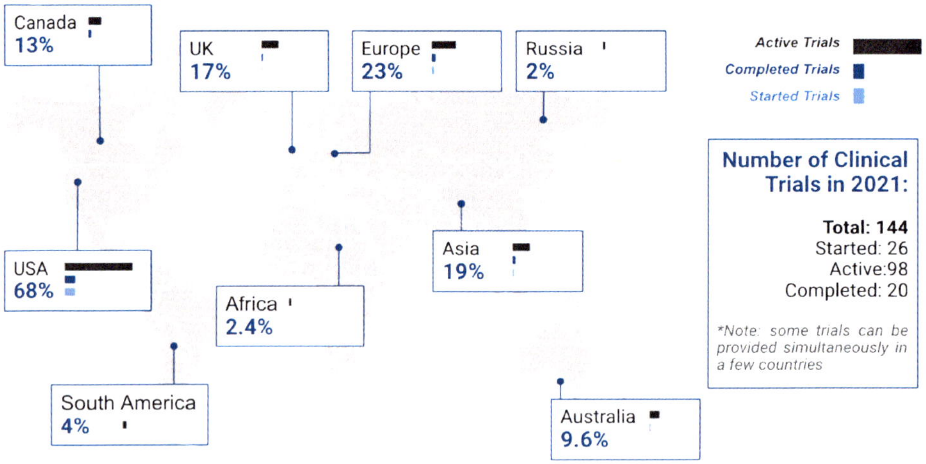

Source: Longevity Dashboard Clinical Trials[1]

The leading countries by research are the USA (26%), France (6.7%), and Canada (4.7%).

[1] www.dashboards.invest-solutions.finance/longevity-investment-clinical-trials

Top Countries by Number of Trials

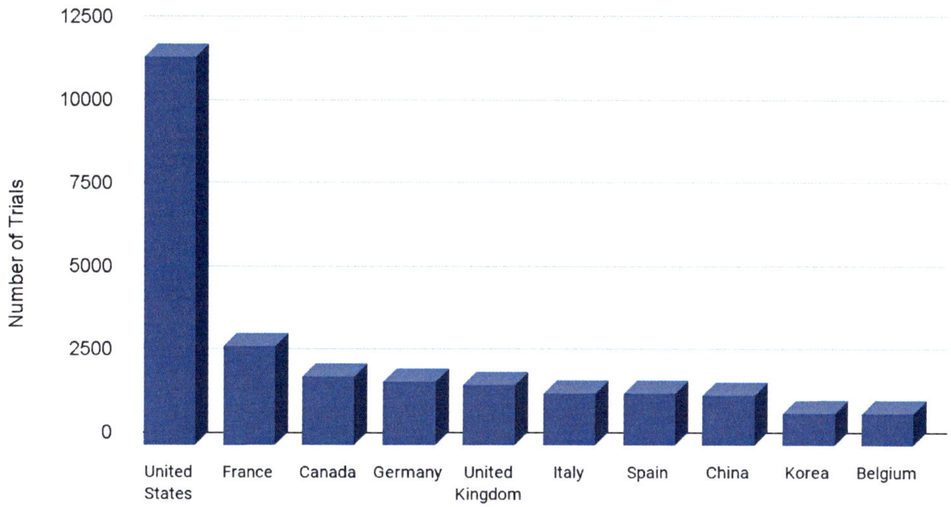

Source: Top Countries by Number of Trials[1]

There are two main types of clinical studies: interventional and observational. The figure below depicts the world's distribution of clinical trials distribution by study type.

Clinical Trials by Study Type, %

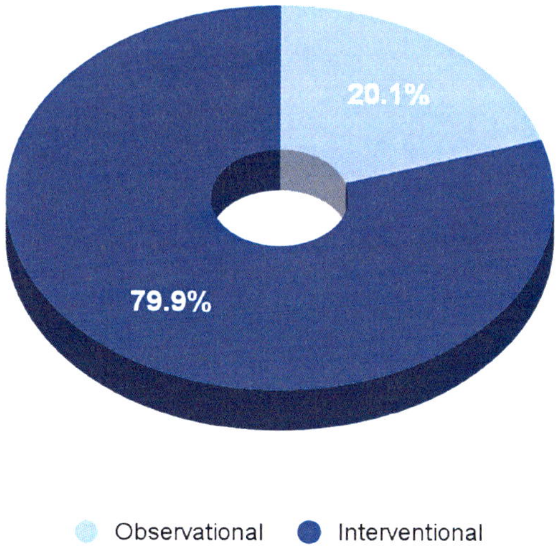

Source: Longevity Dashboard Clinical Trials[1]

[1] www.dashboards.invest-solutions.finance/longevity-investment-clinical-trials

There are a great many age-related conditions, the most common of which are neurodegenerative disorders, cardiovascular diseases, and cancer. Examples of aging-associated diseases are atherosclerosis and cardiovascular disease, carcinoma, arthritis, cataracts, osteoporosis, type 2 diabetes, hypertension, and Alzheimer's disease.

Direct Aging Clinical Trials, %

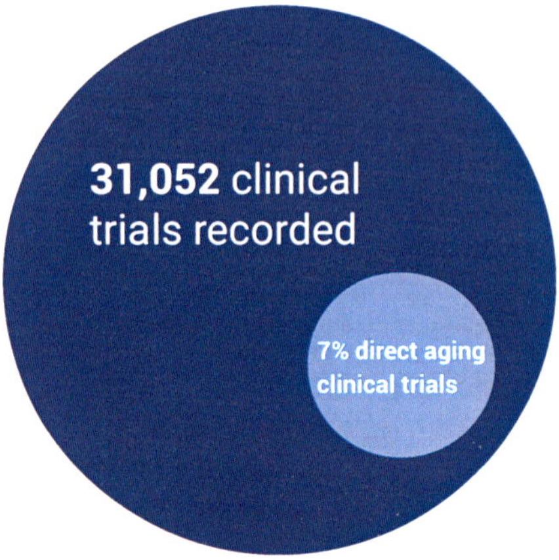

Among the 31,052 clinical trials, 7.5% mention aging as a study condition, resulting in more than 2,000 relevant records. This clearly indicates both an acknowledgment of the importance of such studies and a readiness to execute all related research and investigations. This, in turn, shows investment expediency associated with age-related solutions.

Source: Longevity Dashboard Clinical Trials[1]

Clinical Trials Distribution by Disease Field

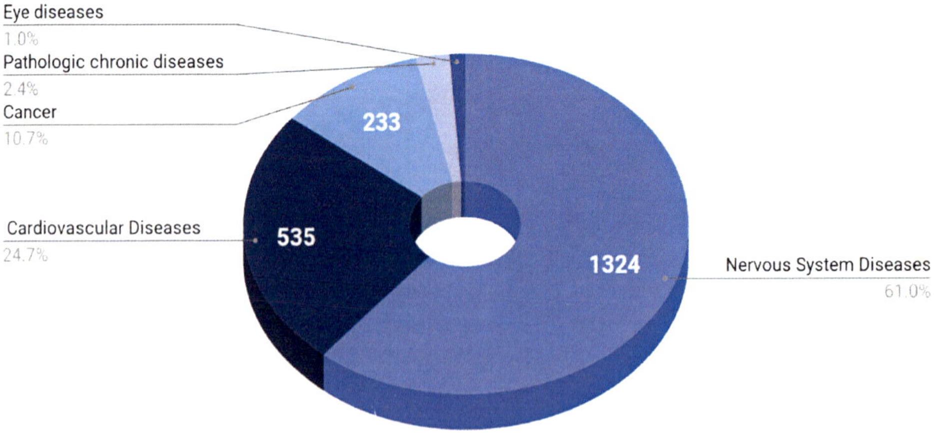

Source: Longevity Dashboard Clinical Trials[1]

[1] www.dashboards.invest-solutions.finance/longevity-investment-clinical-trials

The above-mentioned data corresponds to the clinical trials in certain disease fields connected with the aging condition. Roughly 85% of all directly age-related trials belong to the categories of nervous system disorders and cardiovascular diseases.

Aging is one of the main risk factors for a bunch of neurodegenerative disorders, including Alzheimer's disease and Parkinson's disease. Mitochondrial dysfunction and abnormal protein metabolism in neurons are well-known consequences of aging that lead to neuron damage and the malfunctioning of synapses. Major groups of clinical trials in this category are connected with aging biomarkers in the nervous system, cognitive and behavioral studies (such as sleep, stress, resistance, etc.), and different engineering solutions for disease treatment. This diversity of studies probably explains the huge number of clinical trials in this sphere.

Aging-related cancer clinical trials are focused mainly on early screening for further increase in therapeutic effects on tumors. It is well known that cancer is likely to develop with aging due to genomic instability, mutation accumulation, and epigenetic alterations. Still, the causes are complex, and studies continue to develop in the direction of the tumor microenvironment, oncogenes activation, and other abundantly investigated fields, while the direct research of aging impact doesn't gather that much attention.

Drug development in age-related nervous system disorders is well-founded: antioxidant drugs for mitochondria or lysosome functioning directly affect the causes of disease. Hence, it is less complex and very important to provide effective group therapy for neurodegenerative diseases, while heterogeneous and multiple-cause cancer remains the issue for biomedical scientists to solve.

Longevity Drugs in Phase I Clinical Trials

Clinical Phase I is focused on dose-ranging on healthy volunteers for safety analysis. Due to the aforementioned reasons, the number of drugs being subject to clinical studies is significantly smaller. The distribution of the biological approaches stays close to the one seen in the Preclinical Phase.

Company	Drug Candidate	Company	Drug Candidate
The Sinclair Lab BLAVATNIK INSTITUTE GENETICS	**NMN** boost the ability to repair DNA damage	DENALI THERAPEUTICS	**DNL343** restores RNA and protein function

Company	Drug Candidate	Company	Drug Candidate
NUS National University of Singapore	**Alpha-ketoglutarate (AKG)**	DENALI THERAPEUTICS	**DNL151** to treat Parkinson's disease
Proclara BIOSCIENCES	**General Amyloid Interaction Motif (GAIM)** to disrupt existing amyloid aggregates	salk	**J147** to treat Alzheimer's disease and reverse some processes of aging
DENALI THERAPEUTICS	**DTL201 and DTL151** to restore lysosomal function and treat Parkinson's disease	Amazentis	**Urolithin A** to prevent mitochondrial aging
DENALI THERAPEUTICS	**DNL747** reduces brain inflammation	minovia	Evaluating therapeutic effects of the transplantation of **MNV-BM-BLD**
BlueRock Therapeutics	**CELL+GENE platform** to provide treatment for Parkinson's disease	Intervene Immune	**Human growth hormone (HGH)** and **dehydroepiandrosterone (DHEA)** to regrow the thymus
BlueRock Therapeutics	Surgically injected nerve cells that create dopamine	CohBar INC.	**CB4211** for fatty liver disease treatment
Immunic THERAPEUTICS	Developing **IMU-856**, a compound that targets a transcriptional regulator of immune barrier function	Alterity THERAPEUTICS	**ATH434** to treat Alzheimer's and Parkinson's disease
Epirium	Evaluating the response of biomarkers to **(+)-Epicatechin** in Becker muscular dystrophy	CytoMed Therapeutics 新细胞医学	**MSC infusion** for anti-aging and regenerative therapy
	Umbilical cord mesenchymal **stem cells infusion** for aging frailty	聖釋 SCLnow	**hUC mesenchymal stem cells** for treatment of Cirrhosis
tomorrowlabs	**Investigation of a proprietary cosmeceutical** to reduce the visible signs of cutaneous aging	WISCONSIN UNIVERSITY OF WISCONSIN-MADISON	**Inorganic nitrate supplementation** on cerebrovascular aging and arterial stiffness study

Source: Longevity Clinical Trials[1]

[1] www.aginganalytics.com/longevity-clinical-trials

At the same time, the number of drugs dealing with mitochondrial dysfunction is higher, and, as will be shown later, such drugs tend to go even further, making it through Phase II. It is also clear that it is still hard for genomic-related approaches to pass the Preclinical Phase. The date when these drugs and therapeutic solutions will be as widely accepted as the classical ones is still unclear. On the other hand, the growing popularity of this approach and the genomic breakthrough made in the past 20 years have already created our future, where we only have to accept it.

Longevity Drugs in Phase II Clinical Trials

In some sense, Clinical Phase II is probably one of the most important, since it could propel the drug to Phase III, which is always considered the most complicated and expensive step.

Company	Drug Candidate	Company	Drug Candidate
ChromaDex	**Nicotinamide riboside (NR)** converting the different B3 vitamins into NAD+	LYGENESIS	**The patient's own lymph nodes** for organ replacement
Biophytis LIVE LONGER, LIVE HEALTHIER	**Sarconeos** to treat sarcopenia and age-related loss of muscle mass	LONGEVERON	**Allogenic mesenchymal stem cells (MSCs)** for treating age-related diseases
Stealth BIOTHERAPEUTICS	**Elamipretide** Reduce production of reactive oxygen species	UNITY BIOTECHNOLOGY	**UBX1325** for removing senescent cells
METROBIOTECH	**NAD+** Precursor supplementation for Friedreich's ataxia therapy	ALKAHEST	**Chronokines** (plasma fractions) for rejuvenation
THERAVASC	**Sodium Nitrite** to treat arterial aging	ALKAHEST	Assess the efficacy and safety of **AKST4290**
CBMG Cellular Biomedicine Group	**Mesenchymal progenitor cells** for the treatment of knee osteoarthritis	ALKAHEST	Assess the safety of **GRF6021** infusions
agelessRx	Evaluation of aging with **rapamycin** for Longevity study	LONGEVERON	**Mesenchymal stem cells** to treat aging frailty
Scholar Rock	**SRK-015** for spinal muscular atrophy (SMA) treatment	UT Health San Antonio	**Nicotinamide Riboside** in COVID-19 patients for renal protection

Company	Drug Candidate	Company	Drug Candidate
BIOAGE	**Evaluating BGE-175** in hospitalized adults with COVID-19	UT Health San Antonio	**Metformin** for preventing frailty in high-risk older adults
BIOAGE	**Developing BGE-117,** which inhibits an enzyme responsible for breaking down Hypoxia-Inducible Factor	FREQUENCY THERAPEUTICS	**FX-322** in adults with stable sensorineural hearing loss
Genome Protection	**Entolimod** on immunosenescence in healthy geriatric subjects receiving influenza vaccination	HUMACYTE	**Humacyte Human Acellular Vessel (HAV)** in patients with vascular trauma
Intervene Immune	Thymus regeneration, immunorestoration, and insulin mitigation extension trial **(TRIIM-X)**	LYGENESIS	**Allogenic** hepatocyte transplantation into periduodenal lymph nodes
LINEAGE CELL THERAPEUTICS	Evaluating the safety and tolerability of **OpRegen**	MAYO CLINIC HEALTH SYSTEM	**Dasatinib + Quercetin** for senescence in chronic kidney disease
RETROTOPE	**RT001 in Amyotrophic Lateral Sclerosis**	MAYO CLINIC HEALTH SYSTEM	**Nicotinamide Riboside + Pterostilbene** for protection from acute kidney injury (AKI)
Scripps Research	**Quercetin and Dasatinib** to destroy senescent cells	rejenevie THERAPEUTICS	**Hematopoietic stem cells (HSCs)**
CRATUS	**Mesenchymal stem cell (MSC) therapy** to treat age-related frailty	MAYO CLINIC	**Fisetin** to remove senescent cells

Source: Longevity Clinical Trials[1]

A considerable number of the approaches listed above are closely related to stem cell exhaustion and altered intercellular communication. Since, as explained above, cellular processes are seen as one of the most important and easily affected processes in Longevity, they remain central to a lot of research.

[1] www.aginganalytics.com/longevity-clinical-trials

Longevity Drugs in Phase III Clinical Trials

The third phase is intended to support an eventual New Drug Application and its potential approval.

Lorecivivint (Biosplice Therapeutics) is a potential treatment for osteoarthritis. It entered Phase III of clinical trials, which is designed as a long-term extension to the parent-study SM04690-OA-11. It aims to evaluate the safety and efficacy of the long-term use of lorecivivint (LOR) in subjects with knee osteoarthritis (OA).

SkQ1 (Mitotech) is a drug that has been devised to restore mitochondria function in damaged cells. It demonstrates a statistically significant advantage over placebo, is comfortable and well-tolerated, and causes no unexpected or serious ocular adverse events. It has also entered Phase III and is demonstrating promising results so far.

Elamipretide (Stealth Biotherapeutics) is currently being subject to a randomized, double-blind cross-over trial to evaluate its safety, efficacy, and tolerability in subjects with Barth Syndrome.

RT001 (Retrotope Inc) is an experimental fatty acid that stabilizes the mitochondrial and cellular membranes against attacks and restores cellular health.

Netarsudil (Aerie Pharmaceuticals) is a first-in-class therapy for the treatment of patients with open-angle glaucoma, ocular surface diseases, and retinal diseases. It is now being studied.

Pimavanserin (Acadia Pharmaceuticals) is a drug that should reduce neuropsychiatric symptoms related to neurodegenerative disease in adult and elderly subjects.

Company	Drug Candidate	Company	Drug Candidate
biosplice	**Lorecivivint:** for restoring the Wnt signaling pathway	MITOTECH SKQ	**SkQ1:** a mitochondrial antioxidant
Stealth BIOTHERAPEUTICS	**Elamipretide** in subjects with **Barth Syndrome:** a mito-chond rial antioxidant	RETROTOPE	**RT001** in subjects with **Friedreich's Ataxia**
aerie	**Netarsudil®** reduction of elevated intraocular pressure	ACADIA® Pharmaceuticals	**Pimavanserin** to decrease neurodegenerative symptoms

Source: Longevity Clinical Trials[1]

[1] www.aginganalytics.com/longevity-clinical-trials

Completed Longevity Clinical Trials

Clinical trials that pass Phase III are statistically rare, but they are continuing to grow in number Trials in Aging and Longevity are exceptional and relatively new to the market. Some of them require more than ten years to pass all the stages of clinical research. The novelty of this field is supported by technical and methodological progress, but it also explains why most studies are still within Phases I and II.

Company	Drug Candidate	Company	Drug Candidate
AMGEN	**PCSK9** – downregulation of systemic inflammation	GenSight BIOLOGICS	**GS010** – treating severe inherited retinal diseases.
GRIFOLS	**AMBAR** – plasmapheresis technique that infuses patients with human albumin	Osiris THERAPEUTICS INC	**Prochymal** – allogeneic stem cell therapy based on mesenchymal stem cells
mesoblast	**Rexlemestrocel-L** – Allogeneic STRO-3 immunoselected bone marrow-derived mesenchymal precursor cells	mesoblast	**Allogenic** – mesenchymal precursor stem cell infusion

In the spring of 2019, Amgen completed a Phase III clinical study of proprotein convertase subtilisin/ kexin type 9 (PCSK9). This is a medication that was developed to downregulate systemic inflammation by reducing lipoproteins in the blood. This protein plays a significant regulatory role in cholesterol homeostasis, mainly by reducing the metabolism of LDL-particles. PCSK9 is inactive when first synthesized because a section of peptide chains blocks their activity; proprotein convertases remove that section to activate the enzyme. PCSK9 binds to surface LDL sensing proteins which are responsible for endocytosis. In a further cascade, PCSK9 is activated when entering a lysosome leading to decomposition of LDL and its receptor, whereas during natural endocytosis, receptors are not decomposed. Administration of medicine is performed via injection once a month. The subject of this study was Hyperlipidemia, but potentially this approach can manage other metabolism conditions, obesity, and cardiovascular diseases. On average, PCSK9 treatment for 16 weeks resulted in a 17% reduction of circulating lipids.

GeneSight Biologics is a gene therapy company specializing in treating severe inherited retinal and central nervous system diseases. Its two main products (GS010 and GS030) are designed to treat rare diseases, such as Leber Hereditary Optic Neuropathy (LHON) and Retinitis Pigmentosa (RP). On December 12, 2020, GS010 passed Phase III Clinical Trials. It demonstrated very impressive results in the treatment of Leber Hereditary Optic Neuropathy (LHON), an inherited mitochondrial genetic disease of the eye that causes blindness. In 2020, GeneSight's therapy helped 37 participants. It is an AAV2 gene therapy that contains DNA regions that can encode wild-type ND4 protein and restore an abundance of proteins that underwent mutation in the subject. On June 30, 2021, GenSight reported vital efficacy and safety results at 78 weeks post-treatment in the reflected Phase III clinical trial for LUMEVOQ®. The results show better visual acuity improvements from bilateral intravitreal injections of the gene therapy compared to unilateral injection.

Grifols has developed AMBAR, a plasmapheresis technique that infuses patients with human albumin and intravenous immunoglobulin (IVIG) to treat Alzheimer's disease. The company hypothesized that amyloid-beta, the main driver of Alzheimer's disease, binds to albumin in the brain, which can then be removed from the body. The company's results have confirmed that this approach is valid, although it does not constitute a cure. In a Phase IIb/III trial, Grifols reports that treated patients with moderate Alzheimer's experienced a 61% reduction in disease progression and a 71% reduction of symptoms, with positive results reported for both moderate and mild Alzheimer's. In May 2021, Grifols announced that the first center to offer the AMBAR procedure would open in Barcelona, Spain. New centers in the EU, the USA, and China are planned.

Osiris Therapeutics has developed Prochymal. It is an allogeneic stem cell therapy based on mesenchymal stem cells (MSCs) derived from the bone marrow of donors aged between 18 and 30 years. MSCs are purified from the marrow afterward, cultured, and packaged, with up to 10,000 doses derived from a single donor. The doses are stored frozen until needed. Administration of the product is mainly performed as an infusion. Stem cell therapies have multiple applications, but Prochymal has clinically proven therapeutic effects on Crohn's Disease and Graft-versus-host disease (GvHD). The first clinical approval of this drug happened in 2012. It was the first stem cell drug approved. The trials demonstrated that in 61–64% of children with acute GvHD who were unresponsive to steroids, Prochymal produced a clinically significant response at 28 days after the therapy. Until the approval of Prochymal, there had not been any other effective therapy.

In 2021 Mesoblast completed the Phase III clinical study of Rexlemestrocel-L. This medication is Allogeneic STRO-3 immunoselected bone marrow-derived mesenchymal precursor cells used to treat Degenerative Disc Disease (DDD). It is an age-associated condition that represents back pain caused by worn-down vertebral discs. Disc degeneration and associated low-back pain are significant causes of suffering and disability. Rexlemestrocel-L consists of a unit dose of 6 million mesenchymal precursor cells (MPCs). It is injected by syringe directly into a targeted damaged disc in an outpatient procedure. This clinical trial proved that Rexlemestrocel-L in low doses improves the spine's function and reduces pain through to 24 months post-treatment. Also a different set of stem cells, produced by Mesoblast for the treatment of hematologic malignancies, passed Phase III of clinical trials. This medication differs from Rexlemestrocel-L by its origin of stem cells, but it has the same mechanism of action. Due to a variety of possible stem cell proliferation fates, therapies could have different applications. Thus Mesoblast is running clinical trials of the stem cell medication in such vectors: treatment of Graft-versus-host disease, Myocardial infarction, Osteoarthritis, Rheumatoid arthritis, Spondylosis, and Type 2 diabetes. Clinical trials now in Phase III aim to cure such conditions as acute myeloid leukemia, heart failure, Hodgkin's disease, and non-Hodgkin's lymphoma.

Clinical trials in Longevity have unique features – therapies used to conquer aging have multiple applications. For example, senolytics and stem cell transplantation can be used in treatment for almost any age-associated disease. But, unfortunately, the current paradigm of clinical trials is significantly slowing down the process by separating every possible application of medication into totally different investigations.

Failed Longevity Clinical Trials

Most clinical trials related to Longevity face huge risks of failure. But pioneers of such research are attracting more and more investors and public interest, thus potentially benefiting all upcoming research in this field.

Company	Drug Candidate	Company	Drug Candidate
resTORbio	**RTB101** – a small molecule inhibitor of the rapamycin complex 1	UNITY BIOTECHNOLOGY	**UBX0101** – senolytic drug to eliminate senescent cells
Pfizer	**Capromorelin drugs** – reduces risks of becoming frail.	CONSTANT THERAPEUTICS	**TXA127** – Angiotensin-1-7 – modulates cell signaling

ResTORbio is a clinical-stage biopharmaceutical company focused on the development and commercialization of novel therapeutics and attempts to push the limits of fighting age-related diseases. The main company's goal of clinically applying the geroscience hypothesis has been centered around inhibition of the mechanistic target of rapamycin (mTOR). mTOR has emerged as a critical regulator of aging in laboratory studies, with genetic or pharmacological inhibition of mTOR complex 1 (mTORC1) shown to delay aging in yeast, nematode worms, fruit flies, and several strains of laboratory mice. One of the most promising products was RTB101 – a small molecule inhibitor of the rapamycin complex 1 (TORC1), designed to inhibit the phosphorylation of multiple targets downstream of the TORC1 signaling pathway. RTB101 is the only clinical-stage drug candidate that the company currently has available, although it is developing additional drugs that target the mTOR metabolic pathway just as RTB101 does. The company is planning on expanding its portfolio, but their main focus is on work on clinical trials for RTB101 seeking its approval for clinical use. The first clinical trial was determined to combat respiratory tract infections in older people (65+). The early stages of the study have shown great results, but during Phase III, the cutoff happened. The research did not meet its primary endpoint, and the company has stopped the development of RTB101 in this indication. Yet, multiple preclinical models have demonstrated that the inhibition of TORC1 decreases protein and lipid synthesis, increases lysosomal biogenesis, and stimulates the clearance of misfolded protein aggregates, such as toxic synucleins, that cause neuronal toxicity in Parkinson's disease. In accordance with this data, research was turned toward the field of Parkinson's disease treatment.

RTB101 and other mTOR inhibitors did not demonstrate any results in terms of human life extension. Following the announcement of the clinical trial failures, the company's capitalization fell by 83%. Due to huge investment loss, the company had to merge with Adicet Bio and redesign its whole business strategy, shifting focus from Longevity drugs to treating specific diseases. Again, there was no success in applying Longevity treatments developed for model organisms to humans.

UNITY Biotechnology is a biotechnology business that is developing medicine to target senescent cells. These medicines are known as senolytics. Conquering cellular senescence is seen to be one of the most important steps in combating aging. In addition, cellular senescence is one of the most prominent signs of aging; it is a continual process that causes individual cells to lose function. The company's leading goal is to develop therapeutics to extend the period of life unburdened by disease. UNITY Biotechnologies' initial focus is on delivering localized therapy in ophthalmologic diseases, including diabetic macular edema, age-related macular degeneration, and age-related neurological conditions. Since 2018 they have developed four unique drugs, trials on one of which has already reached Phase II.

UNITY's neurology initiatives draw on a basic cellular senescence research platform to address fundamental characteristics of neurodegenerative illnesses and cognitive impairments. These programs focus on developing senolytic therapies for neurological diseases and complementing the ongoing exploration of novel mechanisms for cognitive benefit, including α-Klotho, UBX2089 as the most recent senolytic products.

UNITY's lead candidate drug, UBX0101, was considered to be a unique small molecule inhibitor of protein interaction.

Unfortunately, UBX0101 failed its Phase II trial of localized senolytics for knee osteoarthritis (UBX0101, a p53/MDM2 interaction inhibitor) in August 2020. When UNITY announced[1] that its lead drug candidate had failed to beat a placebo in reducing joint pain and stiffness, the company's stock plunged more than 60%, and nearly one-third of UNITY staff were laid off.

Despite this initial trial failure, senolytics have strong potential to the broader rejuvenation market. UNITY is developing senolytic candidates for various other age-related diseases.

[1] www.fightaging.org/archives/2020/08/unity-biotechnology-fails-phase-ii-trial-of-localized-senolytics-for-knee-osteoarthritis

Pfizer has been working on *CP-424, 391* – capromorelin drugs drugs to reduce the risk of becoming frail. GHS-1 receptors, for which the natural ligand is the stomach peptide ghrelin, are stimulated by capromorelin, an orally active pyrazolinone-piperidine. It is now proven that GH replacement or stimulation may benefit older persons with a relative GH shortage. Capromorelin has high intestinal absorption in rats; consistent pharmacokinetics in rats and dogs, including a short half-life; and significant GH secretion stimulation in numerous animal species. The availability of an orally given GHS that increases GH pulsatile release allowed researchers to test the idea that sustained stimulation of GH release can improve physical function in older adults. The trial passed Phase I with great results, but not Phase II. Phase II lasted for two years, but did not result in a consistent outcome.

Constant Therapeutics has been trying out its exclusive drug – TXA127 – in the field of Longevity but faced failure during Phase II of clinical trials. TXA127 is pharmaceutically-formulated angiotensin 1-7, a non-hypertensive derivative of angiotensin II. It is a molecule that can be engaged in several immune and cell signaling cascades. Theoretically, it can be used to control immune response and thus inflammatory background. Unfortunately, it turned out that the lowering of inflammatory background in the elderly was not achieved by low doses of medication. Nevertheless, Constant Therapeutics did not stop their range of clinical trial and it is now achieving outstanding results in studies of COVID-19 treatment.

Failures of modern therapeutics in the sphere of Longevity are the normal course of things. Given the low global clinical trials success rate in general, it can be foreseen that this number will stay even lower for Longevity-related research because of its even greater complexity. Nevertheless, all research in this sector is benefiting the whole field due to its overall ideological coherency.

Planned Longevity Clinical Trials

Due to the continuous expansion of research and intensive developments in cellular and molecular biology, more and more approaches and medications are beginning to appear as potential forms of Longevity therapy. The popularity of these approaches is not expected to decrease in the near future.

AgelessRx is a modern company that focuses its resources on developing a Longevity approach and making it accessible. Most of their products are dedicated to treating metabolic conditions; besides that, they also offer bioage tests. One of the main product goals for that company is to start its own anti-aging cosmetic line. These products are already very abundant on the market, but AgelessRx's solution will be unique.

A lot of research is focused on food supplements that theoretically can have a positive effect on metabolism. The idea is that these supplements can allow for a restoration of essential elements of the metabolic system. Disparities in metabolites progressively increase with age. Thus, their normalization can partially restore the normal functions of cells. The most common food supplements are coenzymes and different metabolites.

The National University of Singapore is conducting a human clinical trial of the effects of alpha-ketoglutarate on people aged 45 to 65, testing biomarkers of aging such as epigenetic clocks, pulse wave velocity, and inflammation. Experts believe that a lot of studies will be conducted adopting this idea.

MetroBiotech is planning to conduct substantial clinical research in collaboration with U.S. Special Operations Command. Researchers plan to increase combatants' functional efficiency by altering cell aging cascades, thus improving human performance, such as endurance and recovery from injury. The clinical part of the trial is scheduled for the year 2022. It is also estimated that 3,000+ people will participate. Such an example sets a new field of study – function and regeneration improvement.

Globally, research in this field and the emergence of novel ways of diagnostics are attracting many investments, thus expanding research horizons. Generally, clinical trials focus on three key hallmarks: stem cell exhaustion, cellular senescence and metabolism. Most experts believe that the market interest focus will stick to those Longevity clinical trial groups for a long time.

Metformin

As for drugs that have already completed their path through the Clinical Phases, Metformin is probably one of the most well-known Longevity-related therapeutic solutions. Metformin is a biguanide antihyperglycemic agent. It decreases glucose production by the liver, increasing the insulin sensitivity of body tissues and GDF15 secretion. Metformin was discovered in 1922 but introduced to the market only in 1995 and now it is a part of the World Health Organization's List of Essential Medicines. First studies in humans began in the 1950s and were executed by French physician Jean Sterne. As such, Metformin's real potential value as an anti-aging compound took a long time to become commonly accepted. The MD Anderson Cancer Center has found during research that Metformin could also be used for cancer treatment[1]. Specifically, the pancreatic cancer risk was 62% lower in the group that took Metformin. In addition, for people who took insulin, the pancreatic cancer risk was five-times higher in comparison to the controlled group.

The National Institute on Aging has discovered that the lifespan of the mice that were given 0.1% Metformin in their diet increased by 5.83%; at the same time, the age-related diseases began later for these mice. However, this result has since been called into question after the National Institute on Aging Interventions Testing Program failed to replicate the effects[2] of this study at three different locations using genetically heterogeneous female and male mice.

The man behind the revival of Metformin as a potential anti-aging intervention is Dr. Nir Barzilai. Nir is the founding director of the Institute for Aging Research, the Nathan Shock Center of Excellence in the Basic Biology of Aging, and the Paul F. Glenn Center for the Biology of Human Aging Research at Albert Einstein College of Medicine of Yeshiva University. He also directs the Longevity Genes Project, a genetics study of over 600 families of centenarians and their children. At the same time, Dr. Barzilai is a co-founder of CohBar, Inc., a biotechnology company developing mitochondria-based therapeutics to treat diseases associated with aging.

Barzilai's studies have culminated in a planned series of clinical trials of the anti-aging effects of metformin, titled "TAME" (Targeting of Aging with Metformin[3]), managed by the American Federation for Aging Research. The TAME trial's future status (which was uncertain for a number of years due to issues with financing) was secured in 2019 after closing the final $40 million of its required $75 million budget with a donation from a private source, and the trial design and scope was formally confirmed and officially finalized in 2021.

[1] www.doi.org/10.1053/j.gastro.2009.04.013
[2] www.karger.com/Article/FullText/502257
[3] www.afar.org/tame-trial

While the exact start date of the trial is still uncertain, and probably dependent on the speed at which they progress with trial enrollment, TAME itself consists of a series of nationwide, six-year clinical trials at 14 leading research institutions across the country that will engage over 3,000 individuals between the ages of 65-79. The study's executive committee includes a number of other recognizable Longevity scientists besides Barzilai as well, including David Sinclair, S. Jay Olshansky, Steve Audtad, James Kirkland and Stephanie Lederman.

To date, 14 trial sites have been officially confirmed at its various participating institutions (including Albert Einstein College of Medicine Brigham, the Women's Hospital of Brown University, the HealthPartners Institute, Johns Hopkins University, Northwestern University, the University of Alabama, the University of Connecticut, the University of Florida, the University of Miami and Miami VA, the University of Pittsburgh, the University of Tennessee, the Wake Forest School of Medicine, and Yale University).

Dr. Barzilai's research team at the Institute for Aging Research at Albert Einstein College of Medicine has already conducted a smaller study on metformin, the MILES (Metformin in Longevity Study)[1] study. In the MILES study, half of the subjects (aged 60 and older) were given 1,700 mg of metformin per day, and the other half a placebo. Specifically, they investigated whether metformin treatment would restore the gene expression profile of older adults with elevated blood sugar levels, aka impaired glucose tolerance (IGT), to that of young healthy subjects. They also measured the effect on insulin sensitivity and insulin secretion. This latter outcome is significant, as it has been demonstrated that centenarians (people who are 100 years old or older) compared with "younger aged" subjects have a preserved glucose tolerance and insulin action. However, the study has been criticized by some experts for its small sample size (16 individuals), and the wide-scale applicability of its results for this reason remain uncertain.

Additionally, the limited number of benefits of metformin on health demonstrated in humans comes with a number of known side-effects. Metformin is associated with a higher-than-average risk of vitamin B12 and vitamin B6 deficiency, which is thought to increase the risk of cognitive dysfunction. Metformin also revealed negative effects on cardiorespiratory fitness in some human studies. In one

---

[1] www.clinicaltrials.gov/ct2/show/results/NCT02432287?view=results

placebo-controlled, double-blind, crossover trial using young and healthy adults as study participants, researchers found that metformin caused a small but significant decline in maximal aerobic capacity[1]. Meanwhile, another double-blind, placebo-controlled trial[2], which investigated the dual effect of metformin and 12 weeks of aerobic exercise in older adults having at least one risk factor for developing type 2 diabetes, found that metformin actually decreased the predicted effect of exercise on the normalization of whole-body insulin sensitivity, and increased exercise-associated increases in skeletal muscle mitochondrial respiration."

Metformin also suffers from some issues relating to impurities. In December 2019, the U.S. FDA announced that it had learned that some metformin medicines manufactured outside the United States might contain a nitrosamine impurity called N-nitrosodimethylamine (NDMA), classified as a probable human carcinogen, at low levels. In 2020, some other impurities were found:

February 2020	FDA found **NDMA levels**[3] in some tested metformin samples that did not exceed the acceptable daily intake.
February 2020	**Health Canada**[4] announced a recall of Apotex immediate-release metformin.
March 2020	**Ranbaxy metformin Jamp metformin**[5].
June 2020	In June 2020, the FDA posted its laboratory results showing **NDMA amounts in metformin products**[6] it tested. It found NDMA in certain lots of ER Metformin, and it is recommending companies recall lots with levels of NDMA above the acceptable intake limit of 96 nanograms per day. The FDA is also collaborating with international regulators to share testing results for metformin.
July 2020	**Lupin Pharmaceuticals**[7] pulled all lots (batches) of metformin after discovering unacceptably high levels of NDMA in tested samples.
August 2020	Two lots of tablets were recalled by **Bayshore Pharmaceuticals**[8].

[1] www.karger.com/Article/FullText/502257
[2] www.karger.com/Article/FullText/502257
[3] www.fda.gov/drugs/drug-safety-and-availability/laboratory-tests-metformin
[4] www.healthycanadians.gc.ca/recall-alert-rappel-avis/hc-sc/2020/72281r-eng.php
[5] www.healthycanadians.gc.ca/recall-alert-rappel-avis/hc-sc/2020/72455r-eng.php/
[6] www.fda.gov/drugs/drug-safety-and-availability/laboratory-tests-metformin
[7] www.prnewswire.com/news-releases/lupin-pharmaceuticals-inc-issues-voluntarily-nationwide-recall-of-metformin-hydrochloride-extended-release-tablets-500mg-and-1000mg-due-to-the-detection-of--n-nitrosodimethylamine-ndma-impurity-301089754.html
[8] www.fda.gov/safety/recalls-market-withdrawals-safety-alerts/bayshore-pharmaceuticals-llc-issues-voluntary-nationwide-recall-metformin-hydrochloride-extended

A ***study***[1] published in 2018 has shown that metformin also has an immediate effect on the gut microbiome in humans. This effect is likely to result from an increase in the abundance of opportunistic pathogens and further triggers the occurrence of side effects associated with the observed dysbiosis. Dysbiosis itself can lead to various local gastroenteral (inflammatory bowel disease, irritable bowel syndrome) and systemic diseases (type II diabetes, obesity, atherosclerosis, non-alcoholic fatty liver disease (NAFLD)).

Another study published in the ***BMC Nephrology***[2] journal revealed concerns about metformin usage for patients aged 65+, an age group that suffers disproportionately from chronic kidney diseases (CKD). It is still unknown how, and to what extent, metformin affects kidney functions and the long-term outcomes of people with CKD, but there appears to be a correlation between metformin intake and CKD severity in older adults.

Author's Personal Views on Metformin

While metformin may be generally safe and effective in young, healthy individuals free from other health conditions, its safety, effectiveness and net-benefit in older individuals (especially those with comorbidities) remains uncertain. Yet, despite the lack of research about its safety, metformin is still a commonly prescribed anti-diabetes medication, even to people with CKD. But even this may need to change, and the application of metformin may need to be restricted to relatively young and healthy adults after preliminary doctor's examination. Older adults with dysfunctions or diseases may be risking their health by taking metformin as the anti-aging drug since possibility of severe side effects, though not explicit, remains a definite threat. As such, the safety of metformin for the kidney, its safety for one's overall health, and its overall convenience, are all open questions that still need to be studied thoroughly.

The consensus now seems to be that additional studies are still needed in order to attain efficient, clinically validated metformin that extends healthy lifespan. Granted, we know that metformin slows memory aging, extends nematode lifespan, reduces oxidative stress and inflammation, and extends lifespan and Healthspan in mouse models. But mice are not humans.

[1] www.pubmed.ncbi.nlm.nih.gov/30261008
[2] www.ncbi.nlm.nih.gov/pmc/articles/PMC4057526

The biggest concern is that the impact of metformin on Human Longevity has not been studied and tested on humans in enough depth, and those human studies which were carried out revealed a number of unforeseen side effects. Meanwhile, even if the very modest levels of Healthspan extension demonstrated in model organisms could be assumed to be replicable in humans, these same effects could have been achieved simply by lifestyle interventions (exercise, diet, etc.) without the accompanying potential side effects of metformin use.

However, I believe that the Barzilai TAME studies have used the right approach and, to some extent, set the right MVP standard for future research. Whereas many other studies mentioned in this chapter have used just a few biomarkers, the TAME trial plans on using at least nine (IL-6, TNFα-receptor I or II, CRP, GDF15, insulin, IGF1, cystatin C, NT-proBNP, and hemoglobin A1c). While we would recommend using something closer to 15 or 20 biomarkers, this represents a step in the right direction. Furthermore, the choice of these nine biomarkers was based on an analysis of 258 potential biomarkers of aging[1], analyzed and benchmarked according to (1) measurement reliability and feasibility, (2) relevance to aging, (3) robust and consistent ability to predict all-cause mortality, clinical and functional outcomes, and (4) responsiveness to intervention. This indicates that the choice of biomarkers involved fairly comprehensive consideration of the scope of biomarkers available for use.

Of even greater interest is an adjunct study that will be done in conjunction with the TAME trial, titled TAME BIO[2]. This study seeks to measure the effects of metformin during the TAME trial on a wide range of biomarkers of aging, with the goal of providing "a foundation for future biomarker discovery and validation[3]."

More specifically, TAME BIO will seek to:

- "Establish a repository of biospecimens obtained from TAME participants. This will create a critically needed resource that will store serum, plasma, blood, urine, stool, RNA, and DNA samples that will allow TAME scientists to conduct their trial. Importantly, this repository will also enable the broader scientific community

[1] www.ncbi.nlm.nih.gov/pmc/articles/PMC6294728/#MOESM1
[2] www.afar.org/tame-biomarkers-study
[3] www.afar.org/tame-biomarkers-study

to validate currently known age-related biomarkers and ultimately support the science necessary to identify new biomarkers;

- Compare different biomarker approaches. TAME BIO will assess three separate ways of characterizing biological age and their ability to predict changes in function, multi-morbidity, and subsequent death;
- Evaluate the effects of metformin on TAME trial participants. Using the specimens in the biorepository, TAME scientists will test how metformin has influenced a range of age-related biomarkers among trial participants and help to generate the core data for reporting on the trial's results."

The use of this type of approach – designing an adjunct biomarkers study in coordination with existing clinical trials, in order to use trials themselves as an additive tool to obtain real-world validation on the effectiveness different biomarkers of aging and Longevity – is something to be commended, and something which will hopefully be followed by a larger number of Longevity clinical trials in the future.

The general experience of metformin is complicated by mild side effects such as heartburn, stomach pain, nausea or vomiting, bloating, diarrhea, constipation, weight loss, headache, and abnormal changes to taste. Then there are the multiple concerns raised by "biohackers" who are routine users of metformin (as well as Longevity scientists, entrepreneurs and industry professionals, many of whom have been taking metformin for years), demonstrating a very large volume of evidence that while it may indeed slow down aging or even, theoretically, reverse it in some capacity, comes with many apparent issues as well. There is a range of evidence to suggest that the kind of long-term practical application necessary for it to have a beneficial effect comes with a large number of minor negative side effects, each of which may not be significant or detrimental to overall health but may nonetheless be enough to neutralize the main benefits of the drug. In other words, the cumulative risks and side effects caused by the drug may or may not outweigh its benefits in terms of Healthspan extension.

These risks and side effects can all largely be avoided, but only if the intake of metformin is modified and adapted around certain lifestyle components such as diet, physical exercises, and other factors. It may work, but is such an awkward situation worth the benefits? Is it mass-marketable? It remains an open question whether it can be applied on an industrial scale to entire populations to an extent that impacts the HALY and QALY of entire nations using the existing clinical trials framework. (Note also that such existing frameworks do not take into account individual variability of side effects and co-morbidities. They are also still very granular and not conducted in parallel on highly specific sub-populations according to sex, ethnicity, age, health status and presence of disease).

If these questions can be answered and these problems solved using the current approach of applying large sets of biomarkers at once, this has the potential to lead to a greater degree of industry standardization of particular sets of biomarkers and the general acceleration of research and development.

Rapamycin (Sirolimus)

Sirolimus or Rapamycin is another FDA-approved example of Longevity-defined drugs[1]. Rapamycin is a macrolide compound used to coat coronary stents, prevent organ transplant rejection and treat a rare lung disease called lymphangioleiomyomatosis. It has immunosuppressant functions in humans and is especially useful in preventing the rejection of kidney transplants. It inhibits the activation of T cells and B cells by reducing their sensitivity to interleukin-2 (IL-2) through mTOR inhibition. Sirolimus active compounds are produced by the bacterium Streptomyces hygroscopicus. It was first discovered by Surendra Nath Sehgal and colleagues in 1972. Earlier, in 1964, a Canadian scientific expedition traveled to Easter Island in order to gather plant and soil samples. After the extraction, soil samples were shared with the Microbiology team at Ayerst's research laboratories, where a new biochemical compound with anti-fungal properties[2] was isolated. Suren and his team quickly discovered that the compound also suppressed the immune system. After the test in the National Cancer Institute, it was discovered that this compound is also effective for solid tumor treatment. Unfortunately, due to corporate priorities, management did not want to pursue drugs based on the compound. In 1999, the FDA approved sirolimus under the trade name Rapamune by Wyeth. The pharmaceutical assets of Wyeth, a well-known OTC drugs man-

[1] www.doi.org/10.1016/S1359-6446(05)03395-7
[2] www.sehgal.net/surenshistory.htm

ufacturer, were then acquired by Pfizer in 2009. Since 2012, other Wyeth assets have been purchased by Nestlé.

Sirolimus is used for transplant rejection prevention, Lymphangioleiomyomatosis and Vascular malformations treatment, and other conditions. Sirolimus[1], like metformin, has shown some positive effects in cancer treatment, enhancing the immune response to tumor targeting or promoting tumor regression[2] in clinical trials. As was mentioned before, sirolimus is related to the mTOR inhibition. The mTORC1[3] was first shown to be important in aging in 2003 and then not needed the additional study in 2009. Another study published in 2020 found late-life rapamycin dosing schedules enhanced mouse lifespan in a sex-specific manner where limited rapamycin exposure enhanced male lifespan but not female[4], providing evidence for sex differences in rapamycin response.

Mikhail Blagosklonny is one of the scientists who made a significant contribution to the contemporary aging perception of rapamycin. Blagosklonny is an adjunct faculty member at Roswell Park Comprehensive Cancer Center in Buffalo, New York. He has formulated a TOR signaling hypothesis. According to this hypothesis, despite the functional association of aging with cellular performance decline and the fact that rapamycin is an inhibitor of cellular functions, cell growth, and protein synthesis, it seems to work against aging. This is because aging is not just cellular performance decline but a hyper-activation of cellular signaling pathways and cellular functions, and rapamycin may prevent or slow down such processes.

Author's Personal Views on Rapamycin

In comparison to metformin, sirolimus (rapamycin) seems to avoid any substantial controversies, having a "classical" list of adverse effects, whose presence is conditioned by the chemical structure of our bodies. However, its biggest Longevity controversy lies in the fact that most studies indicating the drug's Longevity-related effectiveness were conducted on mice or other animals rather than humans. This asymmetry has arisen due to the complications – both financial and regulatory – that come with human tests and trials.

[1] www.doi.org/10.4049/jimmunol.1103365
[2] www.doi.org/10.1038%2Fsj.onc.1209886
[3] www.doi.org/10.1093/gerona/glw090
[4] www.nmn.com/news/late-life-rapamycin-regimens-extend-mouse-lifespan-sex-specific-manner

In a similar way to metformin, the clinical use of rapamycin is limited by concerns of side effects associated with the drug. Oncotarget[1] magazine published an article written by Simon C. Johnson[2] and Matt Kaeberlein[3] where the side effects of rapamycin were a significant concern that could limit the drug's application in clinical geroscience. Patients taking rapamycin to prevent organ transplant rejection have experienced adverse effects, including stomatitis, thrombocytopenia, high serum triglycerides and cholesterol, and impaired wound healing.

Doses used to extend lifespan in mice are generally free from these side effects; however, lifespan-prolonging doses of rapamycin may result in a different set of adverse effects in mice, including altered glucose homeostasis (by glucose tolerance test), gonadal atrophy, and increased incidence of cataracts. This is further complicated by the supposition that subtherapeutic doses are more likely to confer beneficial effects in young, healthy individuals, and that there is some evidence to suggest that it will likely be necessary to use more significant therapeutic doses in older individuals to achieve its Healthspan-extending effects.

Moving forward, we believe an emphasis should be placed on understanding how to achieve maximal efficacy from rapamycin treatment while minimizing its side effects. In this context, limited evidence-based consideration of dosing and duration of treatment represents a significant limitation in most preclinical studies.

Meanwhile, rapamycin, much like metformin, has a long history of unofficial use among biohackers and many Longevity scientists, entrepreneurs, and industry professionals, many of whom have reported evidence of personal side effects. While these have not been systematically studied due to it being taken in an unofficial manner, I have personally heard enough direct stories to conclude that rapamycin faces many of the same general concerns as metformin – i.e., that while the number and severity of side effects in young people is likely small enough to be tolerated, its Healthspan-extending effects among the elderly (and those with existing health conditions) may be neutralized by a prevalence of undesired side effects. In addition, neutralizing those side effects would require specific adjustments to

[1] www.ncbi.nlm.nih.gov/pmc/articles/PMC5216691
[2] www.scholar.google.com/citations?user=TT9aI10AAAAJ&hl=en
[3] www.dlmp.uw.edu/faculty/kaeberlein

exercise, diet, and other lifestyle factors in ways that the current one-size-fits-all framework of clinical trials will prove unable to identify in a precise and tangible way.

NMN

Some additional accomplishments could also be traced in nucleotides' research; namely, the aging relatedness of NMN (Nicotinamide Mononucleotide, a precursor to NAD+, Nicotinamide Adenine Dinucleotide) seems to be found. The author of the corresponding results is David A. Sinclair[1], Ph.D., A.O. who is a Professor in the Department of Genetics and co-Director of the Paul F. Glenn Center for Biology of Aging Research at Harvard Medical School. He obtained his Ph.D. in Molecular Genetics at the University of New South Wales, Sydney, in 1995. He worked as a postdoctoral researcher at M.I.T. with Dr. Leonard Guarente, where he co-discovered a cause of aging for yeast as well as the role of Sir2 in epigenetic changes driven by genome instability.

His research has been primarily focused on the sirtuins, protein-modifying enzymes that respond to changing NAD+ levels and to caloric restriction (CR) with associated interests in chromatin, energy metabolism, mitochondria, learning, and memory, neurodegeneration, and cancer. The Sinclair lab was the first one to identify a role for NAD+ biosynthesis in the regulation of lifespan and first showed that sirtuins are involved in CR in mammals. Dr. Sinclair is co-founder of several biotechnology companies (Life Biosciences, Sirtris, Ovascience, Genocea, Cohbar, MetroBiotech, ArcBio, Liberty Biosecurity) and is on the boards of several others.

GSK paid $720 million to acquire Sirtris in April 2008, to get a hold of technology that generated lots of breathless media coverage as a modern-day fountain of youth. The company sought to make drugs that act on sirtuins, a class of proteins that scientists believe play a role in aging, programmed cell death, and other key cell processes. Five years later the company was closed[2] due to the absence of validated results.

NAD+ (of which NMN is a precursor) is a molecule that acts as a coenzyme in many critical processes in our body, including cellular

---

[1] www.sinclair.hms.harvard.edu/people/david-sinclair

[2] www.xconomy.com/boston/2013/03/12/glaxosmithkline-shuts-down-sirtris-five-years-after-720m-buyout/

energy production, DNA repair, and sirtuin activity (enzymes involved in aging). NAD+ levels are known to decline with age — a process that scientists now consider a hallmark of aging. Lower NAD+ levels are also responsible for many age-related conditions, including cardiovascular disease, neurodegenerative diseases, and cancer. Unfortunately, we can't simply boost NAD+ levels by taking NAD+ supplements, because NAD+ in supplemental form has very poor bioavailability; i.e. it won't have a significant effect. However, NAD+ has several precursors or intermediates — molecules that transform into NAD+ through enzymatic reactions.

Several studies[1] indicate that supplementing with precursors[2] like nicotinamide riboside (NR) and nicotinamide mononucleotide (NMN) can increase NAD+ levels[3] and prolong the lifespan of yeast, worms, and mice.

In addition, other animal studies[4] show that raising NAD+ levels may also improve muscle mass, cardiovascular function, and glucose metabolism. But it's important to remember that the results of animal studies do not necessarily apply to humans.

On the other hand, nicotinamide mononucleotide (NMN) research is still in its infancy and requires much more additional investigation concerning how it's addressing the problem of aging. Fortunately, NMN has been gaining a lot of attention lately, especially with the release of David Sinclair's book Lifespan: Why We Age—and Why We Don't Have To[5]. In his book, the scientist discusses his NMN research, especially in mice. He made it clear that he has no opinion on whether anyone else should take the supplement. Phase I of this study was simply evaluating the safety of NMN supplements and therefore NAD+ levels were not even measured. The authors concluded that up to 500 mg of NMN is safe for healthy men. For Phase II, it is reported that researchers will examine the efficacy of NMN[6] as well as the appropriate dosage and frequency. Other ongoing human trials will investigate the safety and effects of NMN in the elderly, and the effect of NMN on cardiovascular and metabolic functions.

---

1 www.doi.org/10.1016/j.cmet.2017.11.002
2 www.doi.org/10.1126/science.aaf2693
3 www.doi.org/10.1111/acel.12461
4 www.doi.org/10.1016/j.cmet.2011.08.014
5 www.amazon.com/Lifespan-Why-Age_and-Dont-Have/dp/1501191977
6 www.blog.insidetracker.com/nr-nmn-longevity-supplements-work-humans

While NMN is expected to be mostly safe, some trials have recorded side effects[1] such as headaches, nausea, diarrhea, and skin flushing, while a 2019 study showed a possible link between elevated levels of NAD and tumor growth in isolated cells and animals[2]. The study, published in the journal Nature Cell Biology, did not find that NAD causes cancer, but elevated levels appeared to accelerate the formation of tumors already emerging.

But above all, concerns remain about whether this animal research will translate to humans. The majority of data we currently have comes from animal model studies, and, while a limited number of trials have demonstrated beneficial health effects on humans, the majority of human studies appear to be too short in duration and include too few study participants[3] to draw definitive conclusions. As such, for a more accurate idea of its safety and efficacy, more robust human studies are needed.

The current state of the science could be summed up by a recent statement by Felipe Sierra[4], the Director of the Aging Biology Division at the National Institute of Aging at NIH: "The bottom line is I don't try any of these things. Why don't I? Because I'm not a mouse."

Author's Personal Views on NMN

It is complicated to study NMN as both its side effects and its Longevity benefits are difficult to determine. High NMN dosages are needed achieve perceptible beneficial effects worth studying and to obtain enough data for tangible analysis, but a recent study has shown that the high doses of Nicotinamide (NAM) – another NAD+ precursor – can cause lethal side-effects for animals[5]. On the other hand, no adverse side effects of the smaller therapeutic dosages have yet appeared. Although headaches, dizziness, and vomiting occurred in healthy humans when NAM at doses up to 6 grammes was ingested on an empty stomach, these adverse effects were minor and resolved upon termination. However, NAM administration above certain levels has shown adverse effects in

[1] www.healthline.com/nutrition/nicotinamide-riboside#risks-and-side-effects
[2] www.nextavenue.org/do-longevity-supplements-work-and-are-they-safe
[3] www.healthline.com/nutrition/nicotinamide-riboside#dosage
[4] www.khn.org/news/a-fountain-of-youth-pill-sure-if-youre-a-mouse
[5] www.ncbi.nlm.nih.gov/pmc/articles/PMC7277745/#:~:text=Animal%20studies%20have%20shown%20the,respectively%2C%20when%20extrapolated%20to%20humans

animals. Moreover, the biochemical nature of NAM raises concerns about the potential adverse effects of administration. Both NAM and its metabolites need to be analyzed in depth for their properties and mechanisms of action. All this leads to the supposition that the larger doses required to replicate its beneficial effects in model organisms in actual human patients may lead to undesired (and as yet unknown) complications.

Unity Biotechnology[1]
(NASDAQ: UBX)

At the same time, other companies are creating a whole class of aging-related drugs[2]. For example, UNITY Biotechnology is the largest of the handful of BioTech companies working on senolytics, therapies capable of selectively destroying a sizable fraction of the senescent cells that accumulate in old tissues.

The company's mission[3] is to develop therapeutics to extend the period of one's life unburdened by disease. The initial focus is on delivering localized therapy in ophthalmologic diseases:

- Diabetic macular edema. Sometimes, tiny bulges (microaneurysms) protrude from the vessel walls, leaking or oozing fluid and blood into the retina. This fluid can cause swelling (edema) in the central part of the retina (macula). This is a serious eye complication, known as diabetic macular edema, that can cause vision problems or blindness;
- Age-related macular degeneration;
- Age-related neurologic conditions.

Unity Biotechnology failed its Phase II trial of localized senolytics for knee osteoarthritis (UBX0101, a p53/MDM2 interaction inhibitor) in August 2020. When Unity announced[4] that its lead drug candidate had failed to beat a placebo in reducing joint pain and stiffness, the com-

[1] www.unitybiotechnology.com/about-us
[2] www.fightaging.org/archives/2020/08/unity-biotechnology-fails-phase-ii-trial-of-localized-senolytics-for-knee-osteoarthritis
[3] www.unitybiotechnology.com
[4] www.fightaging.org/archives/2020/08/unity-biotechnology-fails-phase-ii-trial-of-localized-senolytics-for-knee-osteoarthritis

pany's stock plunged more than 60%, nearly one-third of Unity staff were laid off, and its experimental drug UBX0101 — the first novel "senolytic" agent ever to enter clinical testing — was swiftly abandoned.

That setback threw a pall over the entire senolytic field. A research note from Citi analyst Yigal Nochomovitz[1] cited "a substantially greater risk to the senolytic hypothesis." However, other experts were more positive about a class effect, highlighting issues with trial design, study population, and UBX0101 itself — a small-molecule inhibitor of MDM2 (murine double minute 2) designed to boost the activity of the pro-apoptotic p53 protein — as possible explanations for study failure.

UBX1325, another UNITY product, and a potent Bcl-xL inhibitor, is currently being evaluated for the treatment of age-related diseases of the eye – including diabetic macular edema, diabetic retinopathy, and age-related macular degeneration. A Phase I study is exploring the safety and tolerability of UBX1325 in patients with diabetic macular edema or age-related macular degeneration. The initial assessment of patients in this study did not reveal any dose-limiting toxicities or treatment-emergent adverse events, supporting the initiation of a Phase IIa proof-of-concept study in DME to evaluate the safety and efficacy of UBX1325. Initial data from this Phase IIa study is expected in the first half of 2022. A Phase I study[2] is currently evaluating a single dose of UBX1325 in patients with age-related macular degeneration. A Phase IIa study[3] is evaluating the safety and efficacy of UBX1325 in patients with diabetic macular edema.

Author's Personal Views on Senolytics

Unfortunately, from what we can already discern, senolytic drugs have some severe toxicity and can potentially make things worse.

Senolytics are a field of Longevity therapeutics that has a high likelihood of conferring certain Healthspan-extending effects in specific demographics that can withstand its potential side effects (i.e., young non-immunocompromised individuals), thus serving as an important sector in the overall Longevity Industry. However, their overall levels of media attention and investment prioritization with-

[1] www.doi.org/10.1038/s41587-020-00750-1
[2] www.clinicaltrials.gov/ct2/show/NCT04537884
[3] www.clinicaltrials.gov/ct2/show/NCT04857996

in the industry appear to outweigh the tangible evidence to date of its real-world potential.

In short, senolytics are likely to confer Healthspan benefits in younger people who have enough non-senescent cells to replace the ones cleared away by senolytics. Therefore, the idea of anti-aging therapies that purge the body of dying cells is gaining traction with a raft of start-ups now focused on senescence. Senolyitcs have already been demonstrated many times to have had positive results in mice. And in 2019 a small-scale human trial by Mayo Clinic researchers has shown for the first time that harmful senescent cells that accumulate with age can be removed from the human body by senolytic drugs.

However, when the immune system's functionality is deteriorating, the patient's body cannot produce enough healthy cells to replenish them. Furthermore, older people can experience autoimmune reactions and various side effects that will counteract the Longevity-extending effect of senolytics. All of this raises significant concerns over the effectiveness of senolytics in the elderly, many of whom have some level of immune dysfunction.

According to research published in Fight Aging[1], the senolytic navitoclax, for example, has been shown to have potentially harmful effects on the skeletal-lineage cells of model organisms (mice). Dr. Michael Fossel, in his interview[2] with the life extension science monitoring site Lifespan.io[3], describes the way senolytic drugs work and the fact that senolytics usage can increase the number of senescent cells. In addition, a recent article in The Scientist[4] also reported concerns about the possible adverse consequences of removing senescent cells in the long run. This question needs to be studied further.

The field of senolytics has in recent years been subject to a large degree of media hype and, in my view, a higher degree of interest and prioritization among investors than is warranted by its actual results. In other words, it can be considered as one of the most "over-hyped" sectors of geroscience.

---

[1] www.fightaging.org/archives/2020/06/an-example-to-illustrate-why-navitoclax-is-not-a-suitable-clinical-senolytic-drug

[2] www.lifespan.io/news/are-senolytics-potentially-damaging-for-Longevity

[3] www.lifespan.io

[4] www.the-scientist.com/features/can-destroying-senescent-cells-treat-age-related-disease--67136

We must not allow ourselves to be misled by "group thinking", as many times before some particular technologies have been overly promoted. The most well-known example being "telomerase extension", which just 3-4 years ago was considered as the most hottest area in the aging research field, whereas now many experts are becoming sceptical about whether and when telomerase extension could be used for practical life human extension. The same can happen with the field of senolytics.

It may take many years of hard work and research to truly understand and eliminate or mitigate the possible side effects of senolytics, especially for older people with immune dysfunction, and to transform it into a widely applicable anti-aging drug for mass market.

AgeX Therapeutics[1]
(NYSE: AGE)

Stem cell-derived solutions are probably among the most well-known aging-related technologies. Among a lot of companies using stem cells in their research and therapeutic approaches, AgeX Therapeutics is of great interest to us. It is a biotechnology public company founded in 2017 by Dr. Michael West. Dr. Aubrey de Grey served as the company's Vice President of New Technology Discovery, and the main therapeutic pipeline of the company follows the approach of de Grey's earlier project, the SENS Research Foundation, which is committed exclusively on the repair of aged body tissues and the "repair approach" to aging. AgeX focuses on developing and commercializing innovative therapeutics to treat human diseases to increase Healthspan and combat the effects of aging through its PureStem and induced Tissue Regeneration (iTR) technologies, as well as through its medicines designed to address some of the unsolved problems in aging.

PureStem is a platform for creating pluripotent stem cell-derived cellular (PSC) therapies for use in repairing tissues in the aged human body afflicted with degenerative disease. According to AgeX, PureStem technology is based on the observation that in our early development, tissues in the human body are naturally composed of highly proliferative cells with relatively long cell lifespans. AgeX said that it is possible to generate single cell-derived (clonal) lineages of these cells in the laboratory, a process not easily achievable with cells derived from adult tissues. Pluripotent Stem Cells (PSC) provide the following advantages over adult-derived cells:

[1] www.agexinc.com

- The replicative immortality of PSCs facilitates the indefinite scale-up of PSC master cell banks for the manufacture of the uniform product, as well as an immortal substrate for targeted genetic modifications;
- PureStem technology allows for the clonal expansion of hundreds of purified, identified, and reproducibly scalable cell types that retain regenerative potential (i.e., they have not passed the regeneration limit);
- Since most PSCs maintain long and stable telomere lengths, the replicative capacity of derived differentiated cell types is typically longer.

On the basis of Purstem, the company has two therapeutic candidates – AGEX-VASC1 and AGEX-BAT1 – which are only on to a pre-clinical stage of trials.

Induced Tissue Regeneration (iTR) is another, more innovative area of work of the AgeX research team. The theory behind iTR is that aging and diseases of old age are due to the replicative immortality and inherent regenerative capacity of embryonic cells that are lost with time.

In 2017, AgeX published research[1] showing that the activation and deactivation of certain genes may affect regenerative potential in humans. As a result, based on iTR technology, the AGEX-iTR1547 formulation, which is in the preclinical stage, was developed and patented. The drug showed the possibility of reducing the expression of the marker gene COX7A1, and AgeX is currently performing research to optimize a potential pharmaceutical preparation to reverse the aging of human tissues and thereby restore regenerative potential. When implemented in vivo, this partial reprogramming, or iTR, would be expected to induce tissue regeneration and, when combined with telomerase, could modulate both cellular immortality and regenerative biology for therapeutic effect.

Author's Personal Views on Epigenetic Therapies

In comparison with other therapeutic modalities, epigenetic therapies (which seek to modify gene expression) are still young, and the full

[1] www.doi.org/10.18632/oncotarget.23748

scope of their potential side effects has not yet been robustly identified. iTR involves epigenetic reprogramming relating to genes that confer regenerative potential, and there are complex associations between general regenerative capacity and the regenerative capacity of cancers, while genes, and patterns of gene expression, often have multiple roles within the body. All this leads to the conclusion that the true safety and efficacy of this approach in humans remains uncertain.

Celularity[1]
(NASDAQ: CELU)

Another stem cell-related company is Celularity[2], a clinical-stage biotechnology company founded by Dr. Robert J Hariri, PhD. in May 2016. Dr. Hariri pioneered the use of stem cells to treat a range of life-threatening human diseases. Today, he continues to make transformative contributions in the fields of cell therapeutics and functional regeneration. He is best known as the discoverer of pluripotent stem cells derived from the human placenta and as a member of the team that discovered the physiological activities of tumor necrosis factor (TNF).

The company takes an innovative approach[3] through the use of postpartum placental cells for allogeneic cell therapy. The company's founder, Dr. Hariri, calls the placenta a natural 3D printer. Instead of receiving only a tiny number of stem cells from embryos or abortion material, the placenta gives us thousands to start with. It also allows for huge reserves for further production, as 130 million placentas are disposed of worldwide every year. Placental stem cells are a fundamental part of the ability to respond to illness and disease and, most importantly, respond to the effects of aging.

The company's products can be used without the need to match the patient to the donor, because the placenta normally has a low potential for provoking an immune response. Thanks to advanced technologies, a single placenta is capable of yielding 100–100,000+ doses, depending on the therapeutic, transforming plentiful and normally discarded source material into therapeutics. The company will also be able to quickly scale the therapy due to the large stock of material and make it cost-effective.

---

[1] www.celularity.com
[2] www.celularity.com/company/robert-j-hariri
[3] www.longevity.technology/does-the-placenta-hold-the-key-to-human-Longevity

Currently, the company's pipeline consists of five drugs, two of which are undergoing Phase I clinical trials and three of which are awaiting confirmation from FDE Investigational New Drug in 2021.

- **CYNK-101** is Celularity's lead therapeutic candidate based on its placental-derived genetically modified NK cells. CYNK-101 is an allogeneic, cryopreserved, off-the-shelf investigational cell therapy product with potential applications in hematological cancers and solid tumors as well as infectious diseases. The Phase I/IIa clinical trial will commence in 2021.
- **CyCART-19** is a placental-derived T-cell therapy that is in development initially for the treatment of B-cell malignancies, targeting the CD19 receptor. Placental-derived T-cells are mostly naive and expand readily ex vivo. They express markers of stem cell memory and have lower expression of effector or exhaustion markers, allowing for greater proliferative potential in vivo. Celularity plans to file its initial CyCART-19 IND application and to commence a Phase I/IIa clinical trial in the first half of 2021.
- **CYNK-001** is a non-genetically modified NK cell therapy product being developed from placental hematopoietic stem cells as a potential treatment option for various hematologic cancers, solid tumors, and infectious diseases. CYNK-001 is in development for the treatment of acute myeloid leukemia ("AML"), a blood cancer; for glioblastoma multiforme ("GBM"), a solid tumor cancer; and also for COVID-19. CYNK-001 is currently in a Phase I trial for AML and other blood cancers. It is in Phase I/IIa clinical trials for both GBM and COVID-19.
- **APPL-001** is a placental-derived mesenchymal-like ***ASC***[1] being developed for the treatment of degenerative diseases, including Crohn's disease. Celularity plans to submit an IND in the second half of 2021 and to commence a Phase I/IIa study of APPL-001 for the treatment of Crohn's disease in 2022.

Author's Personal Views on Stem Cells

Despite the enormous potential of stem cells in fighting age-related disorders and advancing Longevity, further research is needed to fully understand the mechanisms of underlying stem cell biology

[1] www.bioinformant.com/celularity-gx-acquisition

and how their regenerative capacity can be enhanced to promote a healthier and longer life before regenerative therapies can enter the Longevity marketplace. Stem cell treatment is considered efficient for the purposes of rejuvenation and Longevity. The FDA approves this method. However, there is a market of unapproved stem cell treatment methods that can lead to possible side effects such as the ability of cells to move from placement sites and change into inappropriate cell types or multiply, the failure of cells to work as expected, and the growth of tumors.

However in general, I would say that in comparison to drug discovery approaches the stem cells area, despite some challenges and concerns, might have more potential to deliver practical applications for Human Longevity in the foreseeable future.

Deep Pharma Intelligence[1], the BioTech focused analytical subsidiary of Deep Knowledge Group, has recently released an analytical project, Cell Therapies In Healthcare Landscape Overview 2021[2], where you can get a better understanding of the scope of this sector, and insights into recent developments and trends.

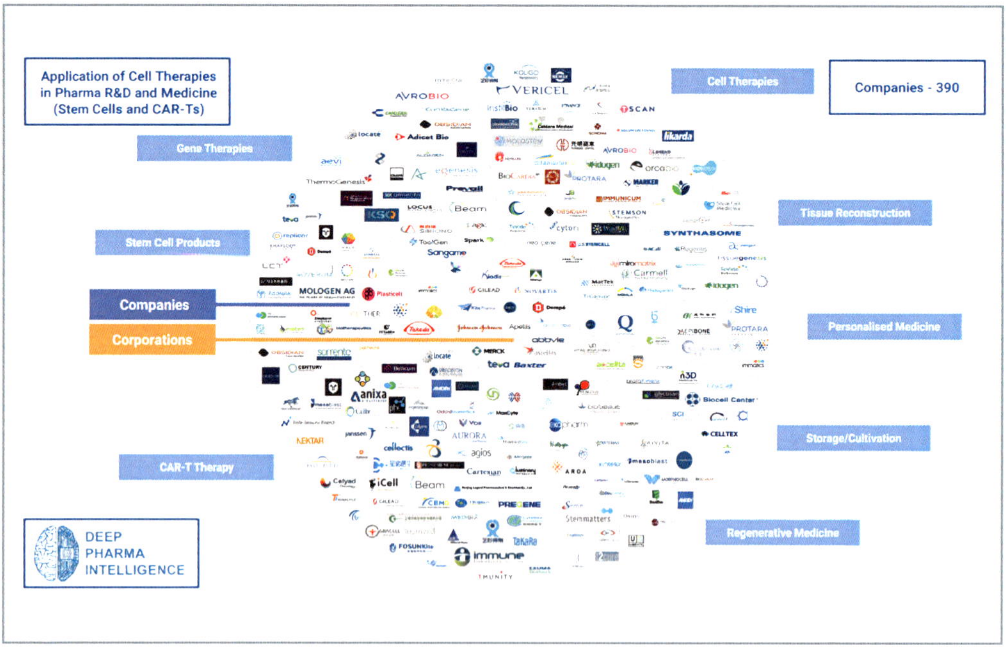

Source: www.deep-pharma.tech/cell-therapies

[1] www.deep-pharma.tech
[2] www.deep-pharma.tech/cell-therapies

The Buck Institute for Research on Aging

The Buck Institute for Research on Aging is a legendary name in the Longevity Industry. It is the most well-funded non-governmental aging research foundation in the US, and describes itself as "the world's only research institution singularly focused on the biology of aging, yielding insights into age-related diseases before they start."

Indeed, the institute can in many respects be considered the North American epicenter of Longevity science, contains many of the best and most well-known scientists and specialists in the entire industry, and houses a great many relevant technologies under one roof. It also has quite substantial budgets driving its activities, exceeding any analogous scientific hubs and epicenters in other regions of the world.

When I was given a tour of the Buck Institute by its current CEO Eric Verdin in 2017, I distinctly remember the clamor and excitement around a new clinic they were building. I thought this was a great idea. However, as the years went by, I expected to hear a steady stream of progress news about in-house clinical research, and I thought that this new clinic would become a powerhouse of Phase III trials. Unfortunately, however, we are not aware of many activities relating directly to Practical Human Longevity at the institute.

It has been home to a substantial number of success stories relating to mice and other model organisms, but we have yet to see a significant proportion of the high-caliber science and R&D that has become a hallmark of the institute actually make the transition into clinical success stories in humans. It is entirely possible that this will change in the years to come (and, indeed, we would be very happy to find this to be the case), and the institute certainly has both the budget and the expertise to make a shift toward a heavier prioritization of human-centric R&D and clinical translation.

We can also see that the burden of clinical research on Buck projects has been taken up by spinouts such as Unity Biotechnology (discussed earlier in this chapter), as well as earlier-stage spinouts including Napa Therapeutics[1] and Gerostate Alpha[2], and this may in part explain the absence of a large number of human-focused clinical success stories coming from the Buck; nonetheless, we would encourage the

[1] www.crunchbase.com/organization/napa-therapeutics
[2] www.gerostatealpha.com/

institute (as we do for all Longevity R&D hubs) to proactively transition to a more human-centric approach, given our concerns regarding the probable success rate of translating positive model organism results into similarly positive clinical results in humans.

Hyperbaric chambers

All the above cases largely relate to drug discovery, and a few to regenerative medicine and stem cell therapies, but it is apparent that there are many other domains, applications, and approaches to Healthspan extension beyond these specific therapeutic modalities, and hyperbaric chambers is one of them.

What all these drugs and stem cell approaches have in common is not that they fail or are destined to fail, but that whatever results they do produce, cannot produce sufficiently strong clinical validation in humans without a broader range of biomarkers. Here we will provide one very notable example of an alternative therapeutic modality (hyperbaric chambers) which, in my view, does have strong clinical evidence. However, there has been a failure to document that evidence in a sufficiently comprehensive and convincing manner due to a lack of consensus frameworks for Biomarkers of Human Longevity.

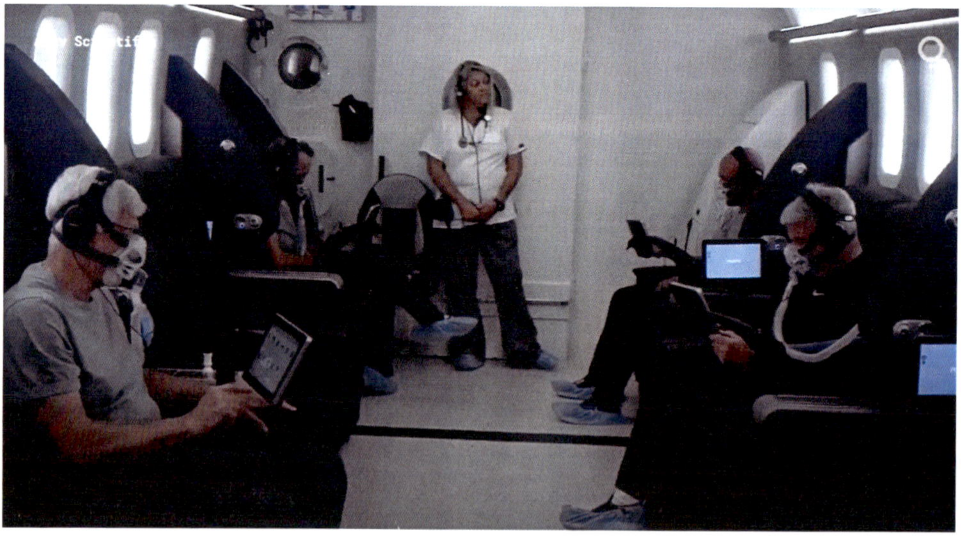

Source: Aviv Clinics[1]

[1] www.aviv-clinics.com

One of the most recent and well-known people in this field is Dr. Shai Efrati (cofounder of Aviv Clinics[1]), who has been developing oxygen therapy programs since 2017 and offering hyperbaric sessions in their high-end clinics in Israel, Florida and Dubai.

Source: Aviv Clinics[1]

Source: ZoomerMedia[2]

In 2020 Efrati recruited 63 people aged 65+, took MRI scans, and tested their cognitive abilities. He then gave some of them a 60-day

[1] www.aviv-clinics.com
[2] www.youtube.com/watch?v=PgSM4TQbKis

course of treatment during which they spent 2-hour stints in a pressurized chamber five times a week, breathing pure oxygen for some of that time. Efrati reported a significant difference in those who received the oxygen therapy – and in tests six months later found the improvements had remained. The MRI scans showed that tissue function in the brain, in areas associated with cognitive decline, had improved. Efrati said that this protocol generates significant fluctuations in blood oxygen, and suggested that the intermittent increase of oxygen concentration induces many of the mediators and cellular mechanisms needed for regeneration. The author decided to try this protocol on healthy adults because, as people age, blood vessels carry less oxygen to the brain, and he thinks that hyperbaric treatment compensates for this. Efrati states that "the occlusion of small blood vessels, similar to the occlusions which may develop in the pipes of an 'aging' home, is a dominant element in the human aging process. This led us to speculate that hyperbaric oxygen therapy may affect brain performance of the aging population."

I am confident that tangible positive results were actually achieved in this clinical study, for reasons elaborated on later in this section. However, this therapeutic approach in general is often met with a large volume of criticisms that we feel are due to the manner in which practitioners and scientists (Aviv Clinic included) have documented their results, and in particular the overall number of biomarkers used to demonstrate their impact on human aging.

For example, O'Rourke, professor at Ben Gurion University's School of Public Health, points out that the sample size isn't huge, although for a first study that's not necessarily problematic. Hillel Aviezer, a professor of psychology at the Hebrew University of Jerusalem, states that there is still room for caution in interpreting the results and more tightly controlled double-blind experimental designs are needed. At the same time, the US Food and Drug Administration[1] warns that the use of the chambers can carry risks of joint pain and even paralysis. It raised concerns that hyperbaric therapy has been "touted" as "universal treatment."

This lack of universal applicability may be linked to a lack of precision. Ilia Stambler, chief scientific officer of Vetek (Seniority) – The Israel Movement for Longevity and Quality of Life, cautions that "the

[1] www.timesofisrael.com/israeli-doctor-claims-to-reverse-aging-with-pure-oxygen-treatment-study

long term effects yet need to be considered" and emphasizes that proper dosage is vital to prevent oxygen poisoning and ensure efficacy and safety. "With the right dosage, you can have some great effects, but not the same dose is appropriate for all, and the elderly may experience more adverse effects of oxidative damage."

Aviv Scientific was the subject of substantial media coverage surrounding the results of clinical study[1] in late 2020, which purported to demonstrate "age-reversing" effects as a result of hyperbaric oxygen therapy. The authors of the study were able to demonstrate telomere elongation in circulating immune cells and a reduction in the number of circulating immunosenescent cells:

"Results: Telomeres length of T helper, T cytotoxic, natural killer and B cells increased significantly by over 20% following HBOT. The most significant change was noticed in B cells which increased at the 30th session, 60th session and post HBOT by 25.68%±40.42 (p=0.007), 29.39%±23.39 (p=0.0001) and 37.63%±52.73 (p=0.007), respectively. There was a significant decrease in the number of senescent T helpers by -37.30%±33.04 post-HBOT (P<0.0001). T-cytotoxic senescent cell percentages decreased significantly by -10.96%±12.59 (p=0.0004) post-HBOT. In conclusion, the study indicates that HBOT may induce significant senolytic effects including significantly increasing telomere length and clearance of senescent cells in the aging populations."

However the media coverage of the study has been accused of overhyping it by many experts in aging research. A closer look at the data in context reveals that, although hyperbaric oxygen therapy did show an increase of ~20% in average telomere length in certain immune cells, the T cells being examined have highly dynamic telomeres, and they can fluctuate considerably, depending on their situation, something that can happen anyway as a result of regular exercise. In other words, the study is less revelatory than it first appears.

On the other hand, during a discussion on this study in an April 2021 webinar hosted by Aging Analytics Agency[2] which I took part in, alongside gerontologist and integrative medicine physician Dr. Arkadi Prokopov, it quickly became evident to me that much of the criticism surrounding the study is the result of measuring just a few biomarkers (i.e., telomere length and the number of circulating

[1] www.aging-us.com/article/202188/text
[2] www.youtu.be/N58v_Wo8wlk

immunosenescent cells). It became clear that **a more diverse and comprehensive set of Biomarkers of Longevity could have produced a more revelatory and evident study.**

Additionally, I believe that in the coming years we will begin to see the use of more and more biomarkers to validate scientific results, and this trend will become increasingly common, and further supplemented via the use of "scientific auditing firms" specializing in biomarkers (and ideally with a focus on the use of AI and data science for precise biomarker analysis). These firms will serve as independent third parties to validate the effects of a given protocol or treatment on a very wide range of biomarkers.

Author's Personal Comments on Hyperbaric Chambers

Despite concerns surrounding hyperbaric oxygen therapy in terms of its actual impact on Longevity, and in contrast to the full scope of other therapeutic modalities described in this chapter, ***this is one particular domain I believe has significant potential for both Practical Human Longevity and performance enhancement. I am 100% certain that the technologies in this domain are working at least in some individuals, because I have seen its potential demonstrated in my own body and on my own biomarkers and many of my acquaintances.***

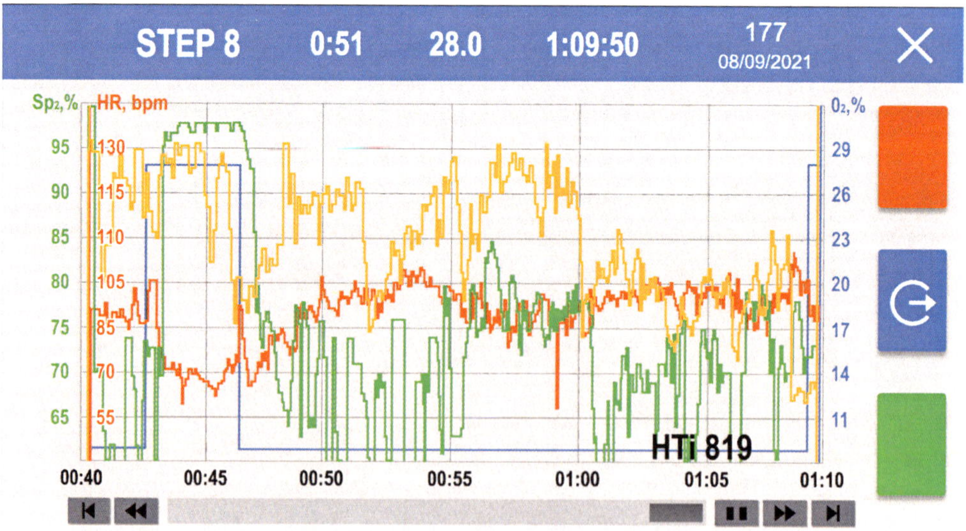

Source: Practical Longevity[1]

[1] www.longevity-book.com/practical-Longevity

For a number of months over the course of 2020-2021, I have been undergoing briefing sessions using a specific and specialized method known as Intermittent Hypoxic-Hyperoxic Intensive Training (IHHIT), and I have seen not only positive changes in my own Longevity Biomarkers, but a significant increase in my physical and cognitive performance and endurance.

Note: The figure above depicts my personal maximum result on Intermittent Hypoxic-Hyperoxic Intensive Training. During a 70 minute session, I achieved an HTI level (***Hypoxia Training Index***[1]) of 819, which may be considered as one of the highest results in the world. Approximately 50% of the training session time was in hibernation mode, whereas during approximately 20 minutes my blood saturation went down to the level of 50-60%.

While my highest result to date was an HTI of 819, typically I am doing intensive training sessions in the range of 300-500 HTI, which is a standard level for professional alpinists, high-performance athletes and marathon runners, and in a more relaxed daily mode I am doing training sessions in the range of 170-250 HTI.

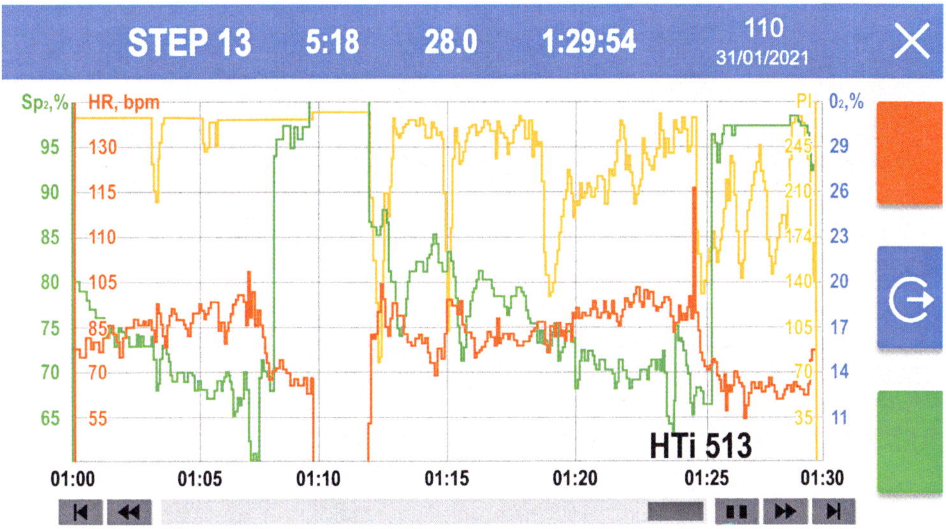

You can read more on this topic, as well as other specific methods I am using to extend my life in my new book Practical Longevity[2].

While the IHHT method utilizes a specific protocol (which enables it to work more safely and effectively than other protocols within

[1] www.en.wikipedia.org/wiki/Hypoxic_Training_Index
[2] www.longevity-book.com/practical-Longevity

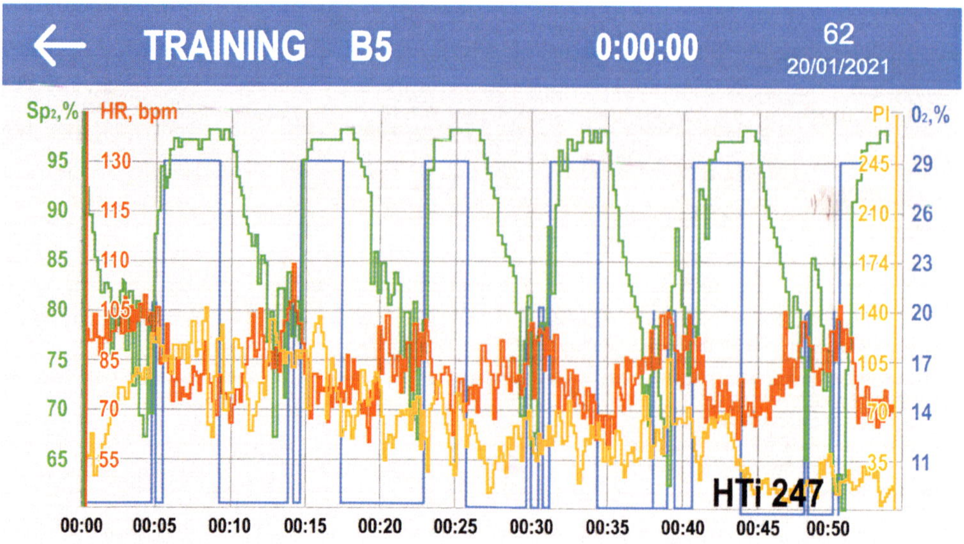

this category of intervention, such as hyperbaric chambers and the HBOT method), the broad domain of therapies encompassed by intermittent hypoxic and hyperoxic stress (and more generally by intermittent breathing techniques) has been used in humans for many decades, creating a large volume of evidence regarding its safety and efficacy in humans. It is well documented in multiple scientific papers. And it is, indeed, uniquely and coincidentally ***a domain of research that accumulated a large amount of human evidence before positive results in model organism studies.***

Prior to formal clinical trials and studies demonstrating the ***safety and therapeutic value in humans***[1] *of intermittent hypoxic/hyperoxic stress for a number of specific diseases and conditions (including* ***cognitive impairment and neurodegenerative disease***[2], ***prediabetes***[3], ***heart disease***[4], ***coronary artery disease***[5] *and* ***cardiorespiratory conditions***[6], ***impaired mobility***[7]*), this broad domain of intermittent breathing techniques received many decades-worth of studies in human patients focused on* ***its effects on intensive athletic perfor-***

[1] www.journals.physiology.org/doi/full/10.1152/japplphysiol.00564.2014
[2] www.ncbi.nlm.nih.gov/pmc/articles/PMC6862463
[3] www.liebertpub.com/doi/10.1089/ham.2019.0053
[4] www.pubmed.ncbi.nlm.nih.gov/30251879
[5] www.pubmed.ncbi.nlm.nih.gov/28323322
[6] www.pubmed.ncbi.nlm.nih.gov/30251879
[7] www.bmcgeriatr.biomedcentral.com/articles/10.1186/s12877-019-1184-1

***mance*[1], *high-altitude (alpine) training*[2]** *(both in young as well as* ***older individuals*[3]**) *and as a core, standard and systematically-applied technique for* ***training cosmonauts for space flight within the scope of USSR-era space medicine*[4]**, *for improving the performance of deep-sea divers, and as a preventive strategy for situations of exposure to extreme environmental hazards (such as clean-up groups working in Chernobyl).*

Today, the broad domain of intermittent breathing techniques enhanced by specialized equipment (which encompass IHHT, HBOT and other variations and specific protocols) are used regularly by high-performance athletes (Olympic champions, long-distance marathon runners, deep divers, alpinists, ***astronauts*[1]**, *etc.), by high-profile businessmen and other intensive-occupation professionals to enhance cognitive performance and endurance, and also as a recovery technique in post-COVID-19 pathologies and other very complex diseases like* ***Lyme Disease*[6]** *and* ***Leigh Syndrome*[7]**.

The specific protocol that I am using personally (IHHT) was developed by Dr. Arkadi Prokopov (mentioned earlier in this chapter), an integrative medicine physician and gerontologist with over 30 years experience in the effects of hypoxia and hyperoxia on Human Longevity, who developed his unique and specialized IHHT protocol through his experience working as part of a government-led Space Medicine program, and the equipment I am using comes directly from Dr. Prokopov's company, ***Athletic HighTech S.L.*[8]** *Arkadi himself uses this technique, and his personal results are also quite compelling – being 72, doing* ***regular athletics and endurance training*[9]**, *and possessing the strength and performance of absolutely healthy people in their 30s-40s, which I have witnessed first-hand during my consultations with him.*

I first came across this technique by accessing analytics produced by a specialized proprietary group ("Practical Longevity"), created

---

[1] www.germanjournalsportsmedicine.com/archive/archiv-2019/issue-5/effects-of-hyperoxic-training-on-human-performance
[2] www.pubmed.ncbi.nlm.nih.gov/7712582
[3] www.pubmed.ncbi.nlm.nih.gov/18331220
[4] www.pubmed.ncbi.nlm.nih.gov/12162864
[1] www.youtube.com/watch?v=BwCMCOhgdX4
[6] www.amazon.ca/Undoing-Lyme-Disease-Mitochondria-Borreliosis-ebook/dp/B085DGS2QB
[7] www.en.wikipedia.org/wiki/Leigh_syndrome
[8] www.a-ht.org/
[9] www.youtube.com/watch?v=FpxNSb8YLZk&t=235s

several years ago within Aging Analytics Agency dedicated to aggregating, analyzing and validating the full scope of technologies, therapies and methods of optimizing Healthspan and extending the healthy period of human life with demonstrable evidence in humans. Being an early founder of Aging Analytics Agency, I myself and my partners at Deep Knowledge Group are having early access to the results of this years-long analytical project.

Meanwhile, the confidence I have in the validity of my own personal results comes not just from obvious day-to-day increases in physical health, endurance and cognitive performance, which may lead to certain subjective psychological biases in regards to my assumptions on this technology's impact on Longevity. Rather, these conclusions are coming from the tangible measurement of my own biomarkers and other quantitative assessments of health and performance. Interestingly, most techniques validated by Aging Analytics Agency's Practical Longevity Proprietary Group (such as intermittent fasting, autologous stem cells, and several dozens of others that I am not at liberty to disclose at this time) can take several months to have a measurable impact on biomarkers of Longevity, health and performance. By contrast, IHHT and similar breathing techniques that induce hypoxia or hyperoxia can be measured in hours and are very evident in days and weeks.

For these reasons, considering both my own personal results and the large scope and depth of human studies published on this technique, I have full confidence that this broad category of intervention, including hyperbaric chambers and intermittent breathing (hypoxic, hyperoxic) methods, are not just working safely and effectively enough to provide health and performance benefits, but are also working to confer some level of rejuvenation and reversal of biological aging in humans.

The majority of the other broad domains of therapies mentioned in this chapter will likely require the development of Digital Human Avatars (introduced in Chapter VI, "A Proposed Solution to the Longevity Industry's Biggest Risk: Modern Tools for Human Validation Results for De-Risking the Investment Decision-Making Processes and IPOs," and expanded upon in-depth in Chapter X, "DeepTech Engineering Toward the Concept of a Comprehensive Digital Human Avatar 1.0, 2.0, 3.0 and 4.0.") to adjust specific therapeutic dosage and cocktail compositions of drugs in humans. In addition to that most likely much more sophisticated technologies of precision delivery will be required to make those drugs actually work. By contrast, Hyperbaric Chambers and IHHT ap-

pear to me to be straightforward (and effective) enough to be applied without the need for complex technologies like Digital Avatars to make them work safely and effectively in practice, as the actual results could be majored even with simple biomarkers systems close in close to real time (typically within days or weeks since the start of procedures).

The possibility of side effects in older adults from geroprotectors and other biopharmaceuticals

Traditional drugs that target particular individual diseases tend to work almost as well on healthy middle-aged patients as on older patients. They are drugs with simple, unidimensional targets and mechanisms of action. However, it is well known that children's and pregnant women's dosages for these drugs vary significantly and that many drugs are even prohibited for them. Furthermore, people with chronic health conditions need to consult with their doctor regarding their dosages and possible side effects. These facts lend support to the thesis that even simple, traditional (legacy system) "blockbuster" drugs trialed at large scales have some side effects, and certain groups of people are potentially more likely to have adverse side effects than others.

Meanwhile, elderly individuals are among the demographics most likely to experience adverse effects, given that they often have multiple existing health conditions, some level of immune system dysfunction, and other chronic age-related diseases. With the knowledge we have about drugs, we can almost undoubtedly expect severe adverse side effects, and we can expect their frequency and severity to grow as the patient gets older. It is, therefore, very likely that Healthspan-extending drugs will have many side effects in older patients – even in the case of simple, single-target drugs (and we can expect this to be multiplied in the case of drugs that target hallmarks of aging, which constitute nodes in a highly multidimensional network of complex biological processes).

However, it is also well known that the prevalence and severity of side effects can be lessened by lowering dosing and through applying so-called subtherapeutic doses. This is, in fact, one way of neutralizing some of these concerns in specific demographics, given that subtherapeutic doses are more likely to confer benefits to younger and middle-aged individuals free from existing health conditions and diseases. However, it is very much likely that subtherapeutic doses will be less effective in the case of older individuals.

Furthermore, the standard recommendation of "consult with your doctor" will not work in this case because only a tiny proportion of doctors practicing today understand the full range of concepts and complexities behind the concept of Longevity. Most of the doctors that individuals have access to today are "sick-care" doctors with no training in the concepts of Healthspan extension and Longevity medicine.

The developers of anti-aging drugs hold fairly well-established notions on how the geroprotectors are supposed to work. And there is a well known concept named "Geroprotectors drug cocktail."

Essentially, geroprotector cocktails consist of micro-doses of several dozens (up to hundreds) of compounds combined into a single mix, and it is likely that the specific dosing and composition of these compounds will need to be adjusted for different demographics according to age, gender, ethnicity and health status (e.g., the presence of existing health conditions).

While this is fairly widely accepted, some specialists (myself included) believe that the dosing and composition of these mixes will have to be adjusted for individual patients on an ongoing basis (potentially even daily). Therefore, in order for this approach to work in a truly safe and effective manner, mechanisms of high-frequency micro-dose adjustment (perhaps even close to real-time fine-tuning) will be required according to a feedback loop that measured the effects of different doses and geroprotectors cocktail compositions upon Biomarkers of Health and Longevity.

Taking all this into account, it is clear to me that the practical implementation of geroprotectors will require near-constant AI supervision coordinated with Longevity Medicine doctors – i.e., medical professionals who have been trained in Healthspan extension rather than "sick-care" (which, unfortunately, represents an exceedingly small proportion of licensed doctors working today). The implementation of geroprotectors will be possible only after establishing the Longevity Biomarkers panels and its deployment so that they will be scalable and usable for people at home, with close to real-time monitoring. Our most optimistic estimate is that this state of Longevity science and technology implementation in real life in regards to personalised high frequency adjustment analytical systems will only be possible by the year 2024-2025.

We can expect that many anti-aging drugs will pass the clinical trials, but aside from them, work-in-practice validation is needed (widespread validation via widespread application) because even if the clinical trials are successfully passed, this does not necessarily translate to drug safety – or effectiveness – for specific demographics. (This remains especially true for older adults with existing health conditions.) In practice, this will require the market-ready use of Digital Human Avatars capable of enabling the sort of high-frequency personalized-experimentation and therapeutic regimen fine-tuning required by this widespread work-in-practice approach, and we can expect the concept of Digital Human Avatars to be sufficiently developed for this purpose between 2024-2025 and adopted by a mass market by 2027-2030. The concept of Digital Human Avatars is discussed more thoroughly in the next chapters, and specifically in Chapter X, "DeepTech Engineering Toward the Concept of a Comprehensive Digital Human Avatar 1.0, 2.0, 3.0 and 4.0."

Concluding Remarks on Longevity Clinical Trials

Based on the above analysis, it should be clear that the well-established paradigm of therapeutic validation that worked for BioTech and BioPharma has proved less applicable when it comes to the Longevity Industry.

There has not yet been any single Longevity drug approved that is completely free from side effects and concerns. In reality, such controversies won't stop either investors or researchers from continuing to use the existing legacy approaches for conducting Longevity clinical trials. So the question remains: how long will we have to wait to see the creation of clinically validated Longevity-related drugs without serious adverse effects and unknown risks?

We forecast that the necessary improvements to the clinical trial frameworks will take place between 2027 and 2030. However, the much much faster pace of evolution is expected to be gained via the AI-enhanced drug discovery platforms, structural chemistry and biology, Big Data-powered ADME-tox analysis, and the rise of

Digital Human Avatar platforms capable of enabling personalized experimentation, validation, and experimental fine-tuning to neutralize negative effects.

In the meantime, we should be seeking to accelerate this timeline by promoting the use of more reliable and relevant methods of conducting Longevity clinical trials, and in general for the validation of Longevity therapies. I will be describing considerations for the development of more progressive frameworks and approaches for this in greater detail in Chapter VI, "A Proposed Solution to The Longevity Industry's Biggest Risk: Modern Tools for Human Validation Results for De-Risking the Investment Decision-Making Processes and IPOs." Meanwhile, this line of work is being simultaneously developed even further by Aging Analytics Agency, which is working on a special case study concerning the new approaches for experiments and clinical trials for the Longevity Industry, which we can expect to receive by Q3 2022.

When I entered the aging research field in 2013, the first group of scientists that I met was preoccupied with geroprotectors, which at the time was one of the most common talking points among geroscientists, just as it remains a frequent topic of conversation today. This was one of the first domains of aging research that I became personally involved in, having provided financing and support in 2013-2014 for the open-source geroprotectors database, **www.geroprotectors.org**. It is now eight years later, however, and we see nothing like the efficient application of geroprotector drugs in humans that many of us back then in 2013 expected to see by now.

Granted, the conversation has advanced since then, but the tangible progress we have seen does not match the conversational hype over the years. While in digital medicine and AI for Healthcare – and many other Tech-driven domains – the tangible progress has kept one step ahead of the predictions, the field of geroprotectors has little to show in the way of validated practical human applications after all these years.

Do not mistake me – I am not a pessimist and I do not regard this as the end of the road for geroprotectors. I am not opposed to the continued pursuit of working geroprotectors and anti-aging drugs, such as senolytics. But after eight years of surveying this field, I see much less in the way of validated, tangible progress in human Healthspan extension than I expected to see by this point, eight years into what became my new life mission in 2013.

I still believe we will see validated geroprotectors slowing down and even reversing aging in humans, but exactly when this will happen remains uncertain. It is my strong expectation that other very sophisticated technologies will need to be developed before we can see real practical applications of anti-aging drugs. Robust panels of biomarkers, personalized real-time AI-empowered systems, Digital Avatars and very advanced systems for precise drug delivery will need to be among these technologies. These systems will need to be capable of measuring the effects of subtherapeutic doses of biopharmaceuticals on biomarkers, such as geroprotectors, in a high-frequency manner close to real time. They will also need to be able to apply precise daily adjustments of both the dose and exact composition of the geroprotector mix based on how they affect personalized Biomarkers of Human Longevity.

Such systems will also rely on other ancillary technological innovations to work in practice by that time. In particular, they will rely on the development of embedded sensors capable of measuring biomarkers in a high-frequency manner, as well as technologies capable of semi-automatically adjusting geroprotector cocktail dosing and composition daily in a very precise, industrial laboratory-grade manner. To put it simply, the equivalent of an AI-driven robotics laboratory will need to be made available to every home, as a consumer device, and with that technology in place, we will always see that some geroprotectors or drugs are working better on some people than on others.

Now, as I am a technological pragmatist, my forecast is that we cannot expect such technologies to be deployed and developed before 2025. By that time we will have seen many failings of Longevity drugs. But we will also be seeing better technologies for more efficient clinical trials, better forecasting of clinical outcomes and neutralizing side effects.

My general prediction that, by 2027, we will have developed anti-aging drugs that really work in the form of precision-made "geroprotector drug cocktails," with AI-driven advanced delivery systems, as opposed to the current perception of a "magic pill" or "blockbuster drug" that works the same way on everybody. And by the year 2030 these systems will be adopted by the mass market.

Summary

In this chapter, I have outlined clinical trials focused on developing aging-prevention drugs, some of which have made remarkable progress. However, we face a lack of human validation when it comes to Longevity drug studies.

- The well-established paradigm of therapeutic validation that worked for the BioTech Industry with 90% failure rate has proved less applicable when it comes to the Longevity Industry. In case of Longevity clinical trials we can expect a much higher failure rate compared to the BioPharma Industry" (note the capital I in Industry.
- The paradigm of therapeutic validation common in BioPharma has led to Longevity company valuation based on the rate of successful preclinical trials on a model organisms. This fact leads to the possibility of overhyping some companies while significantly underfunding the other ones that use in silico human centered experimentation or other progressive drug testing methods.
- Modern progressive companies should consider the severity of the aging issue and the complexity of biological reasons for aging, and they should proceed with scientific research into new approaches to tackle aging.
- There are no approved, concern-free geroprotector anti-aging drugs. Metformin is the closest to this, but there are still concerns, so additional studies are necessary. People have also been taking rapamycin for years, and the risks and benefits are still unclear.
- Working geroprotector anti-aging drugs will first arrive not as a result of the current activities of drug discovery companies working on "blockbuster drug," but in the form of personalized "geroprotector drug cocktails" and AI-driven advanced precision delivery systems.

Chapter IV: The Longevity Validation Gap – Comparative Analysis of How Clinical Trial Results Impacted the Market Capitalization Dynamics and Financial Indicators of 12 Public Longevity Companies

We have established in previous chapters that too many companies are basing their valuations, both pre- and post-IPO, on the results of animal data, and that a more human-centered approach is required for overall industry stabilization.

We have also discussed the ways in which this unmet need poses substantial challenges to the industry's future positive trajectory of growth – and the realization of its practical humanitarian outcomes for global society – by risking a decline in investor sentiment and confidence when positive results in mice fail to be replicated in humans.

While we will explore potential solutions to these issues in greater depth in the next chapter, here we will substantiate this thesis and show how this issue is to some extent already happening within the Longevity Industry. We will do this by analyzing several notable examples of publicly traded Longevity companies whose performance could be divided in three groups:

- Dramatic decline following a failure to replicate the successful results of model organism trials in the case of human clinical trials;
- Fluctuation in capitalization or returns below market average as a result of ambiguous or unclear clinical trials results, even if there are no significant negative events related to R&D, financial activities, or collaborations; and
- Significant growth in the number of successful human clinical trials.

We will see that a significant number of Longevity corporations are characterized by below-average financial performance, and in many cases, Longevity stock returns are too small as for the level of the risk (volatility).

This means that currently investors can select the targets in the other BioTech domains that are both more profitable and less risky. The lack of trust of the investors in the Longevity R&D that is based on model organisms is expressed in stock price dynamics.

Ultimately, this analysis lends real-world validation to the thesis that companies should possess some form of human validation prior to proceeding to go public.

12 Notable Cases of Longevity Publicly Traded Companies

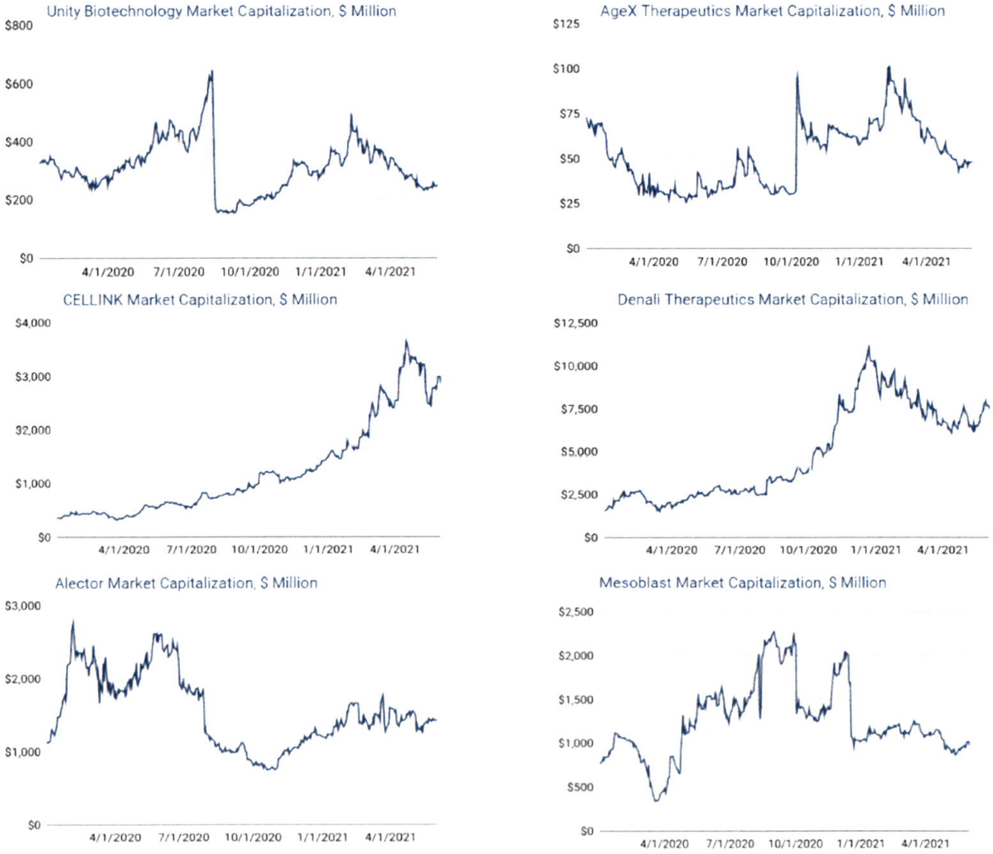

Source: Yahoo Finance[1]

[1] www.finance.yahoo.com/

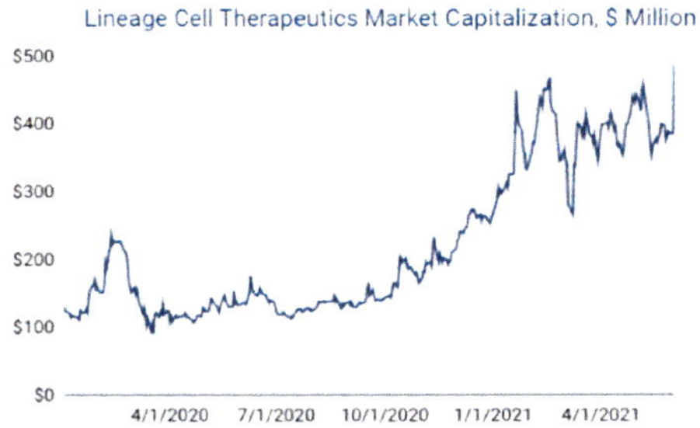

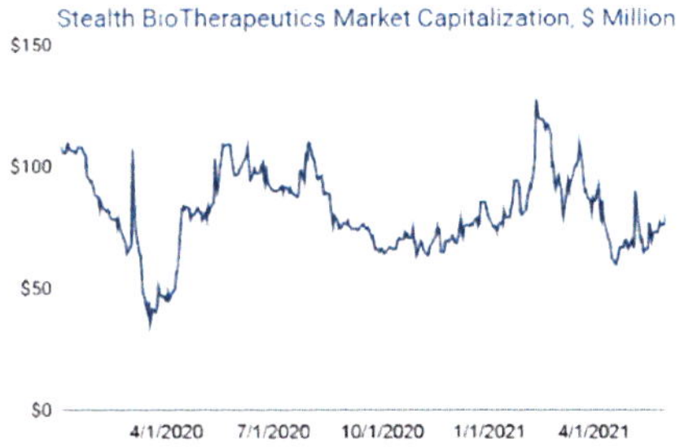

Source: Yahoo Finance[1]

Below, we have provided analytics extracts from various investment digests produced by Aging Analytics Agency and InvesTech Advanced Solutions. You may review the digests using this link: **www.invest-solutions.tech/reports.**

Unity Biotechnology (UBX)

Unity Biotechnology, Inc. is a biotechnology company, engages in the research and development of therapeutics to extend human Healthspan. The company's lead drug candidates include UBX0101, which is in a Phase II clinical trial for musculoskeletal/osteoarthritis disease and a Phase Ib clinical trial for moderate-to-severe osteoarthritis of the knee; UBX1325, which is a Phase I clinical trial

[1] www.finance.yahoo.com/

for the treatment of diabetic macular edema; and UBX1967, for the treatment of ophthalmologic diseases. It is also developing programs in pulmonary disorders and neurological disorders, as well as kidney and liver diseases. The company was formerly known as Forge, Inc. and changed its name to Unity Biotechnology, Inc. in January 2015. Unity Biotechnology, Inc. was founded in 2009 and is headquartered in South San Francisco, California.

On the following chart, the market capitalization of Unity Biotechnology is shown in millions of dollars ($MM). A significant drop in capitalization is gained due to the UBX0101 Phase II failure. More specifically, according to the ***12-week data***[1], no statistically significant difference was found between UBX0101 and placebo groups for a change from baseline in WOMAC-A, which measures pain in OA. The company still has not recovered its financial position and is now waiting for additional R&D breakthroughs in order to outperform its 2020 figures, which is possible only in case of successful human validation of the company R&D developments. But even if we do not take into account the UBX0101 failure, the growing trend is volatile and unsteady.

Source: Yahoo Finance[2]

[1] www.globenewswire.com/en/news-release/2020/11/04/2120416/0/en/UNITY-Biotechnology-Inc-Reports-Third-Quarter-2020-Financial-Results-and-Clinical-Program-Updates.html
[2] www.finance.yahoo.com

AgeX Therapeutics (AGE)

AgeX Therapeutics, Inc., a biotechnology company, focuses on the development and commercialization of novel therapeutics targeting human aging and degenerative diseases in the United States. The company's lead cell-based therapeutic candidates in development include AGEX-BAT1, a cell therapy product candidate for the treatment of various age-related metabolic disorders, such as Type II adult-onset diabetes; and AGEX-VASC1, a cell-based therapy to restore vascular support in aged ischemic tissues, such as the ischemic heart. Its lead drug-based therapeutic candidate in discovery is AGEX-iTR1547, a drug-based formulation to restore regenerative potential in a range of aged tissues afflicted with degenerative diseases. In addition, the company markets human embryonic stem cellsand the GeneCards Database Suite, which includes" genomic interpretation including genomic interpretation algorithms and analysis tools for use by researchers at pharmaceutical and biotechnology companies, and other institutions. AgeX Therapeutics, Inc. has a research collaboration with the University of California, Irvine on neural stem cell research program for Huntington's disease and other neurological disorders; and Sernova Corp. The company was founded in 2017 and is based in Alameda, California.

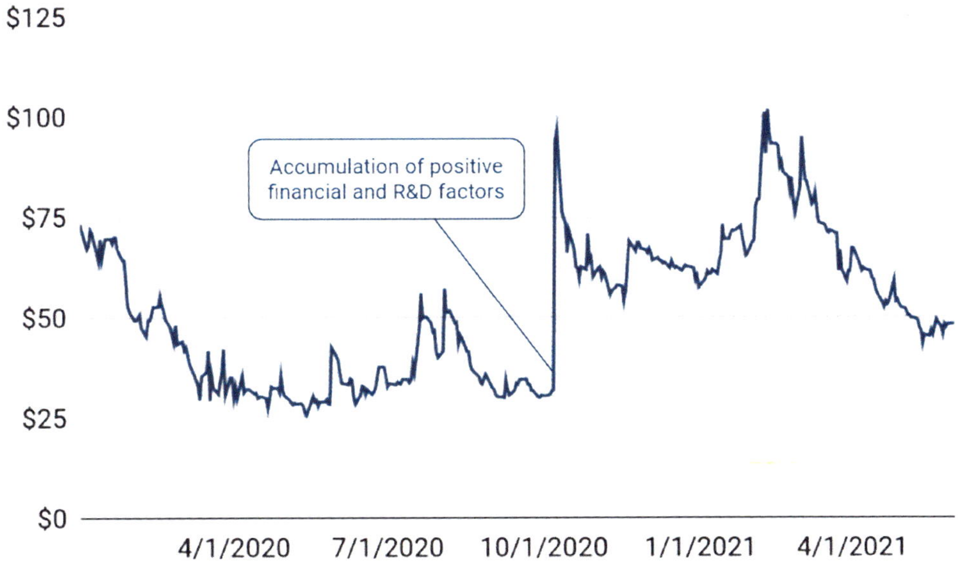

On the following chart, the market capitalization of AgeX Therapeutics is shown in $MM. This market capitalization was due to a natural decline in 2020 that could be seen even in the cumulative capitalization of the Longevity sector. It was also due to the strong financial performance and COVID-19 participation of AgeX Therapeutics, as well as the performance of LifeMap Sciences, the subsidiary of AgeX Therapeutics. For these reasons, the capitalization of AgeX Therapeutics rocketed by the end of 2020, followed by a predicted decline.

Thus, even taking into account breakthroughs in R&D, positive performance of the subsidiaries, and favorable market conditions, we see that market capitalization of the company declined in the last 1.5 years.

CELLINK (BICO.ST)*

Cellink AB designs and develops bioprinting technologies that enable researchers to 3D-print organs and tissue for applications primarily in the pharmaceutical and cosmetic industries. It offers bioprinters, imaging systems, dispensing solutions, and bioinks in the field of academic and clinical medicine for labs and scientists worldwide. The company's bioprinting solutions include BIO X, a bioprinter for life-science companies, researchers, and innovators; BIO X6, a bioprinting platform for printing complex structures; INKREDIBLE and INKREDIBLE+, which are bioprinting solutions for skin, cartilage tissue, and others; Holograph X, a solution that leverages high-resolution holographic stereolithography to bioprint small structures; Lumen X, a solution that offers applications in microfluidics, cell-laden hydrogels, and macroporous structures; and others.

The company also offers biofluids solutions, a single-cell printer, an automated laboratory instrument, and other products. Its customers include universities, hospitals, and public and commercial laboratories. Cellink AB (publ) has a collaboration with MedImmune and a strategic partnership with Made In Space, Inc. It also offers products online. The company was founded in 2016 and is based in Gothenburg, Sweden.

On the following chart, the market capitalization of CELLINK is shown in $MM. In the past two years, CELLINK has become one of the most active players in its market. The impressive financial performance of the company has led to significant acquisition opportunities which the company has made good use of. After the acquisition of Cytena in 2019, CELLINK acquired Scienion along with its subsidiary Cellenion in 2020. The company began 2021 with the acquisition of Nanoscribe for $70.6 million and Visikol, a contract research company. All this led to the skyrocketing of the company's capitalization. The positive performance of the company led to its rebranding under BICO, the world's first Bio-Convergence company. BICO is now the parent company for its 11 subsidiaries, including CELLINK. Applicability of company`s products for practical human Longevity contributes to its outstanding financial performance.

*The stock exchange ticker is provided for BICO Group AB.

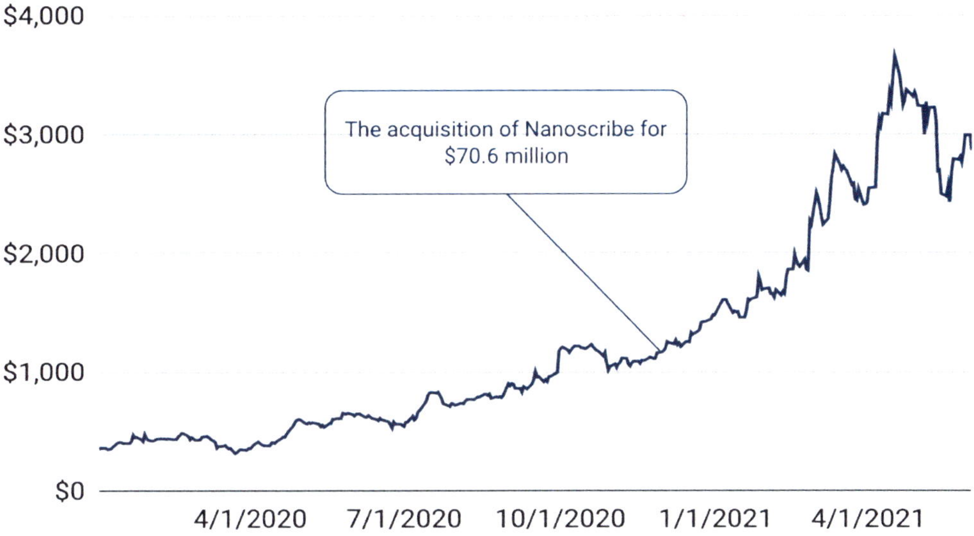

Source: Yahoo Finance[1]

Denali Therapeutics (DNLI)

Denali Therapeutics Inc., a biopharmaceutical company, discovers and develops therapeutic candidates for neurodegenerative diseases in the United States. The company offers leucine-rich repeat kinase 2 (LRRK2) inhibitor product candidates, including DNL201, which has completed Phase Ib clinical trials, and DNL151, which is in Phase I and Phase Ib clinical trials for Parkinson's disease. In addition, the company is developing an enzyme transport vehicle,

[1] www.finance.yahoo.com

as well as an antibody transport vehicle (ATV). Furthermore, it has various seed programs at the research, discovery, and preclinical stages of development. Denali Therapeutics Inc. has a collaboration agreement with Takeda Pharmaceutical Company and Genentech, Inc., Sanofi, F-star Gamma Limited, Harvard University, the Michael J. Fox Foundation, and many other institutions. Denali Therapeutics Inc. was incorporated in 2013 and is headquartered in South San Francisco, California.

On the following chart, the market capitalization of Denali Therapeutics is shown in $MM. The significant spike in the capitalization of the company is due to both its strong financial performance in the past year and its success in Phase I/II study of DNL310 in patients with Hunter syndrome (MPS II) and Phase Ib study of DNL151. In addition, Sanofi, the partner of Denali Therapeutics, submitted an IND application for DNL788, a potent, selective and brain-penetrant small molecule inhibitor of RIPK1 intended to treat patients with Alzheimer's disease, ALS, multiple sclerosis (MS) and potentially other conditions. The company is an example of the impact that human clinical trials results have on the market capitalization.

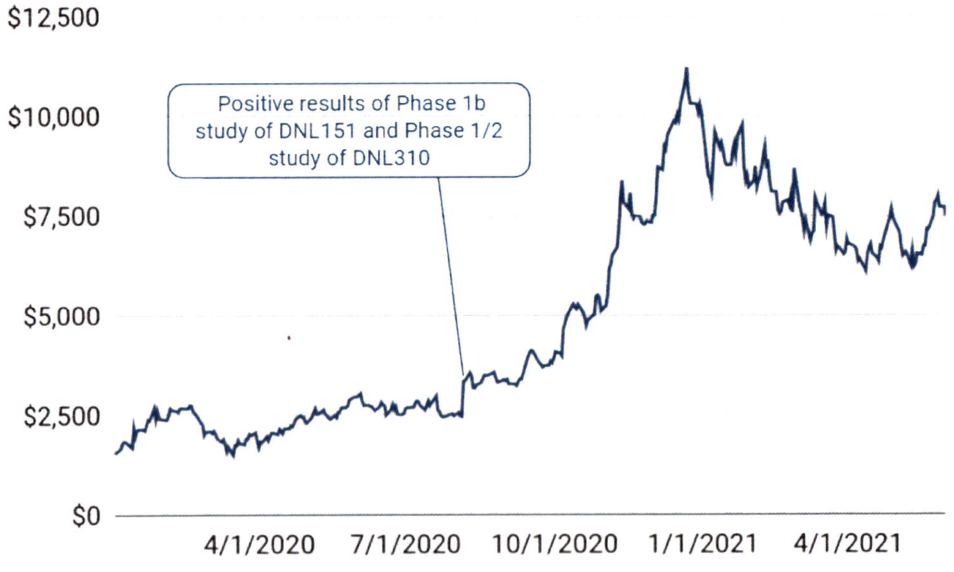

Source: Yahoo Finance[1]

[1] www.finance.yahoo.com

Alector (ALEC)

Alector, Inc., a clinical-stage biopharmaceutical company, develops therapies for the treatment of neurodegeneration diseases. Its products include AL001, a humanized recombinant monoclonal antibody, which is in Phase II clinical trial for the treatment of frontotemporal dementia disease; and AL101 that is in Phase I clinical trial for the treatment of neurodegenerative diseases, including Alzheimer's and Parkinson's diseases. The company also offers AL002 and AL003, which are in Phase Ib clinical trials for the treatment of Alzheimer's disease. In addition, it has 10 programs under research and development. The company has a collaboration agreement with Adimab, LLC for the research and development of antibodies. Alector, Inc. was founded in 2013 and is headquartered in South San Francisco, California.

On the following chart, the market capitalization of Alector is depicted in $MM. The company's financial underperformance ($4.9 million revenue, compared to $6.0 million for the same period in 2019) led to a significant capitalization decline in Q3 2020. On the other hand, since Q1 2021, the company's performance seems to have become more stable despite the departure of the company's CFO. The reason for this performance could be traced back to its corporate portfolio update held back in January 2021. Despite attracting prominent collaborators such as GSK and announcement of a collaboration project in immuno-neurology for two clinical-stage first-in-class monoclonal antibodies for neurodegenerative diseases, stock price returns are still stagnating.

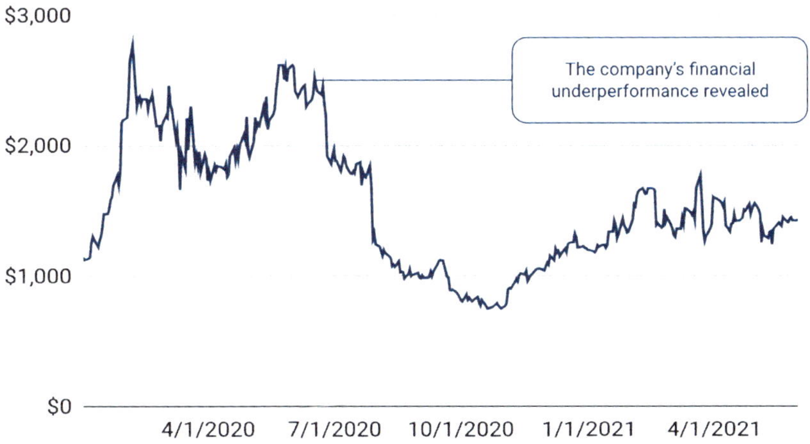

Source: Yahoo Finance[1]

[1] www.finance.yahoo.com

Mesoblast (MESO)

Mesoblast Limited is a biopharmaceutical company that develops and commercializes allogeneic cellular medicines. The company offers products to treat cardiovascular disorders and spine orthopedic disorders, as well as oncological, hematological, and immune-mediated or inflammatory diseases. Its proprietary regenerative medicine technology is based on specialized cells known as mesenchymal lineage adult stem cells. Products by Mesoblast that have completed clinical trials include MSC-100-IV for steroid-refractory acute graft versus host disease; MPC-150-IM for advanced heart failure; and MPC-06-ID for chronic low back pain due to degenerative disc disease. Mesoblast is also developing MPC-300-IV to treat biologic refractory rheumatoid arthritis, diabetic kidney diseases, and type 2 diabetic nephropathy. Mesoblast Limited was founded in 2004 and is headquartered in Melbourne, Australia. The company operates in the United States, Australia, Singapore, the United Kingdom, and Switzerland. The company has strategic partnerships with Tasly Pharmaceutical Group and JCR Pharmaceuticals Co. Ltd.

Source: Yahoo Finance[1]

[1] www.finance.yahoo.com

On the following chart, the market capitalization of Mesoblast is depicted in $MM. Mesoblast's 92% increase in revenues up to US $32.2 million in 2020 and its 127% increase in milestone revenue from strategic partnerships to US $25.0 million, along with strong medical pipeline performance, helped it to achieve impressive capitalization figures. Furthermore, in December 2020 Mesoblast Ltd. saw its stock drop on the news that its allogeneic mesenchymal cell therapy, rexlemestrocel-L (Revascor), failed to meet the primary endpoint of a reduction in hospitalizations in its Phase III advanced chronic heart failure trial. The decline in 2021 is mainly due to the slowing of the company's expansion and entry into the long-term drug approval processes. Such performance, again, indicates the urgent need, the Longevity companies have, in human validation and effective pre-clinical assessment approaches that might differ from the classical ones.

ChromaDex (CDXC)

ChromaDex Corporation operates as a bioscience company focusing on healthy aging. The company operates through three segments: Consumer Products, Ingredients, and Analytical Reference Standards and Services. It conducts research on nicotinamide adenine dinucleotide (NAD+), provides finished dietary supplement products that contain its proprietary ingredients, and develops and commercializes proprietary-based ingredient technologies. The company also holds a patent for NAD+ precursor nicotinamide riboside as the flagship ingredient NIAGEN, which supports human immune function. ChromaDex actively tests the therapeutic effects of nicotinamide riboside on various pathologies. Several of those are classified as Longevity clinical trials, including Nicotinamide Riboside as an Enhancer of Exercise Therapy in Hypertensive Older Adults (The NEET Trial), Nicotinamide Riboside for Treating Elevated Systolic Blood Pressure, and Arterial Stiffness in Middle-Aged and Older Adults. The company is headquartered in Los Angeles, California.

On the following chart, the market capitalization of ChromaDex is depicted in $MM. The astonishing spike is gained due to the following facts: ChromaDex announced the retail distribution of Tru Niagen® in Walmart™ stores across the United States beginning in June 2021 as well as a strategic supply agreement with Health & Happiness Group, a global leader in premium health, human and pet nutrition and personal care brands to sell Niagen® in its exclusively formulated Swisse™ products. The market overreacted in the first few days after the publishing of the Phase III Clinical Study ***results***[1], leading to a sharp increase in market capitalization.

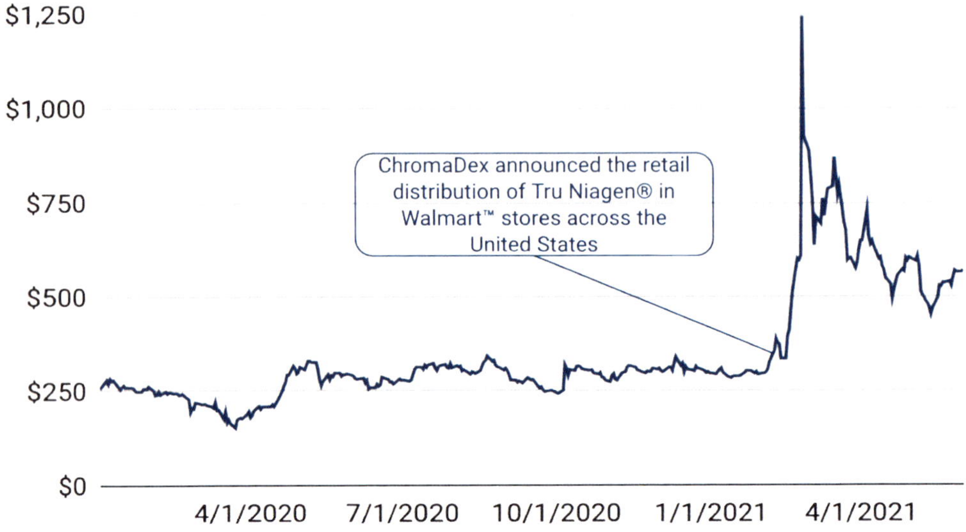

Source: Yahoo Finance[2]

Gensight Biologics (SIGHT.PA)

GenSight Biologics S.A., a clinical-stage biotechnology company, engages in the discovery and development of therapies for mitochondrial and neurodegenerative diseases of the eye and central nervous system. The company develops its products through gene therapy-based mitochondrial target sequence and optogenetics technology platforms. Its lead product candidates include GS010, an AAV2 based gene therapy that is in Phase III clinical trials for the treatment of leber hereditary optic neuropathy caused by mutation of the ND4 gene; and GS030, which is in Phase I/II trial for the treatment of retinitis pigmentosa and geographic atrophy in dry age-related macular degeneration. The company is also developing products

[1] www.investors.chromadex.com/news/news-details/2021/New-Phase-3-Clinical-Study-Finds-Nutritional-Protocol-Including-Nicotinamide-Riboside-Accelerates-Recovery-in-Mild-to-Moderate-COVID-19-Patients/default.aspx
[2] www.finance.yahoo.com

that are in the preclinical stage targeting ophthalmic and neurodegenerative diseases. GenSight Biologics S.A. was founded in 2012 and is headquartered in Paris, France.

On the following chart, the market capitalization of GenSight Biologics is depicted in $MM. The skyrocketing capitalization of the company is based on its strong medical pipeline performance and completion of the Phase III trials of LUMEVOQ® (GS010) for the treatment of Leber Hereditary Optic Neuropathy (LHON). The manufacturing process also seems to be quite stable, since the company's CDMO is Thermo Fischer Scientific.

Source: Yahoo Finance[1]

Lineage Cell Therapeutics (LCTX)

Lineage Cell Therapeutics, Inc., a clinical-stage biotechnology company, focuses on developing and commercializing therapies for the treatment of degenerative diseases in the United States and internationally. The company's lead product candidates include OpRegen, a retinal pigment epithelium cell replacement therapy, which is in Phase I/IIa clinical trial for the treatment of the dry age-related macular degeneration; OPC1, an oligodendrocyte progenitor cell therapy that is in Phase I/IIa multicenter clinical trial for the treatment of acute spinal cord injuries; and VAC2, allogeneic cancer immunotherapy of antigen-presenting den-

[1] www.finance.yahoo.com

dritic cells, which is in Phase I clinical trial to treat non-small cell lung cancer. In addition, the company offers various therapeutic products for the treatment of oncology, neurological diseases and disorders, blood and vascular system diseases and disorders, and blood plasma volume expansion. It also undertakes various research programs for vision restoration and Demyelination. Lineage Cell Therapeutics, Inc. is working in collaboration with Orbit Biomedical, Ltd. It was founded in 1990 and is headquartered in Carlsbad, California.

On the following chart, the market capitalization of Lineage Cell Therapeutics is depicted in $MM. The company's positive capitalization dynamic in 2021 is based on a number of facts. Firstly, the company presented a positive interim clinical update from the ongoing Phase I/IIa study of OpRegen for the treatment of dry-AMD with GA. Secondly, the company also reported the first known finding of retinal tissue restoration in a patient who received a retinal pigment epithelium (RPE) cell transplant and continues to demonstrate areas of retinal restoration as of their last assessment, approximately 3 years after treatment. Thirdly, Lineage Cell Therapeutics has entered into a worldwide license agreement with Immunomic Therapeutics for an allogeneic cell-based cancer immunotherapy based on Lineage's VAC platform with a total of $2 million in upfront payments anticipated in the first year and the potential for $67 million in development and commercial milestones.

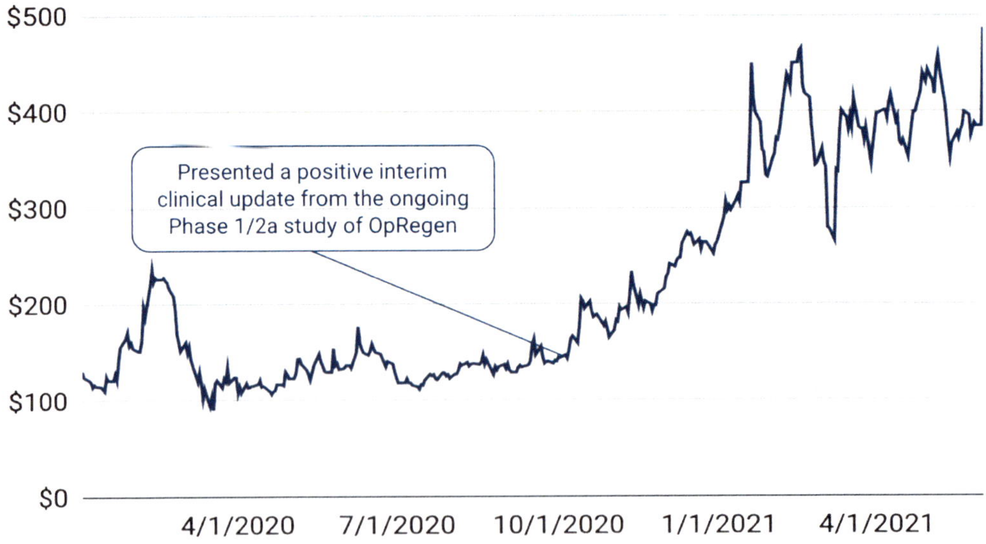

Source: Yahoo Finance[1]

[1] www.finance.yahoo.com

Biophytis (BPTS)

Biophytis SA, a clinical-stage biotechnology company, focuses on the development of therapeutics that slow the degenerative processes and improve functional outcomes for patients suffering from age-related diseases. Its therapeutics focuses on targeting and activating key biological resilience pathways that could protect against and counteract the effects of the multiple biological and environmental stresses that lead to age-related diseases. The company's lead drug candidate, Sarconeos (BIO101), which is orally administered, is in development for the treatment of neuromuscular diseases, including sarcopenia and Duchenne muscular dystrophy (DMD). Its second drug candidate, Macuneos (BIO201), is an orally administered drug in development for the treatment of retinal diseases, including dry age-related macular degeneration (AMD) and Stargardt disease. Biophytis SA has a collaboration agreement with AFM-Telethon for the development of its Sarconeos (BIO101). The company was founded in 2006 and is headquartered in Paris, France.

On the following chart, the market capitalization of Biophytis is depicted in $MM. The company's high capitalization performance is due to its strong financial and medical pipeline performance, including its success in the COVA clinical studies. Additionally, Biophytis receives €980,000 in DeepTech Funding from BPI France for the Development of Macuneos (BIO201) in Dry AMD.

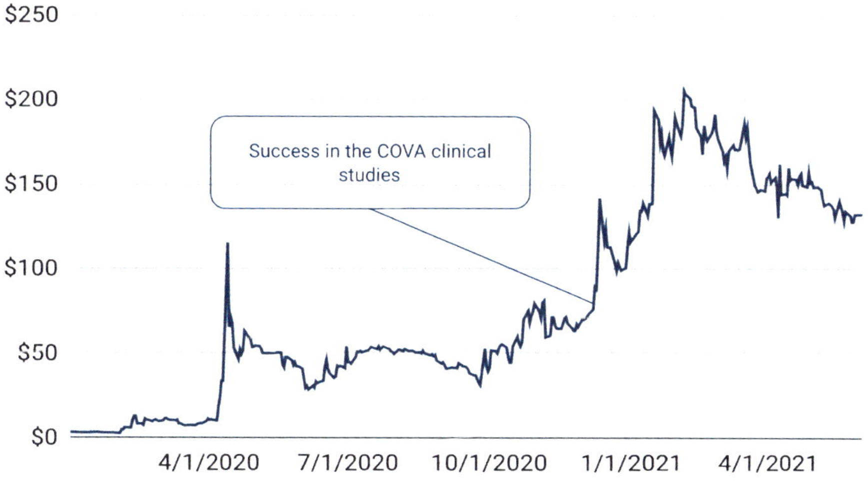

Source: Yahoo Finance[1]

[1] www.finance.yahoo.com

CohBar (CWBR)

CohBar, Inc., a clinical stage biotechnology company, engages in the research and development of mitochondria-based therapeutics (MBTs) for the treatment of chronic and age-related diseases. The company develops MBTs to treat non-alcoholic steatohepatitis (NASH), obesity, fatty liver disease, idiopathic pulmonary fibrosis, type 2 diabetes mellitus, cancer, atherosclerosis, cardiovascular diseases, and neurodegenerative diseases such as Alzheimer's disease. Its lead MBT candidate is CB4211, a novel refined analog of the MOTS-c mitochondrial-derived peptide, which is in Phase Ib stage of a Phase Ia/Ib clinical trial for the treatment of nonalcoholic steatohepatitis (NASH) and obesity. The company's preclinical programs include CB5138 analogs for fibrotic diseases, CB5064 analogs for COVID-19 associated ARDS, MBT5 analogs for CXCR4-related cancer and orphan diseases, and MBT3 analogs for cancer immunotherapy. The company was founded in 2007 and is headquartered in Menlo Park, California.

On the following chart, the market capitalization of CohBar is depicted in $MM. The spikes are due to the fact that the company has completed dosing in its Phase Ib study evaluating CB4211 as potential treatments for NASH and obesity. The CohBar stock price is fluctuating and does not demonstrate a robust positive trend.

Source: Yahoo Finance[1]

[1] www.finance.yahoo.com

Stealth BioTherapeutics Corp (MITO)

Stealth BioTherapeutics Corp, a clinical-stage biotechnology company, focuses on the discovery, development, and commercialization of therapies for diseases involving mitochondrial dysfunction. Its lead product candidate, Elamipretide, has the potential to treat rare genetic and common age-related mitochondrial diseases. The company's product candidates include SBT-20, a small peptide that targets and binds reversibly to cardiolipin, stabilizing mitochondrial structure and function under conditions of oxidative stress; and SBT-272, a lead discovery pipeline compound, evaluating for rare neurodegenerative diseases such as amyotrophic lateral sclerosis. Stealth BioTherapeutics Corp. was founded in 2006 and is based in George Town, Cayman Islands.

On the following chart, the market capitalization of Stealth BioTherapeutics Corp is shown in $MM. MITO shares have fallen 26.40% over the past year. However, they have moved up 9.85% in the beginning of June 2021 after three months of decline in market capitalization. In April 2021, the company delayed an NDA filing for elamipretide for cardiomyopathy in Barth syndrome; this led to a 10% fall in capitalization.

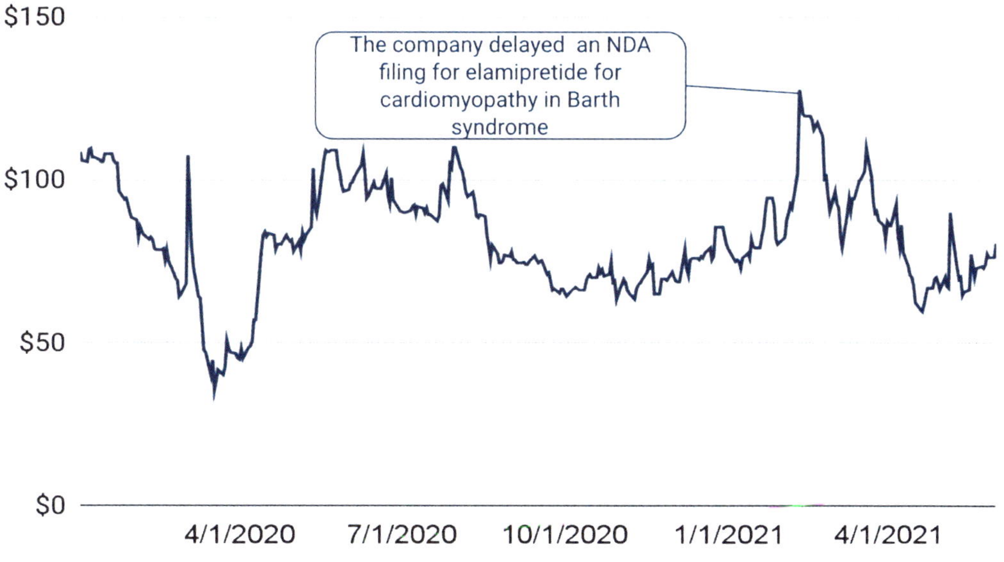

Source: Yahoo Finance[1]

[1] www.finance.yahoo.com

There are three major factors that have the potential to increase the company's future capitalization. The first is recognition by the European Medicines Agency Orphan Drug Designation (ODD) of Barth syndrome and the high unmet need for innovative treatments for this ultra-rare genetic condition. The second is that Stealth BioTherapeutics Corp has reached an agreement with Morningside Ventures for Morningside to provide $30 million of additional payments to the company during 2021. Finally, the performance of the company is highly dependent on the clinical trials of its pipeline, which is now slowed down due to both regulatory and, presumably, technical obstacles.

ResTORbio Case

Early failure is not the worst outcome in the development of clinical therapy. More serious is failure in the final and most expensive clinical trial phase: Phase III. At that point, the treatment is already being tested on many patients, and a lot of money has already been spent on development. This is what happened to RTB101, a drug developed by the company resTORbio. RTB101 – an oral, selective, and potent TORC1 inhibitor – was developed to prevent clinically symptomatic respiratory illness (CSRI) in adults aged 65 and older.

Following the announcement of the clinical trial failures, the company's capitalization fell by 83%, from $260 million to $43.7 million, as shown in the chart below. Dr. Gray believes that the market reaction was too sharp and did not reflect the actual situation with the company, as RTB101 was not the only resTORbio product in clinical trials. In addition, the company had a sufficient stock of cash on the balance sheet.

On September 15, 2020, Adicet Bio, Inc. announced the completion of the merger with resTORbio, Inc. After the merger, prior Adicet equity holders collectively owned approximately 75% of the combined company on a fully-diluted basis, while prior resTORbio equity holders collectively owned about 25% of the combined company on a fully-diluted basis. In connection with the merger, resTORbio implemented a 7-for-1 reverse split of its common shares. Each common stockholder of resTORbio immediately before the merger also received a contingent value right, entitling such holders to receive substantially all of the net proceeds from the commercialization, if any, received from a third party commercial partner of the company's experimental respiratory tract infection drug RTB101. The combined company will be led by resTORbio's co-founder, president and CEO, Chen Schor, with offices in Menlo Park, Calif., and Boston. The board will include Carl Gordon as chairman, with Erez Chimovits, Aya Jakobovits, Yair Schindel coming from Adicet's board, while Jeffery Chodakewitz and Schor have joined from resTORbio's board. Steve Dubin also joined the board at the closing of the deal. Hence, due to the losses in clinical trial failures, the company had to merge with Adicet Bio and redesign its whole business strategy.

Currently, Adicet Bio, Inc. is discovering and developing allogeneic gamma delta T cell therapies to enhance selective tumor targeting, facilitate innate and adaptive anti-tumor immune response, and improve persistence for durable activity in patients.

ResTORbio Market Capitalization, $ Million

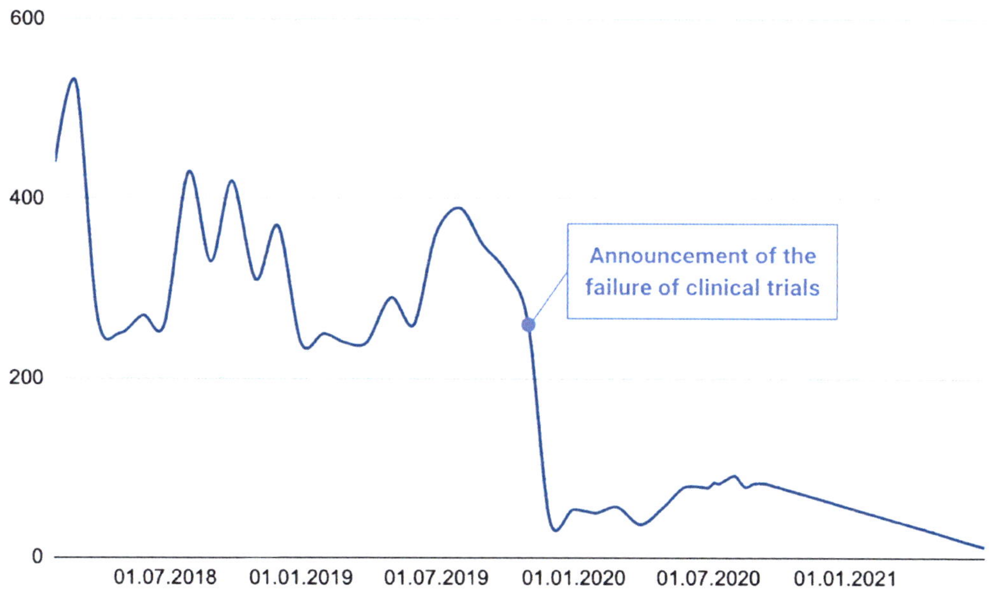

Source: Yahoo Finance[1]

[1] www.finance.yahoo.com

Juvenescence

In 2020, Jim Mellon, the British billionaire, donor and investor in a number of Longevity for-profit and non-profit projects announced plans to make his life extension company Juvenescence public in six to twelve months.

A ***CNBC report***[1] indicated that the IPO was already behind its original schedule as of September 2020, with *The Financial Times* having reported[2] in June 2018 that it would happen at some point in 2019.

London-based Juvenescence is investing in a wide range of relevant therapies mentioned throughout the book, including Insilico Medicine (Deep Knowledge Ventures was the first investor in this company), a leader in the use of next-generation Artificial Intelligence and deep learning for drug discovery. It has also backed Michael West's AgeX Therapeutics, the above-mentioned California-headquartered firm trying to create stem cells that can regenerate aging tissue; Lygenesis, the organ regeneration company; Souvien Therapeutics, which tackles epigenetic mechanisms associated with neurodegeneration; and BYOmass, which develops therapeutics to modulate the central control of metabolism associated with aging and age-related chronic illnesses.

Juvenescence has already privately raised about $170 million, and, according to Mellon, it still has a "fair amount of money left in the bank." Other billionaires – including Mike Cannon-Brookes, the co-founder of Australian software firm Atlassian, and NEX Group founder Michael Spencer – have invested in the company.

Considering the strong R&D, financial, and administrative capabilities of Juvenescence, as well as the general trajectory of its development, the company has potential to provide an example of a successful Longevity investment strategy, although the problem of the absence of clinical results validated on humans may play a negative role here as well. Multiple IPO delays may be a sign of the lack of investors` confidence in the practical applicability of the products developed by the companies comprising the portfolio of Juvenescence.

---

[1] www.cnbc.com/2020/09/29/billionaire-jim-mellon-plans-to-take-life-extension-start-up-public.html

[2] www.ft.com/content/30bb0752-6d5e-11e8-92d3-6c13e5c92914

At the time of writing, Juvenescence's planned IPO has not yet taken place, since the company is aiming to provide positive market indicators in order to gain significant price advantage during the valuation process, which, in turn, is highly dependent on the human validation of their R&D developments.

The Validation Gap

It is clear now that the performance of the Longevity companies is intrinsic to the human validation processes. Although many companies with failed clinical trials generate positive returns, they do not achieve average market results, as measured for example by S&P or NASDAQ indices. Correspondingly, they cannot be attractive for investors who have an opportunity to invest into disproportionately less volatile but more profitable assets. The returns of such Longevity companies, adjusted for overall market growth, are negative. Even in many cases of robust above-average returns, they occur due to the results and products that are not related to Longevity directly, such as chronic diseases or treatments related to COVID-19.

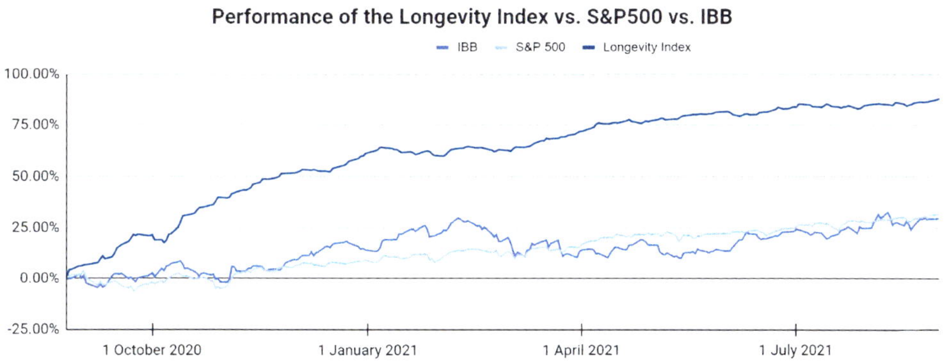

There are, however, Longevity corporations characterized by strongly positive stock price dynamics. The graph below depicts the performance of the returns of the Proprietary Longevity Index developed by ***Longevity.Capital***[1]. It can be seen that the returns of this Index are close to **88%** by the end of the tracking period. On the other hand, the S&P Index, which could be treated as a more reliable, yet more slowly growing one, has experienced growthof around **30%** during the last

[1] www.longevity.capital

year. Some Longevity companies that demonstrate tangible results on humans and are leaders in their subsectors, may achieve 3x or 5x annual returns, such as, for example Biophytis or Gensight Biologics.

At the same time, many Longevity corporations do not manage to achieve above-average returns even compared to the general BioTech market (represented, for example, by the IBB stock index), wihch is more volatile than the overall market and is not characterized by better performance (annual growth is no more than 28%). However, considering that Longevity is more complex, R&D-intensive, and disruptive than general BioTech, we would expect the opposite. It also emphasizes the negativity of the situation that develops around Longevity companies without the results validated on humans.

This implies that a number of companies analysed above are able to enhance their performance in case of overcoming the human validation-related obstacles, since they are not showing investment expediency even in comparison to the S&P 500 Index.

The Increasing Role of SPACs in Longevity Industrialization

While IPOs remain the industry standard for Longevity companies seeking to go public, an increasing number are using an alternative vehicle – a so-called special-purpose acquisition company, or SPAC. This trend has certain specific implications for the ongoing maturation of the Longevity Industry and warrants a dedicated analysis here.

An SPAC is a blank check company with no commercial operations that is formed specifically for the purpose of raising capital through an IPO for the purpose of acquiring an existing company. It is, in essence, a shell company established or sponsored by investors with the single purpose of raising money via an IPO to eventually acquire another company.

SPACs are becoming increasingly attractive to life sciences companies as an alternative to traditional IPOs because they make the task of going public easier. SPACs give companies more time to pitch investors, require a lesser scope of financial statements and disclosures, and need a shorter prospectus, enabling them to be completed in a matter of weeks rather than months. SPACs often result in higher valuations and more price certainty than IPOs because terms are negotiated with just one entity – the SPAC sponsor – rather than several. And, **most importantly, in the eyes of the SEC, SPACs are viewed as a type of M&A deal, which requires much less due diligence than an IPO. As a result, they require less scrutiny and due diligence than traditional IPOs.**

The central theme of this and the preceding chapter has been the need for a higher degree of scrutiny and a more tangible framework of metrics for valuating and performing due diligence on Longevity companies seeking to go public, with the goal of neutralizing the risk of positive results in model organisms failing to be replicated in humans. From this, it should be clear why the rising trend of Longevity SPACs is so concerning.

The first Longevity company to go public via an SPAC was Humacyte, a company developing bioengineered Human Acellular Vessels (HAVs), which provide prosthetic vasculature for the repair and replacement of injured and diseased blood vessels. Humacyte launched their SPAC via a merger with Alpha Healthcare Acquisition Corp, raising $100 million (in addition to a ~$175 million private VC deal) at a $800 million pre-money valuation. This SPAC-mediated merger has created a market capitalization of $1.1 billion. With promising Phase II clinical trial results, three Phase III clinical trials in progress, and plans to file for FDA approval in 2022, Humacyte's choice to go public via an SPAC raises fewer concerns than companies that might seek SPACs as an alternative to IPOs without a high degree of human validation. Humacyte ***successfully completed***[1] the SPAC deal on 27 August 2021.

The second company seeking to acquire Longevity businesses via an SPAC is Frontier Acquisition Corporation, led by Chief Executive Officer Christian Angermayer, Co-Presidents Rick Gerson and Ryan Khoury, and Co-Chairmen of the Board David Sinclair and Peter Attia, which recently announced the successful closing of a $230 million IPO.

While the company's prospectus does not identify Longevity as its focus, it is safe to assume that it will seek to acquire an existing Longevity company, given David Sinclair's status as founder of several private Longevity start-ups, Peter Attia's role as Chief Medical Officer of the fasting app Zero (focused on Longevity-focused fasting), and

---

[1] www.biospace.com/article/humacyte-closes-on-business-combination-with-alpha-healthcare-acquisition-corp-

Christian Angermayer's public statements and general interest in Longevity. Indeed, it may be reasonable to assume that Frontier Acquisition Corp. may seek to acquire Sinclair's primary Longevity-focused company, Life Biosciences, or one of its subsidiaries.

Yet, there is already some ambiguity associated with treatments developed by David Sinclair's Longevity companies, such as, for example, Sirtris which was acquired by GSK.

"We decided to do this because the research at Sirtris exploring the biology of sirtuins has been highly successful and now requires the resources and expertise available from our broader drug discovery organization," a GSK spokesman said. "We're excited about taking sirtuins to the next phase of drug development. Our next steps will be to examine our research against a variety of therapeutic conditions, with the aim of moving potential assets into the clinic within the next three to four years."

Glaxo paid $720 million to acquire Sirtris, David Sinclair's hot BioTech company, in April 2008, generating lots of breathless media coverage. The company sought to make drugs that act on sirtuins, a class of proteins that scientists believe play a role in aging, programmed cell death, and other key cell processes. Five years later the company was closed due to the absence of validated results.

The research into the biological role of sirtuins by David Sinclair attracted its share of skeptics. Sinclair sought to buttress his early work in 2013 with an article in *Science* that stated that resveratrol and related compounds can activate sirtuins.

Life Biosciences, which was established in early 2018, acted as a holding company, acquiring many Longevity start-ups. Some of them are delivering actual results in terms of human Longevity, while the results of other ones are ambiguous and may be overpromoted through the personal brand of David Sinclair. This raises concerns among many scientists and experts in the Longevity field, although David Sinclair himself is definitely one of the leading Longevity researchers in the world. At the time of writing, Frontier Acquisition Corporation has not acquired any Longevity companies, including David Sinclair's ones. Current developments in Longevity SPACs are tracked in the ***Longevity investment digest***[1] by Aging Analytics Agency.

[1] www.aginganalytics.com/investment-digest

It is quite clear that this particular group of companies has very good prospects to achieve significant results; however, potential links with overhyped unvalidated projects may be a point of concern.

Moreover, Longevity companies seeking to go public via SPACs should be treated with some caution given their lower thresholds for due diligence, scrutiny, and company disclosures. There are already existing issues and uncertainties resulting from the growing trend of Longevity companies going public without sufficient human validation and the declines in the market capitalization of public Longevity companies following the failure of clinical trials (and the failure to replicate promising results in model organisms in humans). **The limitations of SPACs represent just one more indication that the future stability of the industry depends upon the creation and adoption of a more reliable and realistic framework of due diligence, combined with the use of valuation metrics that are based on some level of human validation for Longevity companies seeking to go public.**

Author's Personal Views on Longevity IPOs

Deep Knowledge Group's investment arm, Longevity.Capital, is investing in publicly traded companies participating in the Longevity Industry. By our estimates there are more than 600 Longevity-related companies in the global market. As such, our analysts are quite cautious about companies claiming to be involved in Longevity that do not have at least some intermediate results in humans.

There are many instances of financially underperforming Longevity companies and investors' distrust. I am of the opinion that it would be very healthy for the continued systemic growth of the Longevity Industry if companies that want to go public and claim to be Longevity-focused were to use other, more modern, sophisticated, reliable, and validatable approaches for going public.

***Typically, companies that claim to be part of this industry have high valuations during private-stage funding rounds and even higher market capitalizations at the IPO stage. But this also means there ought to be a higher degree of scrutiny as part of going public.** This is an entirely different and complex topic of discussion in need of separate, detailed treatment, and I will be providing some additional consideration on this matter in Chapter VI ("A Proposed Solution to the Longevity Industry's Biggest Risk: Modern Tools for Human Validation and De-Risking In-*

vestment Decision-Making Processes and IPOs"), and in even greater detail in Chapter IX (" Longevity Industry 2.0: Biomarkers of Human Longevity as the Major Catalyst to Increase Investments and Ensure the Sustainable Growth of the Longevity Financial Sector"). **But, in brief, I am of the strong opinion that providing a modern due-diligence framework for investors based on the biomarkers panels will bring clarity and foresight on what they can and should expect from companies aiming to go public and claiming to be focused on Longevity. This framework needs to be based on tangible, reliable, and verifiable metrics for analysis and forecasting of clinical outcomes.** The establishment of such a framework will unleash a literally unlimited source of additional investment capital not only from retail investors but also from institutional investors – on a scale of trillions of euros.

This will serve not just to maintain the Longevity Industry's positive trajectory of growth and provide additional industry stability by protecting and preserving investor sentiment – it will also serve as the catalyst for translating the theoretical humanitarian potential of the industry into real-world practical impacts in terms of human health, HALE and QALY. This is because we estimate that it will take just €1 trillion spent over 10 years on Longevity R&D to ensure that Longevity Escape Velocity will be achievable for the majority of individuals by the year 2030.

Summary

The lack of a tangible framework based on human validation for Longevity company valuation and due diligence, especially in the case of companies seeking to go public, creates significant risks and issues for the continued growth and stability of the industry.

We have argued that this constitutes nothing less than the largest and most fundamental systemic risk faced by the Longevity Industry, posing the threat of a widespread downturn in investor confidence and sentiment if increasing numbers of Longevity companies fail to replicate preliminary positive results in animal models in actual humans via clinical trials. The signs of this problem are already visible: Longevity stock returns are either underperforming compared to market or stagnating in many cases.

In this chapter, we substantiated this thesis with real-world examples, *demonstrating how a number of public Longevity companies experienced declining market capitalization dynamics following clinical trial failures.*

The key points to take away from this chapter are:

- A lack of reliable and verifiable metrics for Longevity companies valuation and due diligence has led, in several real-world cases, to issues with market capitalization.
- A framework is needed for investors to have clarity and reliable foresight on what they can and should expect from companies aiming to go public and claiming to be focused on Longevity. This framework should be based on tangible, reliable and verifiable metrics for the analysis and forecasting of clinical outcomes. The framework will be a necessary tool to neutralize these issues and their accompanying risks for the health of the entire Industry.
- The Longevity companies are struggling when going through the clinical trials process. The companies are facing the same obstacles as ones participating in the pharmaceutical or BioTech industries, yet are sensitive to the complexity of the human-based longitudinal trials, and, thus, are much more vulnerable to the fluctuations gained in the case of failed clinical trials.
- Many Longevity companies are characterized by negative financial performance when compared to such benchmarks as S&P, NASDAQ, or IBB (BioTech) indices. At the same time, their returns are often more risky (volatile) than the returns of the companies in other BioTech sectors.
- The establishment of a standardized validation framework will unleash an unlimited source of additional investment capital not only from retail investors but also from big institutional investors – on a scale of trillions of euros.
- In order to bring closer the golden era of Longevity therapeutics, Aging Analytics Agency is working on a special case study concerning the new approaches for experiments and clinical trials for the Longevity Industry. Aging Analytics Agency is aiming to produce a unique case study on exactly this challenge: safe, modern, Tech-driven approaches for therapeutic validation in humans. We can expect this project to be released in Q3 2022.
- The additional and relevant information about the latest developments of the Longevity publicly traded companies could be found

in the corresponding regularly updated report published by Aging Analytics Agency. For interested readers we also suggest visiting our Public Longevity Companies Investment Big Data Analytics Dashboard, where they can obtain all relevant information about the Longevity-related companies and use uniquely developed analytical tools for assessment of investment expediency.

Longevity Investment Big Data Analytics Dashboard

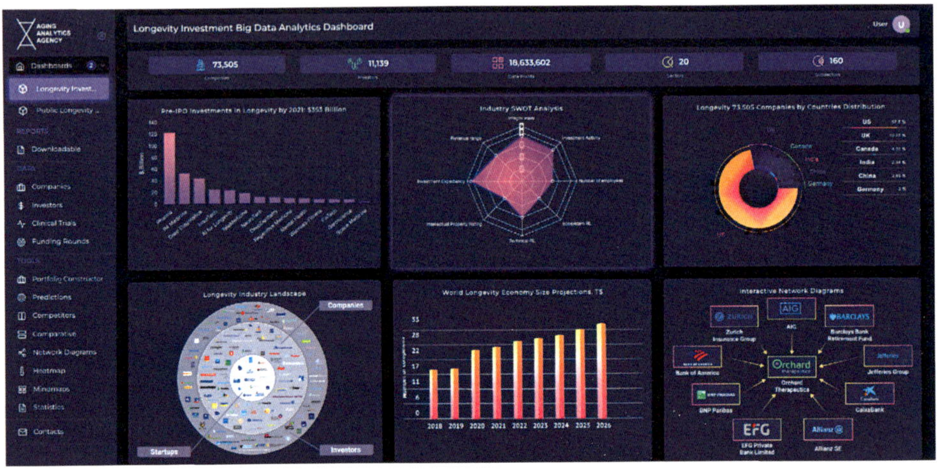

Source: Longevity Investment Big Data Analytics Dashboard[1]

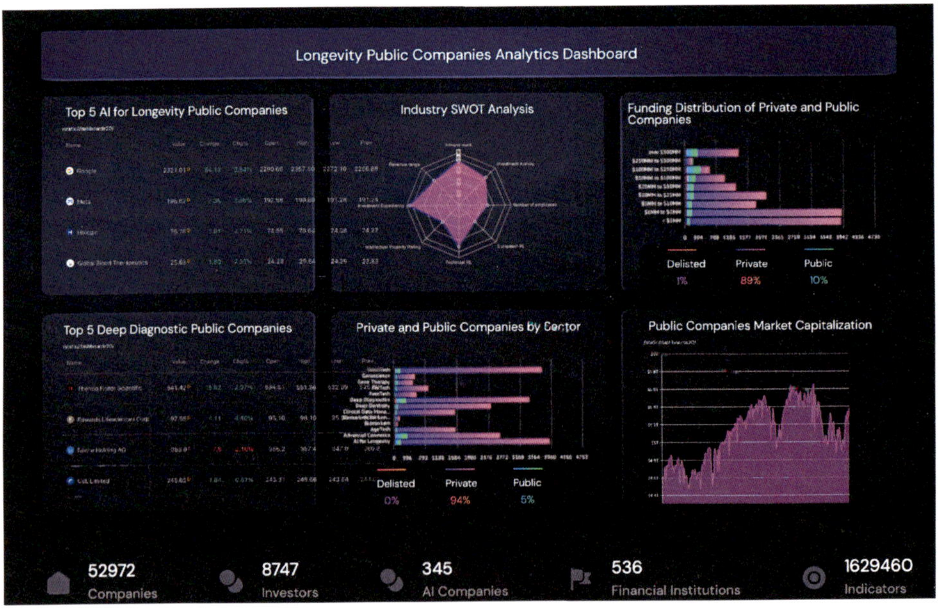

Source: Public Traded Longevity Companies[2]

[1] www.deep-innovation.tech/longevity-investment
[2] www.deep-innovation.tech/public-longevity-investment-dashboard

Chapter V: Biomarkers as the Major De-Risking Tool in Biomedicine: Examples from Oncology and Other Industries

Biomarkers Accelerate the Growth, Maturation, and Stability of the Biomedical Sectors to Which They Are Applied

Biomarkers have come a long way, from simple measurements of clinical diagnosis, to essential tools in R&D and practical applications. The use of biomarkers has been expanding over the last couple of decades due to their potential for disease diagnosis, prognosis, treatment response, pharmacokinetic-pharmacodynamic modeling, and the monitoring of therapies.

Recently, the COVID-19 pandemic boosted R&D for novel diagnostics and led to the approval of many biomarker-based diagnostic tests for the early and rapid detection of the SARS-CoV-2 virus.

Historically, however, pharmaceutical drug developers have often struggled to overcome escalating cost barriers and high drug attrition rates in late-stage clinical trials.

But recent history has shown us that as soon as an industry-wide consensus forms around a relevant panel of biomarkers, the industry reaches a tipping point and experiences runaway success.

The Growing Global Biomarkers Market

The 21st Century Act in the USA allowed the Food and Drug Administration (FDA) to publish guidelines for Biomarker Qualification for use in drug development programs. This paved the way for biomarker inclusion in drug development through either the drug approval process or the Biomarker Qualification Program.

Biomarkers are extremely useful in clinical trials, and they are increasingly used to identify populations for study, monitoring therapeutic responses, and identifying side effects. There is an emerging market of clinical research organizations (CROs) carrying out clinical trial recruitment and other services, while expanding technical expertise in bioanalytical and biomarker development. This enables pharmaceutical clients access to biomarker discovery and development.

The global biomarkers market is growing at a significant pace, driven by an explosion of publications and clinical trials. Enhanced analytical methods and the development of new, sophisticated, sensitive multiplex methods in gene expression analysis, proteomics, metabolomics and transcriptomics bring huge momentum to this market. The development of multi-biomarker assays, novel immunoassays, multi-modal imaging, and mass spectrometry methods further drives market growth.

Collaborations, strategic partnerships, mergers, acquisitions, and other deals between private and public players are on the rise. Companies are strengthening technical know-how and expanding product portfolios in order to offer enhanced services and new offerings to the biomarker research community. Precision medicine, particularly in the field of cancer, has contributed immensely to an interest in biomarkers, leading to the growing adoption of biomarkers in companion diagnostics and in the selection of targeted patient populations for high-value drugs. Other therapeutic areas – such as cardiovascular, neurodegenerative, and autoimmune diseases – are getting noticeable attention in biomarker research.

Challenges for this market remain in the form of disparities between biomarker definitions at an international level and the lack of any defined regulatory guidance for use in R&D. There is still a need to develop sensitive and robust methods of analysis for low concentration analytes via methods that can be validated. Lack of skilled manpower and the high cost of technology are other challenges.

Yet, positive approaches in biomarker research, effective dialogue, and collaborations between all stakeholders are expected to address challenges and take this market forward in coming years.

Personalized medicine runs on "big data," and the data revolution continues. As we unravel individual genomes to look for individualized traits, we are faced not only with just over 3 billion base pairs but also with epigenomic changes, noncoding regions, regulatory sequences, etc., that add layers of complexity. New technologies continue to advance in their ability to identify molecular markers of disease within this information. Although tools have made identifying variants a fashionable pursuit, identifying actionable variants that serve as biomarkers has the potential to revolutionize the way health care is delivered in the future.

The pace at which biomarker discovery is translated into clinical use is often slow and arduous.

This history begins with the Bence Jones protein. In 1847, a peculiar protein was discovered in the urine of a patient with multiple myeloma. More than 100 years later, this was identified as a free antibody light chain produced by the tumor. In 1988, more than 140 years later, a routine diagnostic test for the protein as a biomarker for multiple myeloma was approved by the United States Food and Drug Administration (FDA).

Discovery experiments are often referred to as "fishing expeditions." This often begins with available archived samples followed by investigations such as proteome sampling, which reveals mostly highly expressed proteins such as albumin and immunoglobulins. This contrasts to the identification of a very sensitive and specific biomarker that is diluted by an abundance of peptides. In addition, technical difficulties, simple variations in sex, time of day, and concomitant medical conditions or treatments may alter the proteome within individuals. Reproducibility has been challenging for many as the composition of circulating proteins in the blood represents what is happening to the whole organism at any given time.

It is, therefore, important to understand that the information obtained will be only a snapshot of the organism at a specific time. The added natural complexity of cancer itself limits the insight a single molecular signature provides. Collectively, these elements contribute to the complexity of deriving meaningful data from investigations.

The development of a clinically meaningful diagnostic biomarker best begins with a predefined roadmap. At the beginning of this is a central question: "Is there an unmet clinical need?" As potential candidates develop, a reproducible assay must be available and used to define the markers' distribution in the target population. This should then be tested against the "gold standard" for diagnosis. If added diagnostic accuracy exists, then biomarker performance should be validated.

Central to discovery is study design, part of which includes sample acquisition. Oftentimes, subject selection is driven more by specimen availability than by a stringent study protocol. This retrospective examination is not only subject to immense bias but may also lack thorough patient data for support and be unable to satisfy power calculations based on subject inclusion criteria. This raises concerns of clinical validity as well as future clinical utility.

Given the limitations of prospective studies, it makes practical sense to begin with archived specimens. Although this has not been traditionally accepted as high-quality evidence, some experts have proposed scenarios in which using archived specimens to validate biomarkers may be of considerable value. Emphasis has been placed on "prospective-retrospective" studies, assaying samples from previously conducted prospective trials. For example, genomic analysis of samples collected during Phase II trials can be used to identify a companion diagnostic marker. This can later be applied to a Phase III trial to decide which patients would benefit most from a chosen treatment. With more robust results due to improved patient selection, drug manufacturers, patients, and physicians all stand to benefit from streamlined drug approval, tailored therapies with less side effects, more cost-effective care, and better outcomes.

This elevates the role of the pharmaceutical industry at each step. Large, well-designed prospective trials are costly, and the deep pockets of Big Pharma can support them. The disease process of systemic lupus helps illustrate this. Industry took notice of a medical need for improved therapies and made investments in discovery. There are now a number of new medications in clinical trials as well as newly FDA-approved ones. This created a scenario in which novel biomarkers may help define subsets of patients likely to respond to certain treatments.

More than a century after Bence Jones, there came "Cancer Moonshot 2020." In October 2015, Dr Patrick Soon-Shiong, a billionaire investor and surgeon, met with Vice President Joe Biden and presented a position paper titled, "The Precision Against Cancer PAC: The

Moonshot Program to Develop a Cancer Vaccine for 'I am N = 1.'" This collaborative project aims to enroll 20,000 patients with 20 different tumor types in Phase II cancer vaccine trials. A few weeks later, Vice President Joe Biden gave a speech calling for a "moonshot" to cure cancer. The Cancer Moonshot, managed by a consortium of companies called The National Immunotherapy Coalition, is a public-private partnership consisting of the government (NCI, FDA, the White House, and Congress), the pharmaceutical and BioTech industries, health care providers, and insurance companies, all of which have aligned to compress 10 years of advancement into 5 years. Several pillars of their work include big data analytics, next generation sequencing, and formulating responses to treatments.

Along with unified support comes a very large amount of funding. As part of the 21st Century Cures Act signed into law in December 2016, $1.8 billion was allocated for the Cancer Moonshot and $1.4 billion for the Precision Medicine Initiative aimed at collecting the genetic information of 1 million Americans. If appropriated each year, this funding demonstrates commitment to the importance of finding meaningful biomarkers.

Pharma R&D Has Been Transformed Into a Biomarker-Driven Industry

Biomarkers are becoming an essential part of the drug discovery and development process. A biomarker-driven approach to developing targeted therapies and patient selection strategies has the potential to increase success in the drug development process, decrease costs, and ultimately improve patient outcomes.

The Tufts Center for the Study of Drug Development estimates the cost of developing a new drug at $2.6 billion and suggests that the costs of drug development are rising at a compound annual growth rate of 8.5%.

Additionally, pharmaceutical R&D is moving away from broad-spectrum therapies and toward targeted therapies for treating patients with subtypes of diseases as technologies become more sophisticated.

Technology is fundamentally changing the way that clinical trials are operationalized, says Rick Morrison, CEO of Comprehend. "The excitement around biomarkers is that there are now technology infrastructures that allow researchers to do things that were not previously possible," he commented.

Morrison points out that modern technologies, which are often cloud-based technologies, are purpose-built to manage enormous amounts of data or analysis in ways that were not previously possible.

The global biomarkers market was valued at $24.10 billion in 2015, according to an analysis by MarketsandMarkets, and is expected to grow at a CAGR of 13.58% to reach $45.55 billion by 2020.

An additional example of biomarker market growth may be the biobanks industry. A biobank is a type of biorepository that stores specific types or a variety of types of human or animal samples and, therefore, a huge amount of different biomarkers. Biobanks are the most rapidly growing type of repositories in the biotechnology industry and serve as one of the most frequently-used resources in drug development.

Rising healthcare and R&D expenditures are the key growth drivers in this market. Diagnostics development accounted for the largest share of the global biomarkers market in 2015.

Oncology as the Sector Most Rapidly Adopting Biomarkers

Experts say oncology is the fastest-growing area for the use of biomarkers. In fact, the global cancer biomarkers market, which has experienced substantial growth over the last couple of years, is expected to grow at a CAGR of 11.6% from 2015 to 2020, according to a recent report from MarketsandMarkets.

Factors such as rising incidence of cancer, technological advancements in "omics" technologies, favorable government funding for cancer research, increased demand for personalized medicines in cancer therapies, and increased use of biomarkers in drug discovery development and clinical trials are expected to fuel the growth of the cancer biomarkers market during the forecast period.

Cancer biomarkers are molecules secreted by a tumor or a specific response of the body to the presence of cancer. They can help to enhance cancer detection and facilitate high-speed noninvasive diagnosis using genomics and proteomics tools.

A ***recent study***[1] published in the journal Cancer Medicine provides the first systematic statistical evidence that biomarkers dramatically improve oncology outcomes when it comes to testing new medicines. The new research compared drug approval success rates in biomarker-based clinical trials to outcomes in trials without biomarkers for four types of cancer: breast cancer, melanoma, non-small cell lung cancer, and colorectal cancer.

According to Rachel Laing, Ph.D., manager of Bionest Partners, biomarkers can help better identify patients who will respond to therapy. An increased understanding of disease and technology advancements are creating opportunities in the biomarker space: "We've come a long way in the past five or 10 years to really understand how disease pathways interact with each other," she continues. "And we now have more knowledge and more opportunities for biomarkers, with next-gen sequencing and new platforms."

There have been numerous successes in the biomarker area, especially for companies. One of the major goals of organizations is to bring a product to market more quickly. By incorporating analytics data into their clinical trial programs, companies have found that they can get products to market faster.

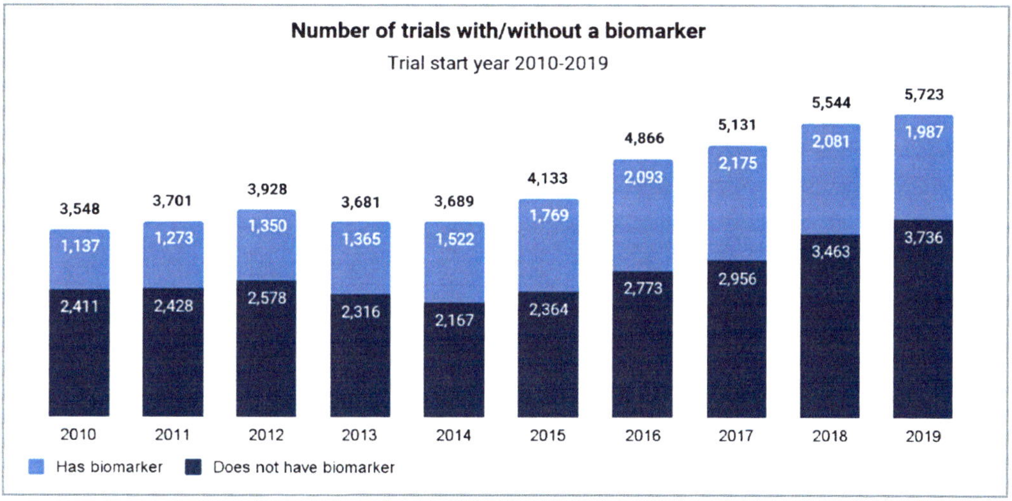

Source: Clinical Leader[2]

This scalability allows researchers and companies to view patients

[1] www.onlinelibrary.wiley.com/doi/10.1002/cam4.3732
[2] www.clinicalleader.com/doc/the-evolution-of-biomarker-driven-approaches-to-clinical-trials-0001

in a much more targeted way. Biomarkers can help identify specific patient populations and streamline the development process. Recently, we've seen this in terms of immuno-oncology, but it could be applicable to other indications as well.

A push to make treatments more personalized and look at the patient as an individual who has specific traits is transpiring in the biomarker space. This could help with the selection of an optimal treatment approach. The best way to do this is to look for genetic biomarkers or proteins that are overexpressed in a particular tumor.

According to Principal Business Insights Analyst Dr. Orestis Mavroudis-Chocholis, this approach indicates that the use of biomarkers in drug development will increase, especially in oncology. The future of oncology involves personalized treatments using biomarkers.

Today and in the next era of drug development, we will see greater personalization. We will see more and more personalized medicines that are designed for highly targeted, highly defined patient populations. This will have a very positive impact on the success of those medicines.

AI for Drug Discovery – Focus on Oncology

In Chapter III ("Reality Check – The Current State of Longevity Clinical Trials"), we provided an overview of the history and current state of the AI in Pharma industry, and the broader potential of AI for accelerating the delivery of practical healthcare results in humans, including its increasingly fundamental role in Longevity R&D and clinical translation. In that section, we also discussed how the use of AI in the oncology industry (from basic research all the way to drug discovery and biomarker development) serves as a leading case study and example of AI's power to rapidly accelerate industry development and stabilization.

This can be clearly seen in the industry statistics and market insights presented by ***Deep Pharma Intelligence's***[1] most recent analytics on this topic: ***Artificial Intelligence in Drug Discovery – Focus on Oncology (Q2 2021)***[2], a follow up to its analogous ***Q1 2021 special case study***[3].

[1] www.deep-pharma.tech
[2] www.deep-pharma.tech/ai-in-drug-discovery-oncology-q2-2021
[3] www.deep-pharma.tech/oncology-report-q1-2021

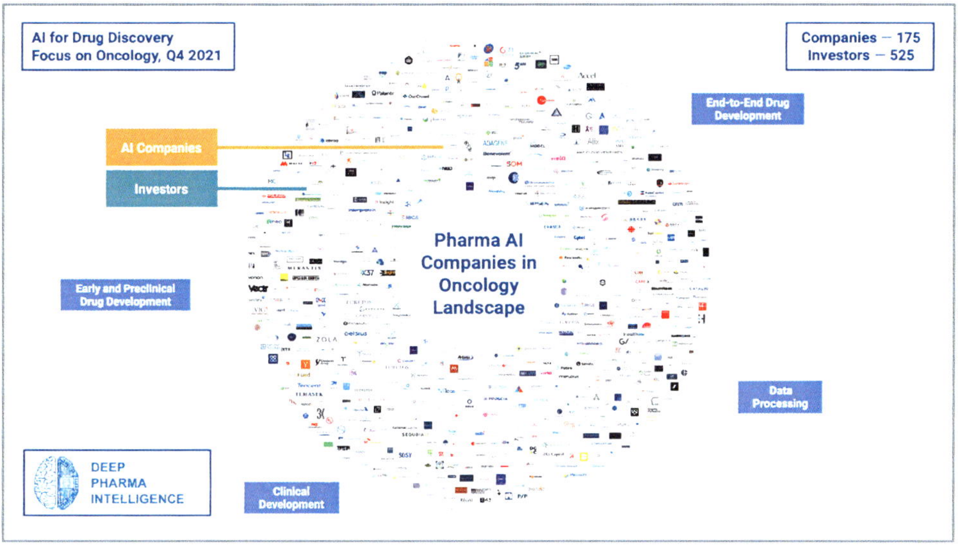

Source: Deep Pharma Intelligence[1]

The AI for Drug Discovery in Oncology special case study highlights leading AI-driven companies, their investors, selected collaborations with pharma corporations, relative expertise of such companies, level of their funding, examples of deals and collaborations, as well as examples of AI applications. The report is based on Deep Pharma Intelligence's ***online cloud-based analytics system (Dashboard)***[2], which features an interactive database of all entities featured in the report, and tens of thousands of additional data points about market trends, key developments, funding, deals, and more.

According to the World Health Organization (WHO), cancer is the leading cause of death globally, accounting for nearly 10 million deaths in 2020. It is the main reason why oncology has traditionally been the major focus of drug developers — oncology therapeutics being the largest category among all FDA-approved drugs last year (35%). It was found that oncology drug discovery has a 3.4% success rate on average; however, it is expected that active implementation of AI in this area would increase this rate.

And, indeed, oncology is (by far) the most prominent therapeutic focus area within the broader AI for Drug Discovery industry landscape, with 74 companies in total working in this domain, and accounting for 23% of the entire company landscape.

[1] www.deep-pharma.tech/ai-in-drug-discovery-focus-on-oncology-2021
[2] www.dashboards.invest-solutions.tech/dashboards/longevity-investment

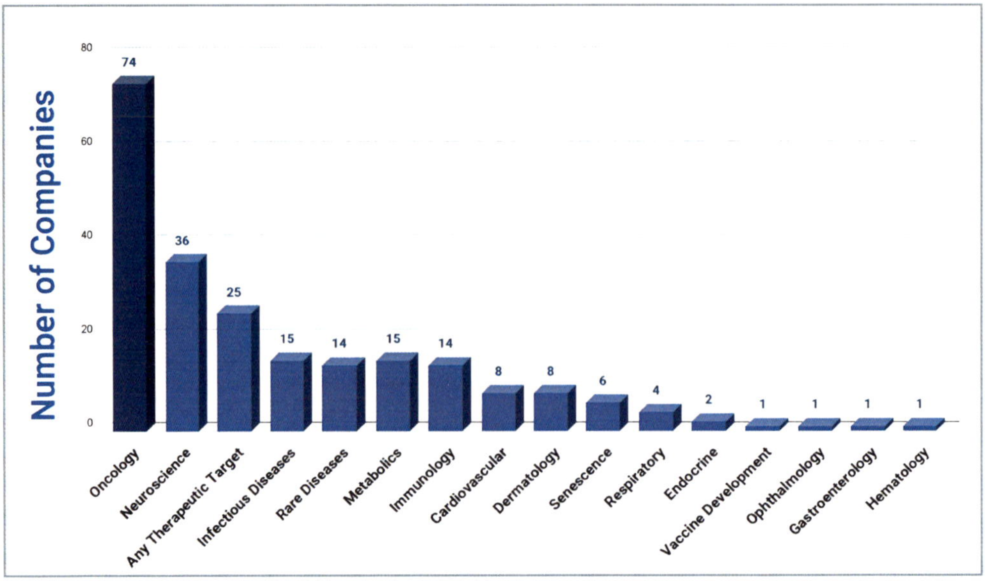

Source: Deep Pharma Intelligence[1]

The leading areas in oncology drug development using AI are hit discovery and target/ID validation. The variety of the data involved in such processes makes AI an irreplaceable tool in prediction of the physicochemical properties, bioactivity and toxicity of small molecules. The third place is taken by preclinical development, where AI is mostly used to predict whether a given molecule will pass preclinical trials considering the genomic profile of the target cell line. The next one is clinical development, where AI can assist in selecting only a specific diseased population for recruitment in Phase II and III of clinical trials.

And while biomarker discovery was, as of Q2 2021, one of the less-prevalent focus areas in terms of discovery stage (with only 14 of 75 AI in Oncology companies specializing in this), it is likely that this is the case because the overall arena of oncology biomarker discovery is already very well developed, having witnessed substantial growth and progress in the past decade.

Considering the rapid pace of industry development in the respective arenas of oncology biomarker discovery and AI for Drug Discovery in oncology in recent years, combined with the established fact of oncology serving as perhaps the best example of effective clinical medicines coming to market much faster than in other pharma sectors, it should be clear that both of these domains (AI for oncology,

[1] www.deep-pharma.tech/oncology-report-q2-2021

and the development of industry consensus biomarkers for oncology) are among the factors most likely to have contributed to oncology's current state of tangible, real-world clinical effectiveness.

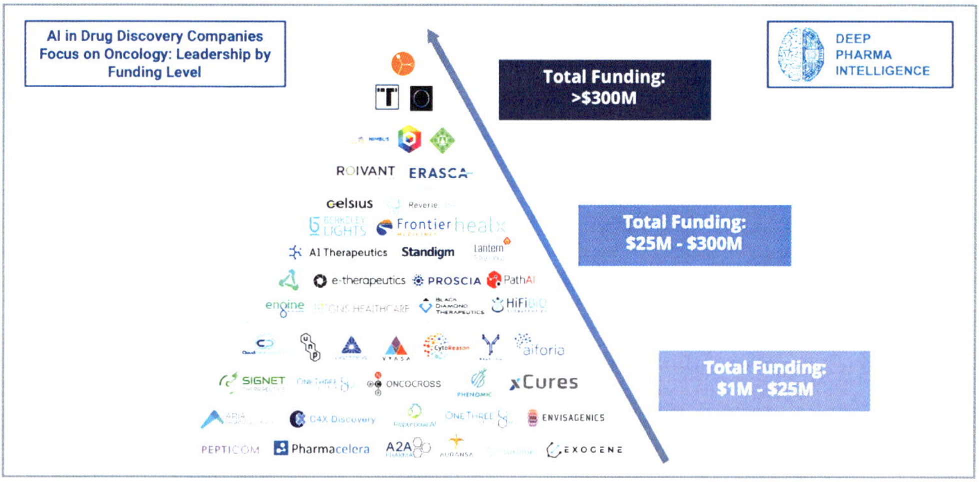

Source: Deep Pharma Intelligence[1]

Biomarkers Will Be the Force Spearheading an Industry that Dwarfs All of BioTech

If those are the accelerative effects that can be achieved by an industry consensus around biomarker panels for Oncology or Pharma R&D, then consider the revolutionary effects of an industry consensus on Aging and Longevity Biomarkers Panels!

This would make biomarkers the driving force that will spearhead the Longevity Industry. And once consensus is achieved, all players in this field will be awash in quantities of capital that will dwarf even the capitalization of oncology.

Why? Because unlike oncology, the current, immediate challenge facing the Longevity Industry is the immense level of complexity and lack of harmonization across disparate sectors. Although cancer, famously, is highly multifactorial, it is an industry comprising relatively few sectors. But both the problems and solutions of aging and Longevity are orders of magnitude more multifactorial and multi-faceted than those of oncology.

[1] www.deep-pharma.tech/ai-in-drug-discovery-focus-on-oncology-2021

As such the Longevity Industry, at the present point in time, attracts a special type of unique talent.

Cancer research includes some of the smartest and most notable scientists on Earth, but Longevity scientists grapple with such exponentially harder challenges that they sometimes exhibit almost religious levels of commitment and personal sacrifice. It is, for example, much more common to find Longevity researchers putting their own health on the line by experimenting on their own bodies.

When the biomarker tipping point arrives for the Longevity Industry, the diligence and commitment of this expertise will have the full power and backing of international finance.

Summary

- Biomarkers, the biological indicators of health and disease, have come a long way, from being used as simple measurements of clinical diagnosis, to becoming essential tools in the clinical space and in drug discovery and development.
- Once a robust, industry-consensus framework of biomarkers is well established for a given biomedical sector, a tipping point is reached, leading to rapid success.
- A biomarker-driven approach to developing targeted therapies and patient selection strategies has the potential to **increase success in the drug development process**, decrease costs, and ultimately improve patient outcomes.
- **Experts say oncology is the fastest-growing area for the use of biomarkers.** In fact, the global cancer biomarkers market, which has experienced substantial growth over the last couple of years, is expected to grow at a CAGR of 11.6% from 2015 to 2020.
- **The Longevity Industry, however, is orders of magnitude more complex than either Oncology or BioPharma**, both in the relationships between its disparate sectors and also the multifaceted nature of the problems it seeks to solve. It, therefore, attracts a unique degree of talent, expertise, diligence, and commitment, which, when the tipping point arrives, will act as an engine of future exponential growth.

Chapter VI: A Proposed Solution to the Longevity Industry's Biggest Risk: Modern Tools for Human Validation and De-Risking Investment Decision-Making Processes and IPOs

Given the enormous differences between the biology of humans and the biology of common model organisms like mice, combined with the higher degree of biological complexity as it pertains to aging vs. single diseases, we can expect the clinical translation failure rate in the Longevity Industry to be much higher than the already-enormous failure rate in traditional BioTech (which estimated to be around 92% in case of drug discovery).

At this point, we at Aging Analytics Agency defined the major problem threatening the stability of the Global Longevity Industry and described why a fundamental shift in the ways that Longevity therapies and technologies are tested and validated (in terms of both safety and efficacy) is needed to safeguard the industry's potential to become the biggest, most stable, and most ethical industry in human history.

The use of Biomarkers of Aging and Longevity constitutes the most market-ready and validated means of proving efficacy in humans and can serve as the basis for demonstrating human-validated results by Longevity companies and start-ups. A wide array of single biomarkers and panels of biomarkers of Longevity exist in market-ready form and should be adopted into due-diligence practice by Longevity investors to create a more modern, sophisticated, and robust method of preliminary validation of therapeutic safety and efficacy.

Besides biomarkers, there are a number of other ways to provide preliminary indications of human validation, which could be used as the basis for a coordinated framework to provide investors with greater confidence in the likelihood of clinical translatability. These include:

- **In silico human modelling**
- **In vitro tests using human cells and tissues**
- **Human-animal chimeras (e.g., human-mouse)** for safety, toxicity and efficacy testing. This approach is already common in immuno-oncology research, and a wide array of validated approaches can be applied for testing of aging-focused interventions.
- **In vivo administration of subtherapeutic doses using microfluidic chips** (i.e., in vitro "skin-on-a-chip" testing)

Below are several of the companies that we feel may be capable of delivering prototypes of these solutions, including in silico human modeling solutions, in vitro tests using human cells and tissues, panels of biomarkers, and more.

For the relevant assessment of technology or treatment claimed by any Longevity company in the best-case scenario, their projects should be cross-checked by at least two independent Longevity Biomarkers companies, and if both are coming to favorable decisions, that is an excellent mark to proceed further. If both are making an adverse decision, there is a very high likelihood that the projects should not proceed further.

Any decision-makers – investors, entrepreneurs, policymakers – are advised to engage with the entities presented on this ***Longevity Biomarkers Platform***[1]. There are already 85 companies, 40 research labs and 7 non-profits. And in the coming years, we can expect that there will be more players in this field, and the market leaders will enhance, improve, and update the precision and accuracy of their biomarkers panels. Right now, the TRL of such panels is around 7-8; next year, it will be 8-9; in 2-3 years, it will be 9-10.

[1] www.longevity.international/biomarkers

AI Platforms (digital biomarkers)

Source: The Rising Wave Of Human Biomarkers Longevity Landscape Overview Q2 2021[1]

Nowadays, if relevant decision-makers fail to use these biomarkers panels, ***it may be considered as unreasonable***; in 2-3 years from now, ***it will be considered as unthinkable and irrational*** not to use them.

I would highly recommend investors take this into account already now because there are currently existing solutions, and some of them are already on the market.

[1] www.docs.google.com/presentation/d/1gLp5ChlGJt2DUY_Zn23mR5c6rc5y7lBu/edit#slide=id.ge00030d1ab_0_410

Modern and Historical Precedents for Human Validation

While the notion of human validation may seem controversial to many readers, the truth is that we now have a complete set of modern tools and approaches for safely demonstrating varying levels of human validation of Longevity therapies and technologies. Furthermore, the notion of human validation is not modern but very ancient and predates the present-day system of model organism studies.

In the early centuries and decades of biomedical research, it was common for doctors to engage in self-experimentation and to test their therapies on their own bodies. The most famous example is Edward Jenner (inventor of the vaccine), who inoculated himself with his smallpox vaccine in the late 18th century. This practice is more common among doctors and biomedical researchers leading up to the 20th century than many people realize. This is not to say that the same simplistic approaches should be revived, but the mindset of practical self-experimentation and human validation is more ancient than its opposite.

These practices were rightly replaced by paradigms and systems of model organism experimentation and validation in the 19th and 20th centuries. The transition secured the safety of doctors and scientists and proved effective for many decades when the complexity of biomedical treatments was low enough not to become fundamentally stalled by differences in the biology of humans and model organisms. However, today this entire mindset serves as a significant bottleneck for scientific and industrial progress, posing massive bureaucratic hurdles that prevent us from adopting and developing to scale modern technologies that can ensure safe human validation and the equipment to handle the increasing complexity of biomedical research and translation.

Another historical example is warfare medicine, which by necessity historically focused on human validation rather than model organism experiments precisely because of its urgency – i.e., the recipients of these techniques need immediate care and would be likely to die if not treated, while the sheer scale of those needing care (especially in the cases of World Wars I and II) was enormous. While this seems like an isolated and unrelated example, in many ways, we are in a similar war with aging, and each of us is assigned a death sentence if we do not get treatment. There are approximately 170,000 people dying every day globally on average, and we can consider that at least 100,000 of them are dying because of aging. Likewise, the scale of the need for the urgent anti-aging treatments is even more pressing, affecting all of humanity.

There are also modern-day examples, the most well-known of which are the biohacking and Quantified Self movements, where individuals attempt to push the limits of the Longevity medicine methods and technologies available to them to the extreme, utilizing the full arsenal of tools at their disposal (including those outside the scope of FDA-approved treatments) to optimize their state of physical, biological, and cognitive health and performance. California is home to a community of specialists who practice self-experimentation and so-called "Quantified Self" methods by using advanced approaches and techniques related to precise diagnostic and monitoring and implementing safe applications of geroprotectors. This approach translates to a more balanced and pragmatic alternative to biohacking as the latter is perceived as a more radical option.

Meanwhile, I know several medical doctors who pride themselves on the development, refinement, and validation of their Longevity-focused lifestyle practices and even technologies on their bodies. And while I am not in a position to reveal specifics in this book, I can attest to several technologies (not just lifestyle and behavioral practices, diets, and supplements, but actual technologies) that I am personally using to optimize my Healthy Longevity and performance.

It is vital that the modern set of tools, technologies, and approaches that make up the proposed framework for safe human experimentation and validation – as described earlier in this chapter – be adopted not only by the Longevity Industry but also the underlying drivers: Longevity scientists, doctors, and practitioners.

While the number of scientists and practitioners experimenting on themselves is increasing, it remains a core taboo of the scientific community. More work is needed to create institutional and infrastructural framework that normalize these activities, such as establishing a peer-reviewed journal dedicated to exactly this – publishing the results of personal Longevity experimentation by scientists, doctors, and practitioners.

We also have industry examples of biomedical sectors that emphasize human validation more than model organism studies; this serves to strengthen the notion that there is nothing fundamentally wrong or misguided in human validation if done safely. One of the most immediately obvious sectors is the NeuroTech Industry, which focuses on human validation primarily out of necessity. The size and functionality of mouse brains are so vastly different from human brains that there could be no valuable extrapolation from experimental results in mice. And, indeed, while the difference between the genomes, physiological complexity, and the nature of the biology of aging in mice and humans is arguably just as large, if not larger, than the difference in the size and capacity of brains between mice and humans, it is less recognized by conservative scientists and regulators because the difference is not as immediately understandable.

Finally, the domain of Space Medicine, in particular, is an area that is not only overwhelmingly focused on human-centered validation but, interestingly, has many technical and scientific convergences with Longevity research and clinical translation as well, so much so that it warrants being analyzed and discussed as an exemplary case study of this larger phenomenon. As such, we will cover them in-depth in the next chapter.

Deep Knowledge Group Holds Human Validation as a Fundamental Due Diligence Requirement

Deep Knowledge Group and its venture investment arms, Deep Knowledge Ventures and Longevity.Capital, have been warning of the fundamental false assumption at the very bottom of the Longevity Industry for several years now. They have been working to develop modern approaches to combat this systemic problem, both to optimize their internal investment strategies and to promote the adoption of similar approaches by other investors to help secure the long-term stability of the Longevity Industry more generally, as I have explained in detail in my first book, "***Longevity Industry 1.0: Defining the Biggest and Most Complex Industry in Human History***[1]."

On the one hand, Deep Knowledge Group has been hedging against this fundamental and paradigmatic risk by strategically

[1] www.longevity-book.com

prioritizing other sectors with high degrees of market-readiness which are not exclusively focused on biomedicine – such as AI for Longevity, Age- Tech and Longevity FinTech – while also developing and deploying sophisticated and modern approaches to investment analytics, due diligence, and company valuation that are equal to the complexity, intersectionality, and multidimensionality of the Longevity Industry itself. These approaches have been developed over the period of 2013-2021 by Deep Knowledge Group's Longevity-focused analytical subsidiary, Aging Analytics Agency.

However, Deep Knowledge Group has also been working to combat this insidious threat to the stability of the Global Longevity Industry more directly and tactically; this has not yet been publicly disclosed. It is a strict internal mandate that Deep Knowledge Group's venture arms do not make any investments into biomedical Longevity companies without some form of human validation.

What is more, in many cases, Deep Knowledge Group uses as a component of its scientific due diligence process something that at first glance may seem controversial, but which, in my belief, will come to be recognized as an increasingly common, pertinent, and sensible approach within the Longevity investment community: **namely, by having Deep Knowledge Group partners validate the safety and efficacy of the Longevity technology or therapy offered by a given investment prospect through their own bodies and biomarkers.**

This specific process of technological and therapeutic due diligence is guided by a special proprietary group ('Practical Longevity') assembled several years ago within Aging Analytics Agency, which is dedicated to aggregating, analyzing and validating the full scope of technologies, therapies and methods of optimizing healthspan and extending the healthy period of human life with demonstrable evidence in humans, as well as for evaluating in a safe manner new upcoming techniques and proposals. Being an early founder of Aging Analytics Agency, I myself and my partners at Deep Knowledge Group are having early access to the results of this years-long analytical project. You can read more about the major outputs of this special proprietary group here: '***Practical Longevity***[1]'.

[1] www.longevity-book.com/practical-Longevity

I am the foremost practitioner of this mandate. I am the Deep Knowledge Group partner who most frequently partakes in the practical, real-world use of Longevity technologies as a component of our internal due diligence process (though I am certainly not the only partner to do this). Having pushed the consortium to make its first investments in the Longevity space in 2014 – and having founded its Longevity-focused analytical subsidiary in 2013 precisely as a conveyor-belt and gigafactory of analytics, research, market intelligence, and industry understanding of Longevity science, business, finance, and policy – I consider this quite natural. In addition, I have been developing and refining my own personal, practical Longevity regime for almost a decade, and I am very well equipped – both psychologically and practically applicable expertise – to understand what has a high likelihood of being safe before applying it and assessing its effects on my own body. While the work with the extended panel of my own biomarker is at the core of this process.

I believe that as such approaches for varying levels of safe human validation in the Longevity Industry become more common, accepted, and demanded as industry-standard components for scientific and technological due diligence by investors, an increasing number of Longevity entrepreneurs and even scientists will begin to adopt the mindset that the only reliable form of validation – and the most definitive proof of safety – is in their bodies.

Due to the Longevity Industry's overcomplexity and the difficulties in distinguishing hype from reality in this industry, Deep Knowledge Group's Longevity-focused analytical subsidiary Aging Analytics Agency has developed an AI-driven Big Data Analytical System[1] for more tangible, multiparametric comparative analytics, which is necessary to reliably provide the metrics this industry needs.

The analytics dashboard developed by Aging Analytics Agency aims to deliver to strategic decision-makers across the private sector, global investment community, Financial Industry, governance, and policy ecosystem access to Big Data analytics and visualization, market intelligence, competitive analysis, technology and policy benchmarking, SWOT analysis, practical recommendations, and other strategic toolsets capable of handling the unprecedented complexity and multidimensionality of the full-scope Global Longevity public and private sector ecosystem.

[1] www.aginganalytics.com/big-data-analytics-dashboard

Big Data Analytics Platform at a Glance

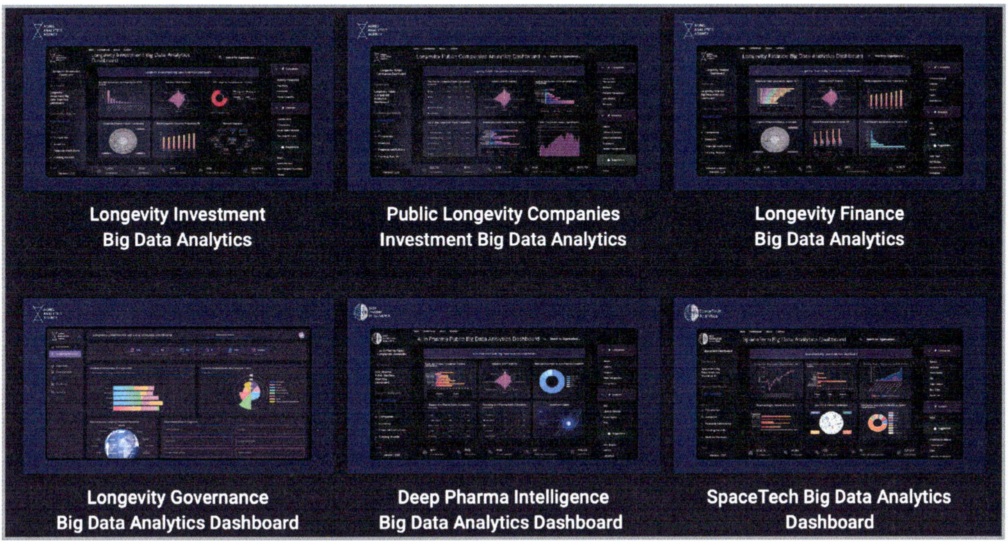

Source: Big Data Analytics Dashboard[1]

AI + Consumer-Grade Biomarkers = the Natural Evolution of Digital Human Avatars

At this point in the book, it should be clear that the true power of Biomarkers of Human Longevity lies in the increased level of precision and personalization – and the advanced levels of predictive forecasting and high-frequency adaptability – afforded by the application of AI and data science techniques to very high volumes of biological and non-biological data, and that these trends are leading toward a systematically increasing capacity to utilize Biomarkers of Human Longevity as a means of more reliable mechanism for forecasting clinical outcomes and for real-world clinical validation.

To give greater substance to this thesis, we will briefly explore the realm of digital biomarkers, and the ways in which the increasingly prominent and fundamental role of AI in biomarker development and real-world applications is already now converging toward the creation of powerful predictive models of individual humans – or what could be named Digital Human Avatars.

[1] www.deep-innovation.tech/longevity-dashboard

Once more, the use of AI to execute, manage, and optimize this process at every stage — from data aggregation to validation, model construction, analysis, predictive forecasting, and real-time adaptation — is not just a complementary and accelerative factor but a critical component of the entire process. It is necessary for handling the extremely high volumes of data involved; for performing high-frequency (and, eventually, real-time) joint analysis and readjustment of biomarkers and external factors; and for achieving the levels of precision required by the highly personalized nature of individual health, behavior, and performance. Just as the nature of aging varies markedly between mice and humans, so too does it vary quite substantially between different demographic groups (based on age, sex, ethnicity, health status, and environment), and even between the biomarkers of a single individual at different periods of time over their lifecycle.

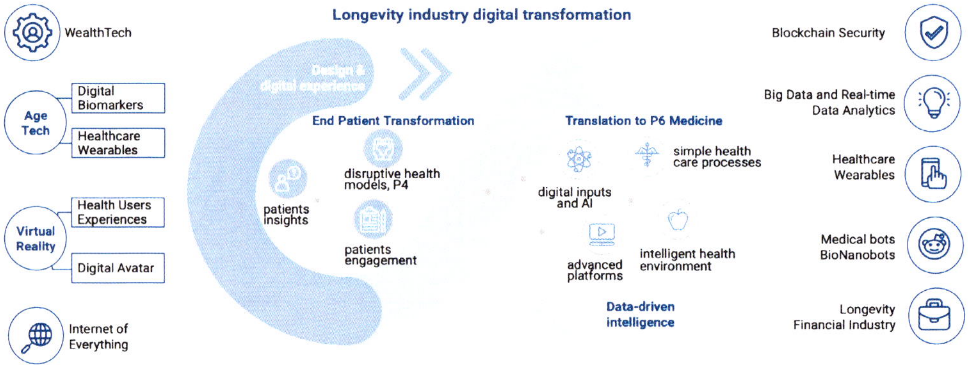

Source: Longevity Biomarkers Landscape Overview Q4 2021[1]

Taking into account this general trend, in combination with the increasing power and sophistication of AI, data science, and other modern IT technologies – as well as the exponentially increasing amount of biological and non-biological data points being created and collected on the individual and population level – it is clear that these developments are quickly moving toward a future that would enable the synergetic and comprehensive integration of all data points impacting the quality of life, performance, health, psychology and wealth. These data points would be amalgamated into a Digital Human Avatar – a full-scope, full-system model of the human body and mind that could serve as the ideal tool for the optimization and

[1] www.longevity.international/biomarkers

predictive forecasting of health, and performance. It would be utilized for the high-frequency validation, adjustment, and returning of all internal and external factors behind the length and quality of life.

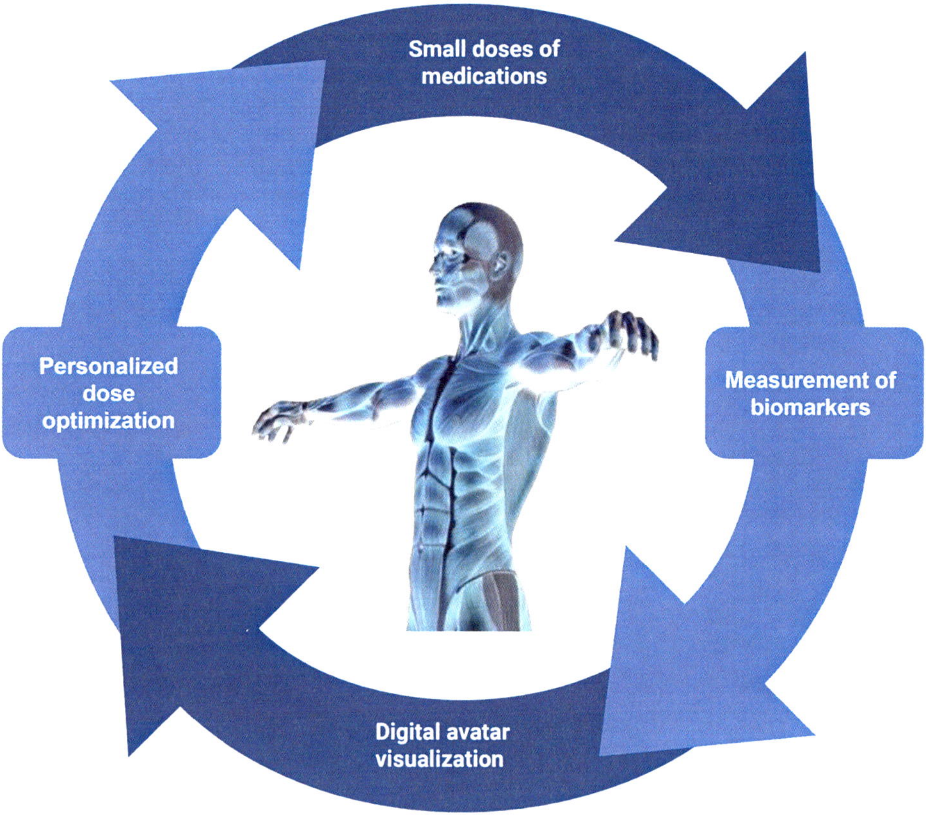

Source: Longevity Biomarkers Landscape Overview Q4 2021[1]

Perhaps unsurprisingly, given that Deep Knowledge Group's definition and conception of the Longevity Industry is broad, encompassing not just biology and biomedicine but also non-biological approaches to maximizing wellness and performance (e.g., AgeTech), and because the concept of Healthy Human Longevity is about much more than simply biological age and health, our conception of Biomarkers of Human Longevity is also broad, encompassing non-biological markets such as lifestyle, behavior, psychology, social activity, finance, and the environment.

[1] www.longevity.international/biomarkers

Based on this framework, it is clear to us that, as the volumes and types of data being generated and collected increase – and as the capacity for AI and data science to synchronize, orchestrate, and integrate those data types together into a fully harmonized whole also increases – global society is poised to witness the ongoing evolution of increasingly powerful, precise, and functional Digital Human Avatars.

The Increasing Role of Digital Biomarkers

Quantified in ones and zeros, Digital Biomarkers and their related Digital Panel Platforms can support continuous measurements outside the physical confines of clinical or hospital environments by using home-based, connected products. These products have created new opportunities, enabling remote monitoring for biomedical research, decentralized clinical trial designs, and routine patient care. These components are already well-integrated in the market, but they are still waiting to be used to good advantage by the world's health systems.

The increasing capacity for inexpensive, consumer-grade devices to be used as tools for the aggregation, storage, and analysis of high volumes of individual biological and non-biological data, is a critical component in the development of Digital Human Avatars.

Today, Digital Biomarkers are widely studied in order to reveal the broad spectrum of possible uses and to revolutionize current methods of monitoring patients' state of health and predicting the outcomes of diseases. According to the Digital Biomarkers Journal[1] – a multidisciplinary-by-design open access journal that bridges the disciplines of computer science, engineering, biomedicine, regulatory science, and informatics – Digital Biomarkers represent an opportunity to capture clinically meaningful and objective data.

Digital Biomarkers represent a breakthrough tool for Precision Health, which is why not only individuals and healthcare providers but also many companies have started to onboard this technology into their day-to-day practices. Breathometer, Xsensio, Scailyte AG, Nightingale

[1] www.karger.com/Journal/Home/271954

Health, FEET ME, xbird, Mindstrong Health, Serimmune, IXICO, etc., are top private companies that successfully carry out the mission of Digital Biomarker popularization. They are known for the development of unique sensing platforms and chips, human liquid testing systems, devices for health monitoring, and single-cell profiling devices. They are also known for the provision of unique information that fuels the development of new diagnostics, vaccines, and therapeutics. All of these enable novel approaches for preventing and treating a great many diseases – not only pathologies and conditions of aging but the vast majority associated with age and Longevity.

"Digital Biomarkers" covers a broad range of preventive, diagnostic, and prognostic measurements, with opportunities and challenges related to their use. Some Biomarkers are immediately familiar to patients or physicians as they are digitized versions of well-established metrics — such as glucometer readings transmitted by Bluetooth or the timed six-minute walk test measured with a smartphone's built-in gyroscope and accelerometer. Others, such as the smartphone-derived tapping test for Parkinson's disease severity, are entirely novel and evolving. They can detect eye diseases from scans as accurately as experts can; they can also predict patient deterioration up to 48 hours earlier than is currently possible – this is a development carried out by DeepMind Technologies, now belonging to Alphabet. Also, Digital Biomarkers can be components in autoregulated closed-loop systems; for instance, a continuous glucose sensor linked to an insulin pump in the pancreas can automatically dose or micro-dose insulin to patients with diabetes.

The Increasing Need for AI-Mediated Digital Biomarkers

A large part of health information is digitized. This allows us to compile vast amounts of data, access global servers, and compare patient information; it is a dynamic repository of information that is constantly being updated. This is a massive advance as these databases help doctors with their diagnostic process, improving their ability to measure, analyze, compare patients, and produce medical reports that are more accurate and personalized. This will, in turn, lead to the best available therapy or treatment of the time.

AI-Driven Biomarkers and P4 Medicine

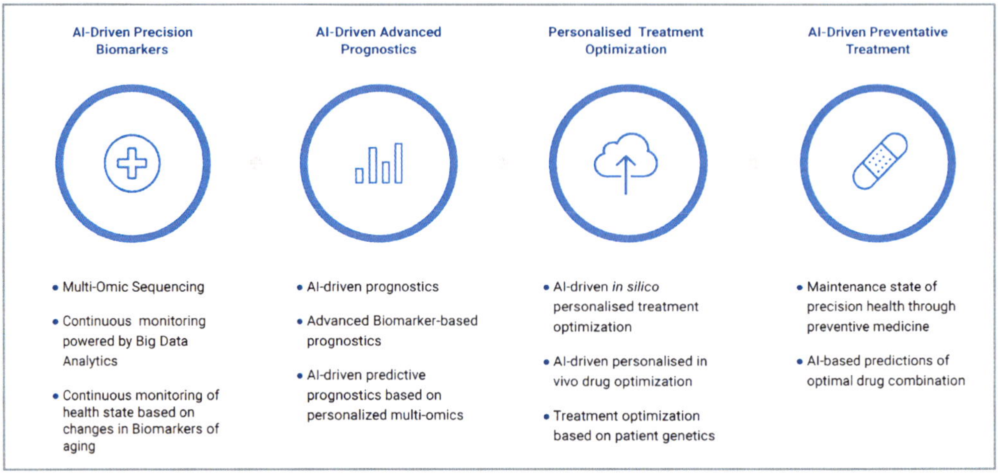

Source: Longevity Biomarkers Landscape Overview Q4 2021[1]

Aggregation of Biomarkers of Longevity, rather than biomarkers of disease only, and from healthy populations – among the young and the even younger, rather than bedside data from the hospital populations, will be part of everyday life due to the novel Digital Health platforms capable of extracting genuinely massive amounts of clinically relevant data from a single patient through electronic layers. These transformations already underway in biomarkers assessment modality will allow moving from the conventional therapeutic approach toward large-scale precision preventive medicine, the necessary vehicle to build both individual and national Healthy Longevity, and this will also be the foundation for the achievement of a unified theory of the root causes of aging and Longevity. The one plus the other, these two emerging together, will provide a framework to intervene in aging in a somewhat accelerated way rather than gradual. Tech companies rather than healthcare companies will play the leading role in this process; the threshold is already blurring. The use of AI will give sense to the thumping amount of digital biomarkers data.

Furthermore, it will allow the development of undoubtedly multiple optimal biomarker panels for the monitoring of aging and its correction, as well as countless other health biomarkers (digital and non-digital) for AI-driven analysis of each person's deep health status, allowing to orchestrate extremely personalized therapeutic interventions in

[1] www.longevity.international/biomarkers

response to minimal and currently imperceptible fluctuations in all those biomarkers.

As the number of data points increases, AI-driven analysis will be strictly necessary for the Longevity Industry and the health industry as a whole. At this point, it is impossible to determine whether biotechnologies for Longevity have been successful if we cannot tell how advanced the aging process is in any given individual. However, at the same time, the latter will not be feasible until successfully achieving highly actionable Panels or "Aging Clocks" that allow evaluating the aging process in broad healthy and less healthy differentiated ranges of the population spectrum.

The intensive application of AI to all stages of Longevity and Preventive Medicine R&D and healthcare has the potential to rapidly accelerate the clinical translation of experimental biomarkers toward practical applications for diagnostics, prognostics, and therapeutics. It can empower patients to ultimately become the ***'CEOs of Their Own Health'*** through continuous AI-driven monitoring of minor fluctuations in biomarkers, and it allows for the rapid development of the Global Longevity Industry to scale.

AI Biomarkers Conditioning Stage

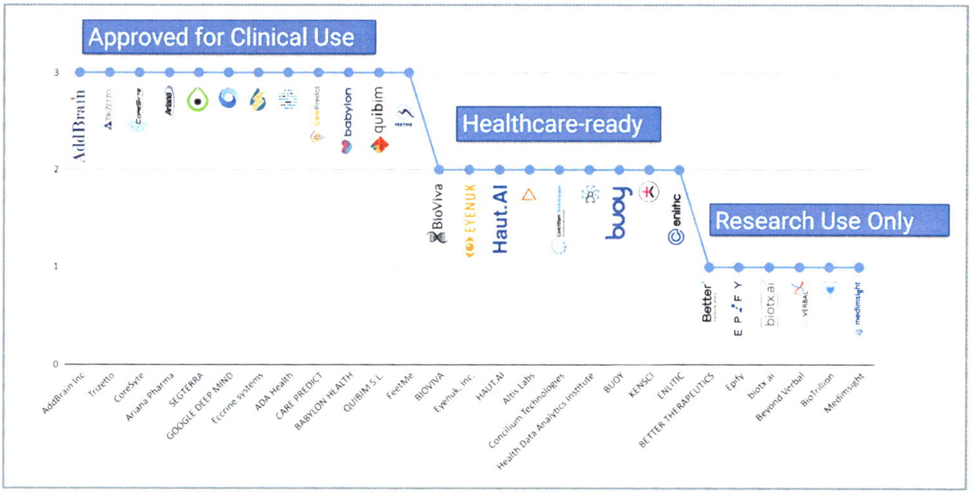

Source: Longevity Biomarkers Landscape Overview Q4 2021[1]

Artificial Intelligence is the next frontier of Longevity industrialization because it can integrate multimodal data from different types of measurements and observations, identifying global patterns across

[1] www.longevity.international/biomarkers

all data sets simultaneously. This ability of AI (mostly deep neural networks) can be used to realistically integrate data and to allow the most accurate prediction of aging and Longevity risks and dynamics. Recent advances in the understanding of the aging process have led to the creation of next-generation aging clocks (such as "epigenetic clocks") and advances in the area of deep learning — all this makes the creation of **Digital Avatars** a reality for the foreseeable future.

This "**digital twin**" approach could become the most disruptive episode for aging research and the implementation of public policies to improve individual and national health metrics, solving the current need for a biomarkers panel as a predictor of biological age and a measure for testing the effectiveness of Healthspan-extending interventions. The proof of concept for this approach already exists in the manufacturing arena, where the "digital twin" approach has already become one of the most interesting technologies for modelling and predictions. A digital twin is a simulated replica of a physical object that preserves a number of its key features. The replica is built in a digital environment that allows researchers to manipulate its features by changing its parameters or altering other objects in its surrounding simulated environment. The name "digital twin" was proposed by John Vickers of NASA in a **2010 Materials, Structures, Mechanical Systems, and Manufacturing Roadmap Report**[1]. This technology is already widely used by Siemens, IBM, Microsoft, General Electric, PTC and others.

AI and Digital Platforms: SWOT Analysis

S — Strengths

- Non-invasive.
- Collecting vast amounts of data from the daily lives of people not in a hospital.
- Use powerful statistical algorithms for data operating.
- Individual patient level.
- Multi-Omic Sequencing Data approach.
- Increasing percentage of deep diagnostic and preventive medicine.
- Maintenance state of precision health through preventive medicine.
- Advanced Biomarker-based Prognostics.
- Continuous monitoring of health state based on changes in Biomarkers of ageing.

W — Weaknesses

- Use only the data from mobile apps, or devices, not medical equipment.
- A lack of digital biomarkers popularization.
- Lack of accurate measurement of some physical biomarkers by mobile devices.

O — Opportunities

- Possibility to create new apps, mobile and wireless devices or human chips for collecting and integrating data and using Big data.
- Digital biomarkers can be used to improve clinical trials of drugs.
- Design of the life insurance apps.
- Aggregation of photographs (imaging technologies) into the analysis accelerated developments in Digital Imaging Biomarkers.
- Continuous monitoring powered by Big Data Analytics.
- Advanced Biomarker-based prognostics.
- AI-driven in silico personalized treatment optimization.
- AI-driven personalized in vivo drug optimization.
- Treatment optimization based on patient genetics.

T — Threats

- Possibility to hack collected data or use in criminal cases or in geopolitical migration processes.
- Risks associated with the microchips technologies.
- Decentralized clinical trials designs, and routine patient care.
- The problem of tests standardization.
- The problem of the data storage technologies.
- The productivity of machine and computer analysis.
- The problem of integrating data from molecules, cells, organs, the individual, families, communities, and the natural and man-made environment.

Source: The Rising Wave Of Human Biomarkers Longevity Landscape Overview 2021[2]

[1] www.nasa.gov/pdf/501625main_TA12-MSMSM-DRAFT-Nov2010-A.pdf
[2] www.data.longevity.international/biomarkers-report-2021.pdf

In the coming years, we can expect the convergence of AI, novel aging biomarkers R&D, and digital biomarker applications to accelerate the emergence of AI-driven biomarkers panels of aging and Longevity.

The Current State of Digital Human Avatars in Practice

A virtual profile of all health data can be generated by collecting multiple types of data, some of which are also visualized in 3D through devices or augmented reality. Biomarkers serve to diagnose issues and evaluate the overall health status and predict the aging rates of each individual. Gathering more of this type of data, e.g., periodic blood tests, will enable a complex, highly personal picture of each person, whose predictive power will be proportional to the quantities of input classes and the AI-driven analysis's inherent capabilities of recreating biomarker networks.

The added accuracy and size of the data sets means it will inevitably provide an invaluable source for AI to pick out critical trends for each kind of patient cohort or indeed work out the cohorts themselves. Despite being a trove of knowledge for health practitioners and researchers, individuals will still be the sole proprietors of their health data, allowing them to control who sees what and how it is used for various monetization purposes. The individual's property rights on data are enabled by blockchain storage of patient data.

We are already beginning to see the first real-world applications of this emerging trend (i.e., secure blockchain-based platforms for personal health data storage, or what could collectively be titled a 'Health Data Wallet') on the market today. These applications allow individuals to store their personal medical and health records on blockchain, which gives them both individual ownership over that data, as well as the capacity to selectively share their data in a cyber-secure manner with a number of third parties including doctors and personal physicians.

Besides giving individuals greater control over their own health data, and storing it in ways that enable that health data to be predictively analyzed in a more consistent and machine-readable fashion, this trend also opens up the possibility of individuals monetizing their own health data to third-parties in an anonymous fashion, thereby giving drug-makers, healthcare and healthtech companies, and governmental healthcare departments and ministries access to larger volumes of population-level health data (for their own R&D purposes) than they have ever had access to before.

One such company developing innovative solutions within this emerging domain is ***Longenesis***[1], a spin-out of Insilico Medicine, which has created an end-to-end blockchain-based environment for biomedical institutions, patient organizations and research partners and sponsors to communicate directly, enabling both safe data curation and compliant, consent-enabled biomedical data utilization for research.

It is fortunate, in terms of the overall development of Digital Human Avatars to allow more people to take control of their own healthspans, that one of the leaders of this emerging domain is a spin-out of a company with an existing focus and prioritization of Human Longevity. However, they are not the only players, with a number of other companies developing similar blockchain-based solutions for personalized health data storage, management, analysis and anonymized monetization, including Medical Chain[2], Factom[3], DreamzTech Solutions[4], BurstIQ[5], and others.

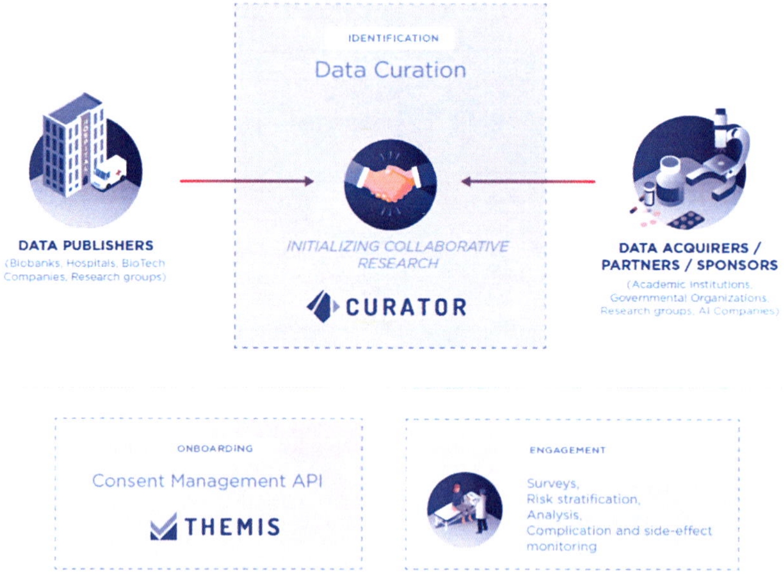

Source: Longenesis[6]

[1] www.longenesis.com
[2] www.medicalchain.com/en
[3] www.factom.com
[4] www.dreamztechusa.com/health-wallet-solution
[5] www.burstiq.com
[6] www.longenesis.com

In addition to biomarkers, all other data types relevant to healthcare will be integrated into the digital avatar. Health records, medication, lifestyle will be fed into the profile. The ability to process patient data on a bigger scale than traditional medicine enables genuinely personalized healthcare and removes the difficulties of identifying individual patient backgrounds and needs that doctors may fail to obtain in time. With the technologies of healthcare advancing, the digital avatar will evolve from a data collection and disease-focused tool to a truly Longevity-focused tool. Instead of looking at unidimensional, disease-linked biomarkers, it will look at the whole organism from an overarching health perspective and focus mainly on prevention and extending patient Healthspan.

There are at least six companies already establishing themselves as dominant players in this space.

Deep Longevity, for example, is a biotechnology company spun out of Insilico Medicine with the aim of transforming Longevity R&D through AI-discovered biomarkers of aging. They are the original inventors of "deep aging clocks," an extremely promising form of aging biomarker discussed in greater detail later in the book, and they have several granted patents on the technology. The company providing mobile app for individuals and platform solution for medical businesses – ***Young.AI***[1]

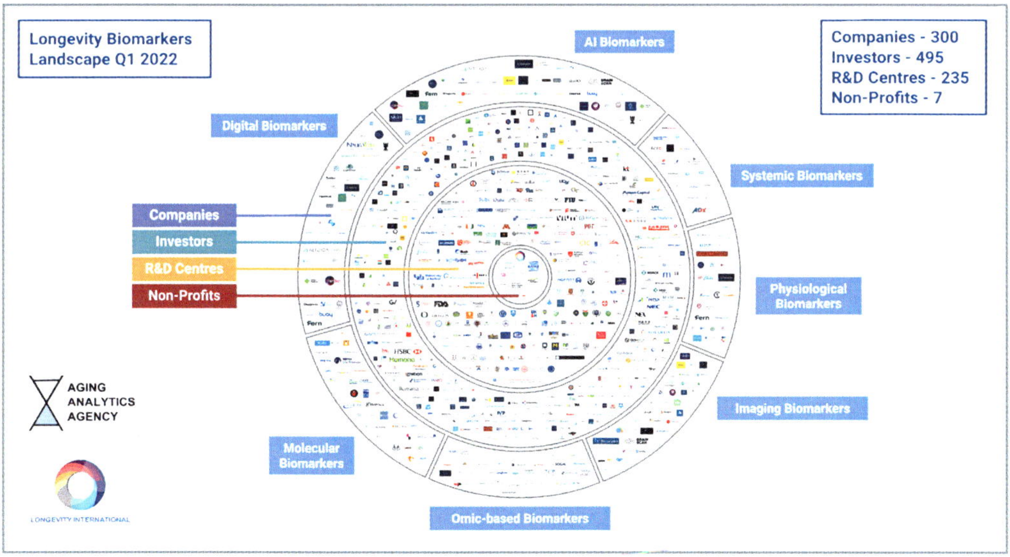

Source: Longevity Biomarkers Ecosystem[2]

[1] www.young.ai
[2] www.longevity.international/biomarkers

Seqster is a San Diego-based Software-as-a-Service(SaaS) solution that can be seamlessly integrated into any payer, provider, or clinical research enterprise. Seqster provides instant interoperability to retrieve and harmonize electronic health records (EHR), genetic, and continuous monitoring data from different sources. Users create their own matched, longitudinal health data profile across all of their US-based healthcare data sources through person-centric interoperability. The company serves healthcare enterprises such as pharmaceutical companies, payers, life insurance, and providers. In 2020 they launched the Seqster 5.0 system.

122 Labs has developed what they refer to as their **Life Ecosystem *app***[1], a set of wireless health monitoring devices connected to the application that detects anomalies, threats, and potential diseases by using an interactive machine learning platform. The company also provides Rehabilitation Circle, which is its proprietary medical and rehabilitation product for the elderly.

Biocanic is a **health data platform**[2] that streamlines the functional health business to an improved scale, achieving better outcomes for clients. Biocanic centralizes all information from intake forms, assessments, and lab results and supplements into a simple and easy-to-understand interface. Using proven clinical analysis, they highlight clients' potential health issues and put them on the quickest path to resolution.

LifeOmic is a software company that offers for individuals and corporate clients **innovative platform solutions**[3] to healthcare providers, researchers, health IT companies, and patients. It allows government agencies to integrate biomedical data management and informatics via their Precision Wellness and the Precision Health cloud-based platform, LIFE mobile apps, in use at major medical and cancer centers.

Onegevity is an AI and health intelligence engine of Thorne HealthTech, which combines proprietary AI models with a massive multi-omics database to develop new nutritional and pharmaceutical products and enhance or repurpose existing compounds for innovative uses with higher efficiency.

---

[1] www.122labs.com/lifeecosystemapp
[2] www.biocanic.com/individual
[3] www.lifeomic.com

The current state of biomarker development is not yet sufficient to constitute a full-fledged Digital Human Avatar at this point in time (as of Q3 2021). As such, in the taxonomy proposed in Chapter X (which charts the future evolution of such avatars), I would classify the current state of biomarker development as appropriate for "a type 0.5 Avatar." Nonetheless, given the productivity of cutting-edge companies already achieved this level, and the pace of development in the field generally, we can expect that the breadth and depth of players focusing on creating predictive models of individual humans of sufficient power and comprehensiveness to be called Digital Human Avatars will grow exponentially. We can, therefore, estimate that the realization of a full-fledged Digital Human Avatar MVP, or what could be considered as a Digital Avatar 1.0, is possible within the next year.

While this book's final chapter outlines my personal thoughts on the future evolution of the concept of Digital Human Avatars more thoroughly, the above spotlight on the concept – and its present state in Q3 2021 – should be sufficient to lend additional real-world validity to this chapter's core thesis that the scope of market-ready approaches for safe human experimentation and for tangible, reliable methods of forecasting clinical outcomes is more than sufficient to begin being on-boarded in a truly systemic manner by the majority of Longevity Industry stakeholders.

Summary

In this chapter, I have outlined the pressing need for modern, relevant, and reliable human-centered approaches to company and technology validation, valuation, and due diligence capable of overcoming the translational gap between results in model organisms and humans, and of facilitating reliable forecasting of clinical outcomes. The chapter also outlined a framework of approaches that can be used to facilitate this transition from overreliance on results in model organisms toward a more dependable, human-centered approach, using the concept of Digital Human Avatars as a showcase and practical example of the market-readiness of Biomarkers of Human Longevity for this purpose. The key points to take away from this chapter include:

- In the coming years, we can expect the convergence of AI, novel aging biomarkers R&D, and digital biomarkers applications to accelerate the emergence of AI-based Biomarkers of Aging and Longevity. This is likely given the emergence of Longevity Bio-Technology as a standalone industry and the many players who are entering the field, from universities and non-profits to large corporations, investment funds, and start-ups.
- As the volume of investments and IPOs in the Longevity Industry continue to rise, the threat posed by the potential failure of well-funded companies to translate their results into humans, leading to a downturn in overall investor sentiment toward this industry, becomes bigger and bigger.
- The use of Biomarkers of Aging and Longevity constitutes the most market-ready and validated means of proving efficacy in humans and can serve as the basis for demonstrating human-validated results by Longevity companies and start-ups. A wide array of single biomarkers and panels of biomarkers of Longevity exist in market-ready form and should be adopted into due-diligence practice by Longevity investors to create a more modern, sophisticated, and robust method of preliminary validation of therapeutic safety and efficacy.
- Deep Knowledge Group and its investment arms Deep Knowledge Ventures and Longevity.Capital have an internal rule to only invest in companies that possess some form of human validation. The scope of approaches available to companies and investors for human-focused validation and reliable forecasting of clinical outcomes is more than sufficient at this stage to begin the transition toward making these approaches the industry standard for due diligence, valuation and investment decision-making.
- The increasing capacity for inexpensive, consumer-grade devices to be used as tools for the aggregation, storage, and analysis of high volumes of individual biological and non-biological data, is a critical component in the development of Digital Human Avatars, which is an excellent example of the real-world capacity of Biomarkers of Human Longevity to be used for the reliable forecasting of clinical outcomes, and for safe human validation of Longevity therapies and technologies.

Chapter VII: SpaceTech and Space Medicine as a Modern Precedent for Safe and Effective Human Experimentation and Therapeutic Validation in the Longevity Industry

The previous chapter described and defined what Deep Knowledge Group believes to be the largest and most fundamental source of risk threatening the continued stability of the Longevity Industry (the assumption that positive results in model organisms will translate to humans successfully). It outlined an integrated set of modern technological and scientific approaches plus the framework for a proposed solution that can allow for the safe and effective human-centered validation of Longevity therapies and technologies at the pre-clinical trial phase. Investors can use the solution to de-risk their decision-making in this sector; Longevity companies can use it to reliably validate the safety and efficacy of their therapeutic pipelines. Longevity start-ups preparing to launch IPOs can use it to prevent dramatic declines in their market capitalization following failures of their model organism results to translate to humans.

In this chapter, we will turn to a specific modern precedent of safe and effective human experimentation and validation within the realms of SpaceTech and Space Medicine. We will explore the lessons that can be learned and applied by the Longevity Industry to facilitate a paradigm shift away from almost complete reliance on the results of model organism studies for due diligence, company valuation, and investment decision-making – and toward a more realistic, relevant, and modern human-centered approach. We will also address some interesting scientific and technological convergences between accelerated aging and the negative effects of spaceflight, as well as the ways in which the specific therapeutic approaches used to protect and preserve the health of astronauts intersect with Practical Healthy Human Longevity.

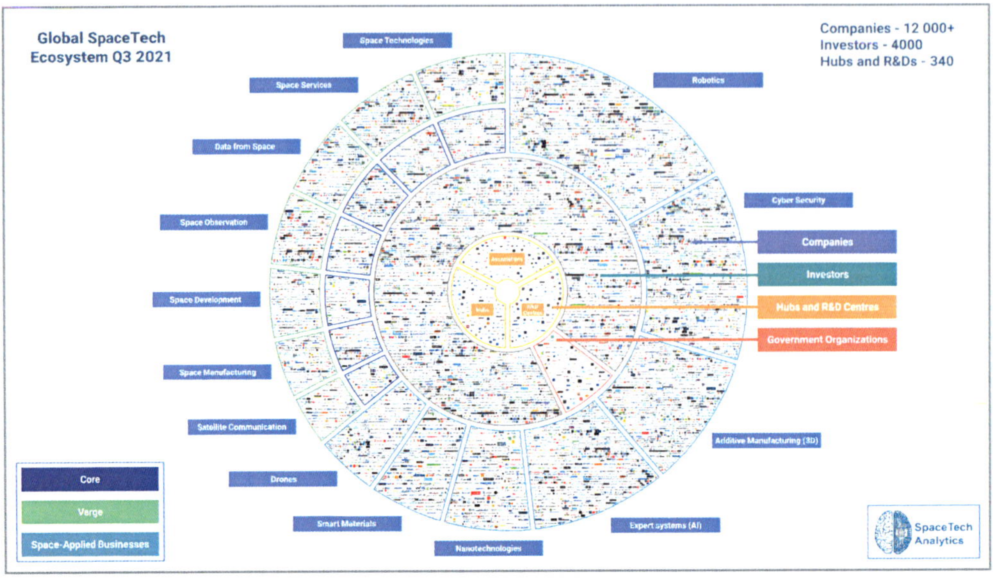

Source: SpaceTech Industry Landscape[1]

Human-Centered Validation in Space Medicine and its Convergences with Longevity Science and Clinical Translation

There are several lessons to be learned on the topic of Longevity Industry bottlenecks and the challenges for tangible and relevant Longevity investment decision-making posed by the sector's focus on results in model organisms (predominantly mice) as opposed to humans. These lessons are from the arena of SpaceTech and the approaches used in validating methods of preserving and protecting the health of astronauts during spaceflight.

In early 2018, I participated as a co-author on a scientific paper that explored convergences between SpaceTech, Space Medicine, and Longevity in a concrete form in the scientific journal Oncotarget, titled "Vive la radiorésistance!: converging research in radiobiol-

[1] www.spacetech.global/mindmaps-q2

ogy and biogerontology to enhance human radioresistance for deep space exploration and colonization.[1]"

The paper contained over 30 articles, including some from the Biogerontology Research Foundation[2] (the UK's oldest Longevity-focused charity, for which I serve as Managing Trustee), as well as others written by scientists from NASA Ames Research Center, the Environmental and Radiation Health Sciences Directorate at Health Canada, the University of Oxford, Canadian Nuclear Laboratories, the Belgian Nuclear Research Centre, Insilico Medicine, the Biogerontology Research Center, Boston University, Johns Hopkins University, the University of Lethbridge, Ghent University, the Center for Healthy Aging, and others.

In the news   Media coverage of health and science topics

The Guardian   APRIL 7, 2018

Gene therapy may help astronauts going to Mars resist deadly radiation

by Sarah Betancourt

Researchers and scientists say new discoveries and drug creation could be beneficial to future astronauts on deep space missions

Source: The Guardian[3]

As I stated in the press release associated with that paper: "In linking aging and radioresistance and tying together research into enhancing the radioresistance of astronauts with the extension of healthy Longevity, we hope to have shown how aerospace research could be used to leapfrog the massive funding deficit surrounding the clinical translation of Healthspan-extending interventions, in order to brave the storm of the oncoming Silver Tsunami and prevent the looming economic crisis posed by demographic aging."

The scope of the paper was also summarized by Afshin Beheshti, another co-author of the paper, and a bioinformatician at NASA Ames Research Center: "Our recent manuscript provides a comprehensive review of radioresistance for space radiation. Currently, there is minimal research being done for radioresistance against HZE irradiation.

1 www.ncbi.nlm.nih.gov/pmc/articles/PMC5865701
2 www.bg-rf.org.uk
3 www.theguardian.com/science/2018/apr/07/gene-therapy-may-help-astronauts-going-to-mars-resist-deadly-radiation#%3A~%3Atext%3DGene%20therapy%20may%20help%20astronauts%20going%20to%20Mars%20resist%20deadly%20radiation%2C-This%20article%20is%26text%3DAn%20international%20group%20of%20researchers%2Cthree%20years%20of%20space%20travel

The importance of these types of studies will be to reduce the associated health risks for long-term space exploration and allow for the development of potential countermeasures against space radiation. In addition, the synergy between understanding aging with radioresistance will allow for further benefits for humans in long-term space missions and allow for reduced health risk. This review sets the stage for the potential research the scientific community can do to allow for safe long term space exploration."

The roadmap outlines future research directions toward the goal of enhancing human radioresistance, including upregulation of endogenous repair and radioprotective mechanisms; possible routes into gene therapy in order to enhance radioresistance via the translation of exogenous and engineered DNA repair and radioprotective mechanisms; the substitution of organic molecules with fortified isoforms; the coordination of regenerative and ablative technologies; and methods of slowing metabolic activity while preserving the cognitive function.

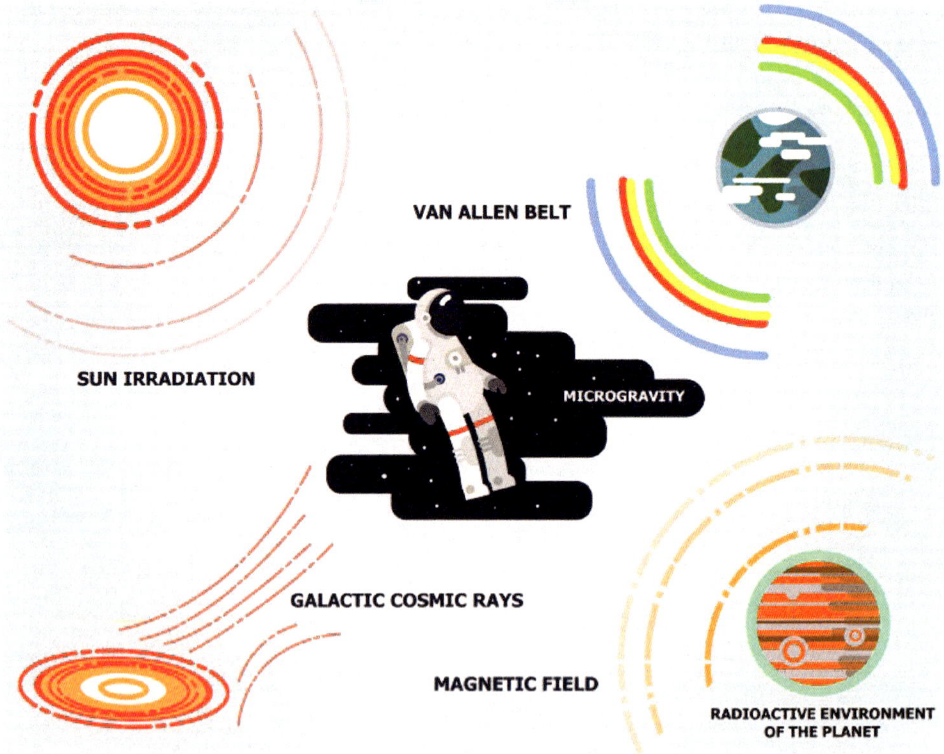

Source: Vive la radiorésistance!: converging research in radiobiology and biogerontology to enhance human radioresistance for deep space exploration and colonization[1]

[1] www.ncbi.nlm.nih.gov/pmc/articles/PMC5865701

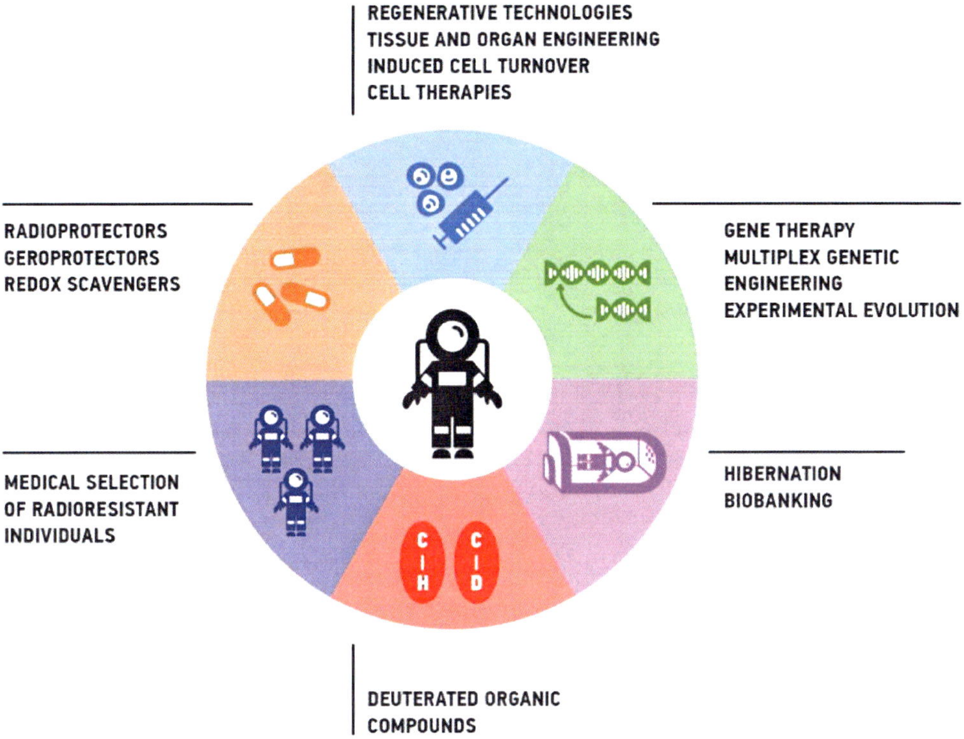

Source: Vive la radiorésistance!: converging research in radiobiology and biogerontology to enhance human radioresistance for deep space exploration and colonization[1]

The paper concludes by presenting the known associations between radioresistance and Longevity and articulating the position that enhancing human radioresistance is likely to extend the Healthspan of human spacefarers as well.

While this paper was published back in early 2018, the role that SpaceTech example can play in developing a more robust and relevant framework for validating the results of Longevity companies – to serve as the major tool for tangible investment decision-making in this sector – has been on the radar of Deep Knowledge Group for the past several years.

SpaceTech is known for its emphasis on human validation, and necessarily so, given the need to equip astronauts with the most advanced methods of health preservation and protection. As such, we can expect more robust framework for the safe, human-centered validation of Longevity therapies and technologies to learn from both the testing and validation approaches used and developed within the

[1] www.ncbi.nlm.nih.gov/pmc/articles/PMC5865701

scope of SpaceTech, as well as its general mindset that the only valid proof of safety and efficacy can be found in humans and not in mice.

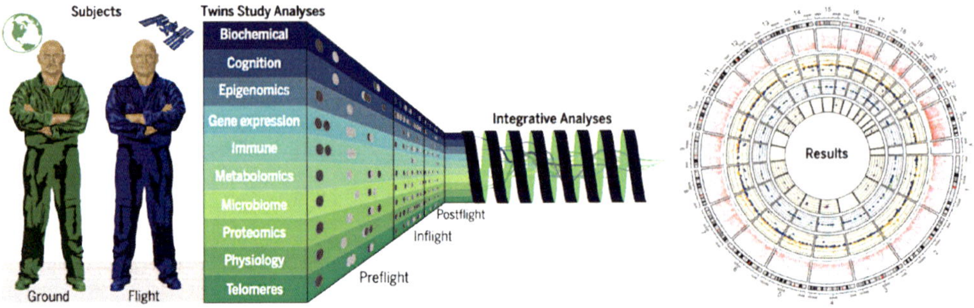

Source: Science[1]

The identical twin subjects were assessed according to 10 generalized biomedical modalities before (preflight), during (inflight), and after the flight (postflight) for a total of 25 months (circles indicate time points at which data were collected) (Right). Data were integrated to guide biomedical metrics across various "-omics" for future missions (concentric circles indicate, from inner to outer, cytokines, proteome, transcriptome, and methylome).

Scientific, Technological, and Medical Convergences Between Space Medicine and Longevity Medicine

Interestingly, the convergences between the nature of SpaceTech health optimization and Longevity research are quite numerous, and each warrants a brief exploration in its own right.

One of the most far-sighted Longevity studies is related to NASA astronaut studies. It is devoted to the multidimensional analysis of a year-long human spaceflight. But it could also be described as a study of accelerated aging.

The NASA Twins Study represents an integrated, multi-omics, molecular, physiological, and cognitive portrait of an astronaut and

[1] www.science.sciencemag.org/content/364/6436/eaau8650

reveals the biomedical responses of the human body during a year-long spaceflight as a model of accelerated aging.

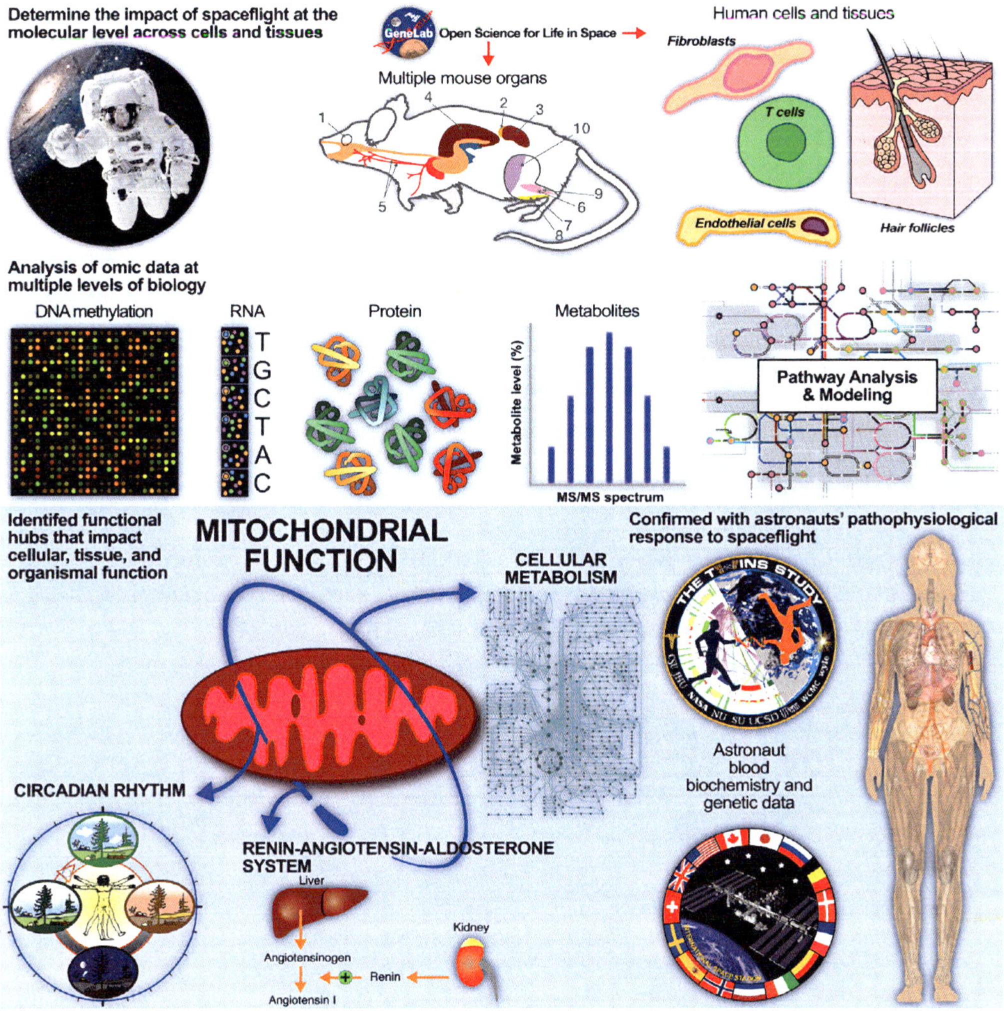

Source: Comprehensive Multiomics Analysis Reveals Mitochondrial Stress as a Central Biological Hub for Spaceflight Impact[1]

Given NASA's results, it is expected that astronauts conducting exploration-class missions could experience risks (related to hallmarks of aging) from mitochondrial dysfunction, immunological stress, vascular changes, fluid shifts, and cognitive performance decline, as well as alterations in telomere length, gene regulation, and genome integrity.

[1] www.www.cell.com/cell/fulltext/S0092-8674(20)31461-6?_returnURL=https%3A%2F%2Flinkinghub.elsevier.com%2Fretrieve%2Fpii%2FS0092867420314616%3Fshowall%3Dtrue

Given the limitations of studying a single spaceflight subject, studies of additional astronauts on long-haul missions are needed to confirm these findings and resolve outstanding questions.

Long-haul missions that will take humans to Mars and beyond are planned by public and private entities for the end of the 2020s and 2030s; therefore, comprehensive studies are underway to assess the impact of long-duration spaceflight on the human body, brain, and overall physiology.

Multiple factors combine to make the space environment harsh and challenging, including confinement, isolation, and exposure to environmental stressors such as microgravity and radiation. The selection of a pair of monozygotic (identical) twin astronauts for NASA's first one-year mission allowed for a study comparing the impact of the space-flight environment on one twin to the simultaneous impact of the Earth environment on the other.

The known impacts of the spaceflight environment on human health and performance, physiology, and cellular and molecular processes are numerous and include bone density loss, effects on cognitive performance, microbial shifts, and alterations in gene regulation.

Physiological, telomeric, transcriptomic, epigenetic, proteomic, metabolomic, immune, microbiome, cardiovascular, vision-related, and cognitive data were collected over 25 months. Some biological functions were not significantly affected by spaceflight, including the immune response (T-receptor repertoire) to the first test of vaccination in flight. However, significant changes in multiple data types were observed in association with the spaceflight period; the majority of these eventually returned to a preflight state within the time period of the study.

These included changes in telomere length; gene regulation measured in both epigenetic and transcriptional data; gut microbiome composition; body weight; carotid artery dimensions; subfoveal choroidal thickness and peripapillary total retinal thickness; and serum metabolites. In addition, some factors were significantly affected by the stress of returning to Earth, including inflammation cytokines and immune response gene networks, as well as cognitive performance. For a few measures, persistent changes were observed even after six months on Earth, including some genes' expression levels, increased DNA damage from chromosomal inversions, increased numbers of short telomeres, and attenuated cognitive function.

Telomere length shortens with cell division and thus with age, as well as with a variety of lifestyle factors such as stress and environmental exposures, including air pollution and radiation. Most notable was a significant increase in telomere length during flight for TW (14.5%), as compared with his preflight and postflight measures. TW's increased telomere length was observed in sorted CD4, CD8, and LD cells but not in CD19 cells. Notably, telomere length shortened rapidly within ~48 hours of TW's return to Earth but stabilized to near-preflight averages within months.

Spaceflight effects were more evident in isolated cells than whole organs, suggesting that tissue complexity plays an essential role in response to space-related stress. The liver undergoes more differential gene and protein expression changes than other organs, consistent with the role the liver plays as a dynamic and critical hub in sensing changes in blood composition and maintaining homeostasis. Space-flight impacts mitochondrial function at the genetic, protein, and metabolite levels of cellular, tissue, and organismal biology.

Additionally, spaceflight causes a universal change in gene expression related to energy generation. Observed changes include altered mitochondria-associated metabolites and modified nuclear DNA (nDNA) and mtDNA OXPHOS gene expression; reduced antioxidant defenses and increased urinary markers of oxidative stress; and altered integrated stress response (ISR) gene expression. These and other related observations suggest that mitochondrial dysfunction may alter metabolic flux through mitochondrial pathways, perturb mitochondrial gene expression, and activate the ISR. Spaceflight suppresses mitochondrial OXPHOS gene expression predominantly in oxidative tissues, and the induction of the mtDNA genes partially compensates for the diminished mitochondrial oxidative metabolism.

Thus, it should be clear that future success in long-duration space exploration requires a comprehensive understanding of the impact of spaceflight on human biology, and such knowledge could be used to design efficient countermeasures that would benefit astronauts and the health and lifespan of people on Earth. The domain of Longevity science – and its translation from theory into practice in the Global Longevity Industry – has a lot to learn from SpaceTech and Space Medicine, not only in its approach to an emphasis on human-centered research and validation but also in many of its specific approaches to protecting astronauts from the physiologically detrimental effects of spaceflight, which share many fundamental consistencies with the nature of biological aging in humans.

This was the subject of the conference Practical Applications of Mitochondria Research for Human Longevity and Space Medicine[1] held by Aging Analytics Agency, where a broad array of speakers – including Practical Longevity Experts and Space Medicine Experts shared their insights, findings, and personal experiences about the role of mitochondria in Practical Human Longevity and Space Medicine.

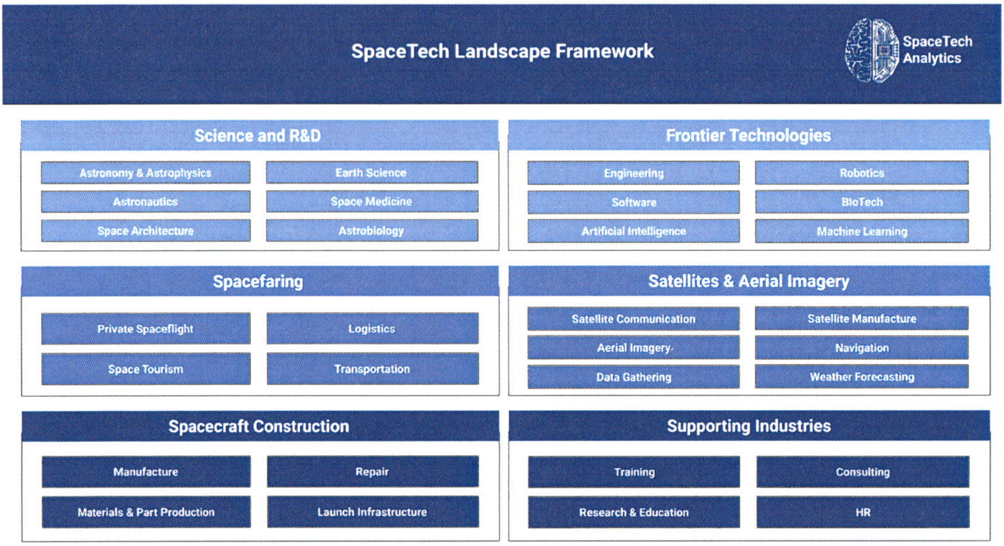

Source: SpaceTech Industry Landscape[2]

Due to the rising importance of SpaceTech, especially Space Medicine, that this chapter has described, Deep Knowledge Group has launched in Q2 2021 a new analytical subsidiary called SpaceTech Analytics, which is focused on the markets related to Space Exploration. Its range of activities includes research and analysis on major areas of high potential in the SpaceTech Industry, profiling of companies and government agencies based on their innovation potential and business activities, and the provision of consulting and analytical services to advance the SpaceTech sector.

SpaceTech Analytics has launched an ***IT-Platform***[3] and produced an analytical report, "***SpaceTech Industry Q2 2021 – Landscape Overview***,[4]" which summarizes key observations in the SpaceTech ecosystem, a

[1] www.biogerontology.info/longevity-mitochondria-conference
[2] www.spacetech.global/mindmaps-q2
[3] www.spacetech.global/dashboard
[4] www.spacetech.global/report

rapidly evolving and exponentially growing industry. Specific emphasis was given to Space Medicine. The IT-Platform contains a database of more than 10,000 SpaceTech related companies, 5,000 investors, 150 R&D Hubs and Associations, and 130 governmental organizations.

Summary

In this chapter, we have provided an overview of, perhaps, the strongest and most relevant modern precedent for safe and effective human experimentation and validation within the realm of SpaceTech and Space Medicine. We have explored the lessons that can be learned and applied by the Longevity Industry to facilitate a paradigm shift away from almost complete reliance on the results of model organism studies for due diligence, company valuation, and investment decision-making – and toward a more realistic, relevant and modern human-centered approach. Against this background, we have also given a brief overview of several interesting scientific and technological convergences between aging and the negative effects of spaceflight, as well as the ways in which the specific therapeutic approaches that are used to protect and preserve the health of astronauts intersect with Practical Healthy Human Longevity. The key points of this chapter are the following:

- There are known associations between radioresistance and Longevity – enhancing human radioresistance is likely to extend the Healthspan of human spacefarers as well.
- The convergences between the nature of SpaceTech, health optimization, and Longevity research are quite numerous.
- NASA's twin experiment showed changes between the two subjects in telomere length; gene regulation, measured in both epigenetic and transcriptional data; gut microbiome composition; body weight; carotid artery dimensions; subfoveal choroidal thickness and peripapillary total retinal thickness; and serum metabolites.
- It should be clear that future success in long-duration space exploration requires a comprehensive understanding of the impact of spaceflight on human biology; this knowledge could be used to design efficient countermeasures that would benefit astronauts, and if these measures will work on humans in hazardous space environments, they definitely will work even better for people on Earth.

Chapter VIII: NeuroTech and the Role of Cognitive, Neurological and Psychological Biomarkers of Human Longevity

If we aim to speak on the topic of Practical Human Longevity, we must also, necessarily, speak about cognitive health. Without the Healthy Longevity of the brain, the Healthy Longevity of the body is practically meaningless and, to a certain extent, impossible.

Whereas healthy brains even with an unhealthy body may mean a lot. Example of Steven Hawking is remarkable for this case.

In other words, among all tissues in the body the focus on health and Longevity of brains should be the #1 priority.

Our thoughts, feelings, beliefs, and attitudes positively or negatively impact our biological functioning. Our minds affect how healthy our bodies are. Similarly, the health and performance of the rest of our physical bodies can shape our mental states either positively or negatively. A sub-optimal state of health can exacerbate both age-related cognitive decline and psychological aging, and psychological aging (loneliness, social isolation, stress, etc.) can impair the rest of the body's state of health. This results in a complex, cyclical, and interdependent interrelationship between our minds and bodies.

As such, developing and applying biomarkers of cognitive and psychological Longevity to optimize cognitive health and performance is a fundamental requirement for the overall success of the Longevity Industry, for the translation of its potential into real-world humanitarian impacts, and for an increase in the QALY of global society.

Fortunately, cognitive Biomarkers of Longevity are in a similarly advanced state of development as other types of Biomarkers of Human Longevity are – they are steadily growing in scope and market-readiness but have yet to be assembled into an industry standardized framework for the purpose of accelerated translation of theory into practice.

Neurological Biomarkers: TRL Analysis

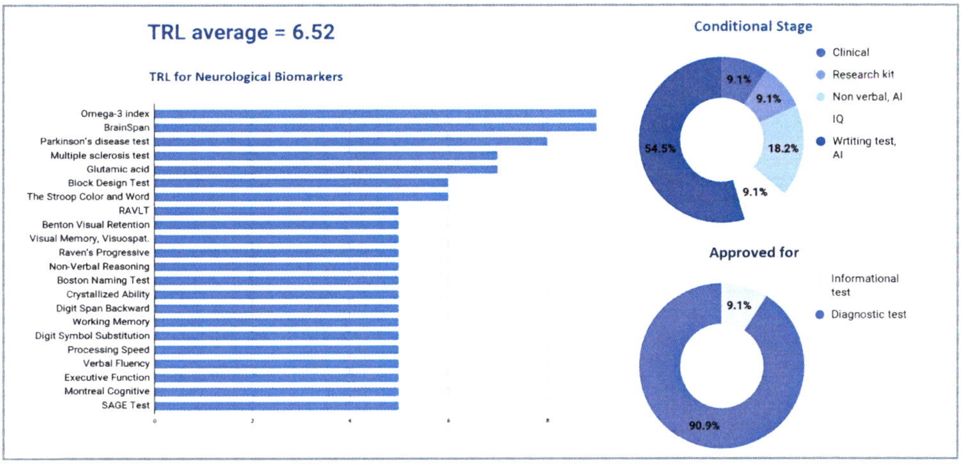

Source: The Rising Wave of Human Biomarkers of Longevity Landscape Overview Q2 2021[1]

Similarly, the overall number of cognitive biomarkers in use is large. Examples of market-ready, commonly-used biomarkers of cognitive performance and psychological aging include Executive Function, Verbal Fluency, Processing Speed, Digit-Symbol Coding, Working Memory, Digit Span Backward, Crystallized Ability, Boston Naming Test, Attention, Stroop Task, Nonverbal Reasoning, Raven's Progressive Matrices, Visual Memory and Visuospatial Ability, Benton Visual Retention Test and Block Design Test, Verbal Memory & Learning Rey, and Auditory Verbal Learning Test.

In addition, as was briefly explained in the preceding chapter, the broader domain of NeuroTech, much like Space Medicine already, uses a more human-centered approach to R&D and clinical translation than most other sectors of biomedicine, and serves as a modern precedent for safe, tangible, human-focused therapeutic validation. As such, the practically-oriented engineering mindset behind the creation of consensus framework of cognitive and psychological Biomarkers of Human Longevity should already be much more palatable to NeuroTech industry professionals than in other sectors.

In the present chapter, we will explore the current state of cognitive, neurological, and psychological Biomarkers of Human Longevity, including a specific spotlight on the role of digital cognitive biomarkers. We will also discuss the convergences and intersections between NeuroTech and Space Medicine beyond their dual relationship as modern precedents for safe human experimentation and therapeutic validation.

[1] www.longevity.international/biomarkers

Impact of Cognitive Decline and Global Neurological Biomarkers

Cognitive decline may limit independence and signal dementia. Although debated, evidence indicates that the onset of cognitive decline is detectable relatively early in adulthood, for example, from around ***45 years of age or earlier***[1] in some functions. Executive function, processing speed, and episodic memory are a possible minimum set of domains to be assessed in aging studies. If assessment time allows, tests of crystallized cognitive ability and nonverbal reasoning would be useful additions. Executive function is markedly affected by aging, exhibiting an inverted U-shape pattern across the lifespan. Processing speed declines progressively with age and is associated with greater mortality risk, cardiovascular, and respiratory diseases. In addition, episodic memory is sensitive to brain aging and declines in individuals with mild cognitive impairment and neurodegenerative diseases. A standard deviation advantage in memory is associated with a 21% reduction in mortality risk among older individuals.

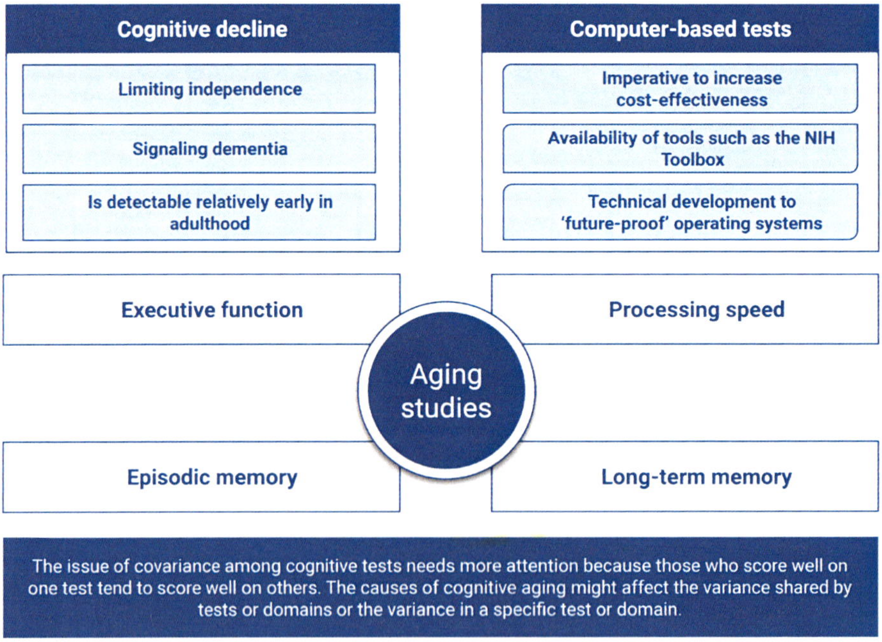

Source: Longevity Biomarkers Landscape Overview Q4 2021[2]

[1] www.alz-journals.onlinelibrary.wiley.com/doi/full/10.1002/alz.12068
[2] www.longevity.international/biomarkers

To date, computer-based tests are not widely used in major cohorts; the availability of tools such as the NIH Toolbox and the imperative to increase cost-effectiveness are likely to drive the migration to digital methodologies. This will require that tests are supported by ongoing technical development to "future-proof" operating systems and hardware. Where tests are administered repeatedly in the same individuals, problems associated with practice and familiarity need to be addressed. The issue of covariance among cognitive tests needs more attention because those who score well on one test tend to score well on others. Timothy A. Salthouse and others have highlighted that the causes of cognitive aging might affect the variance shared by tests or domains or the variance in a specific test or domain.

Prospects of being able to monitor a person's health status non-invasively without the participant having to engage in specific tasks is a hot topic now. To facilitate this, large datasets are required with extensive validation in healthy and clinical populations. Data sources can be combined to create complex phenotypic signatures or digital avatars which relate to a person's health or disease status. These can be enhanced by coupling data sources with active cognitive assessments to determine relationships with health. Cognitive biomarkers serve as both the immediate gold-standard endpoints for cognitive function, but also as a benchmark for training and validating novel measurements.

The global neurological biomarkers market size was estimated at $5,564.8 million in 2018 and is expected to exhibit a CAGR of 14.5% during the forecast period. The rising prevalence of neurological diseases and an increasing emphasis on early diagnosis and treatments are expected to drive the market growth.

Some notable commercial companies developing and utilizing neurological and cognitive Biomarkers of Human Longevity include the following:

Olink Proteomics (Sweden, United States): Olink offers an unmatched high-multiplex technique to identify actionable biomarkers, with a strong focus on the human plasma proteome.

Myriad RBM, Inc. (United States): Myriad RBM, Inc. is the multiplexed immunoassay testing laboratory that solves complex drug development challenges with innovative biomarker services and a novel whole blood immunophenotyping device (TruCulture®).

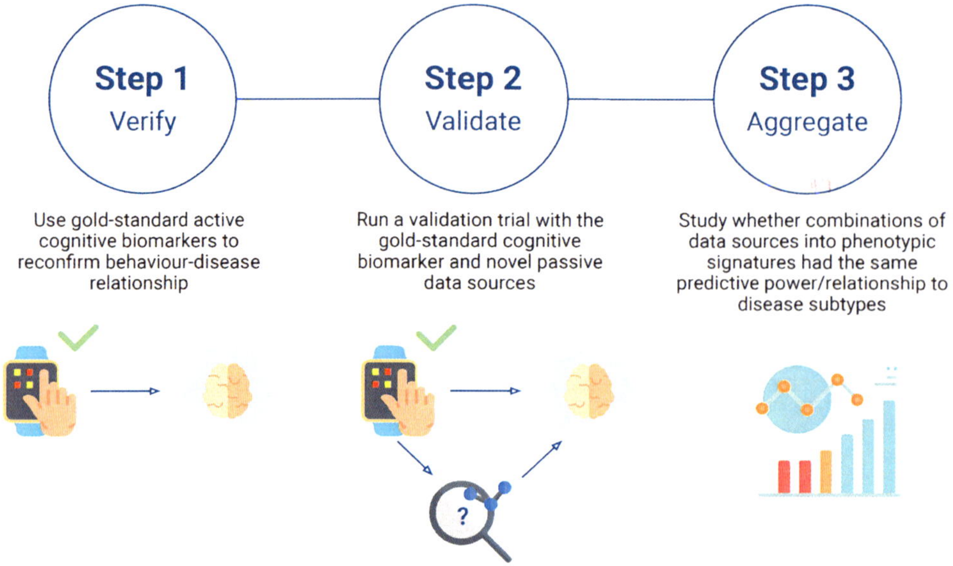

Source: Cambridge Cognition | Jose Lara. 2015[1]

BEYOND VERBAL (Israel): Uses Voice and AI to extract human emotions and reveal Vocal Biomarkers – acoustic features indicating various health conditions. In collaboration with leading medical institutions, Beyond Verbal performs multiple clinical trials and conducts Big Data projects aimed at revealing signs of cardiovascular, pulmonary, neurological, and other diseases and chronic medical conditions in the voice – allowing continuous patient monitoring and predictive analytics. Products/Services: Vocalis Health platform. Research Use Only. Pilot, validation. Analytics, Artificial Intelligence, Big Data, Call Center, Cloud Computing, Health Care, Internet of Things, Machine Learning, Market Research, Medical, Mobile, Personal Health, Wellness.

COMPANION MX (United States): A digital health technology company with a proven platform for proactive mobile mental health monitoring for better clinical outcomes. Born out of the Massachusetts Institute of Technology Media Lab and clinically validated at Harvard Medical School teaching hospitals and multiple Veterans Affairs clinics, the Companion™ system uses active monitoring of voice and passive monitoring of other smartphone metadata to continuously produce acoustic and behavioral biomarkers that predict core symptoms of mood and

[1] www.cambridgecognition.com/blog/entry/optimal-approach-for-using-digital-cognitive-biomarkers-in-clinical-trials

anxiety disorders. Products/Services: Companion™ mental health monitoring system, AI Platform. Clinically validated, Mobile, health care.

Cognitive and Neurological Biomarkers of Human Longevity SWOT Analysis

Neurological and cognitive biomarkers have a number of strengths, weaknesses, opportunities, and threats compared to the broader Biomarkers of Human Longevity landscape. These include:

Strengths	Weaknesses
• Good support: test is widely used by clinicians and researchers. It is also used extensively to aid in the selection of employees for jobs that require fine and gross motor dexterity and coordination; due to this, there is a considerable amount of research available in support of its use in clinical practice and scientific research. • Strong reliability as an aging marker. There are a number of studies that provide age-valued norms. • Good operability. Low complexity level. No special training is required for use and interpretation, although it is recommended. The manual provides detailed directions, scoring, and referenced norms across the lifespan. • Good inter-patient adaptability. The ranges are optimally standardized; thus, it can be used in a wide variety of patient populations where dexterity is of concern. • Good transferability. The mental tests would have a similar value in a variety of different health-related implementations, including measurements of functional age in panels.	• Studies have suggested that group administration may lead to increased competitiveness and knowledge of scores can create a situation where subjects try to beat previous scores, thus relativizing the results. • No fully integrated approach that can easily collect a variety of different data points from human participants.

Opportunities	Threats
• Identifying the biological correlates of late-life cognitive function is important for developing treatments to help reduce age-related cognitive decline. • The chance to develop a low-cost, modular touch screen device for integrating multiple cognitive and biometric assessment technologies. • The chance to develop statistical models and functional age calculation algorithms.	• Expensiveness and complexity of MRI analysis • Non-accurate writing tests

Source: The Rising Wave of Human Biomarkers of Longevity Landscape Overview Q2 2021[1]

In recent years, undeniable successes have been achieved in studying the genetic, molecular, and physiological mechanisms that contribute to the functioning of a brain. NeuroTech is one of the most promising and prospective sectors of modern biotechnology. Corporations, investors, and other players who will utilize this growth and development will gain essential competitive advantages in their struggle for new markets, audiences, and funding.

An excellent example of a high level of development is brain-imaging and neuromonitoring technology — machines for MRI and computed tomography (CT) were designed in the 1970s. These devices allow medical professionals and researchers to help diagnose and treat injuries and diseases and better understand how the brain works. As mentioned above, one of the challenges related to the Cognitive and Neurological Biomarkers of Human Longevity is MRI's expensiveness and complexity. As an alternative to MRI, contrast-enhanced ultrasound (CEUS) could be used. There are two types or generations of ultrasound contrast agents used in this technology. First-generation agents are presented in the bloodstream for a limited time, since they contain microbubbles of air that were dissolved in blood when exposed to acoustic pressure during the analysis. Second-generation contrast agents are distinguished by the presence of microbubbles of perfluorocarbon, nitrogen gas, or sulfur hexafluoride stabilized in a phospholipid membrane. When the bubbles are subjected to the ultrasound beam, they oscillate (they are being compressed by the effect of positive pressure

[1] www.longevity.international/biomarkers

created by the ultrasound waves, and they expand in the negative pressure phase). The compression of the gas is greater than the expansion, resulting in a non-linear reaction (echo). This has a significant impact on ultrasonic backscatter and enhances vascular contrast, much to the intravenous contrast material used in CT and MRI 2. Commercially available ultrasonography contrast agents include SonoVue/Lumason® (a trademark of SonoVue/Lumason® [Bracco]) and Sonazoid® (authorized in Japan, Korea, and Taiwan as of 2020).

Another common and well-known example is pharmaceuticals that alter brain chemistry to change human behavior. Drugs such as selective serotonin reuptake inhibitors (SSRIs) and medications used to treat attention deficit disorder (ADD) and attention deficit hyperactivity disorder (ADHD) are widely used, as they specifically change the way the brain functions. Neuroprosthetics is mainly associated with mixing human nervous tissue with artificial implants, for example, through a brain-computer interface or other devices. Implants can be used to treat non-congenital blindness and the causes of hearing loss.

Neural implants are a highly effective treatment for sleep apnea and assist the cardiovascular system after a stroke. Many technologies can potentially revolutionize neuroscience and health care over the next decade. They provide doctors and patients with a new sense of optimism about the future diagnosis and treatment of neurological and psychiatric diseases and contribute to future medical innovations.

The Role of Digital Cognitive Biomarkers

Similar to physiological biomarkers, cognitive biomarkers are objective measurements that can be used to track the progression of a disease or the outcome of a treatment. The defining difference is that cognitive biomarkers use non-invasive behavioral performance on objective cognitive tasks as an indicator or predictor of disease status.

The digital age recognizes the value of combining multiple complex data sources into a phenotypic signature used to monitor health on an individual and population-based level. These phenotypic signatures rely on numerous, interacting data sources to evaluate the risk status for the manifestation, progression, or relapse of disease. This is in contrast to the historical method of using a single independent biomarker for risk identification.

Within pharmaceutical trials, cognitive biomarkers have traditionally been used to determine the safety and efficacy of novel compounds. This application continues to grow and develop, with many trials now including high-frequency, daily assessments of cognition to evaluate real-world data from patients rather than infrequent in-clinic tests every 3-6 months.

More recently, the industry has recognized the added utility of digital cognitive biomarkers to help enrich patient recruitment and identify relevant candidates using remote, non-invasive methodologies which can be deployed at scale with demonstrable cost-effectiveness.

In addition, digital cognitive biomarkers hold promise for better patient stratification within-trial populations for more informed subgroup analyses to better understand the characteristics of drug responders versus non-responders. For example, stratifying participants by cognitive ability provides visibility into whether those who respond have a specific cognitive signature. This can lead to more specific drug approval for a targeted sub-population rather than a generalized late-stage drug failure.

Biomarkers enable us to study the incidence and to recruit specific patient populations and stratify sub-groups. Until recently, most biomarkers were invasive and overly reliant on biological processes. The intersection between healthcare and technology has made viable the creation and adoption of digital biomarkers: the collection of non-invasive behavioral data from multiple sources to create a phenotypic signature.

The complexity and heterogeneity of psychiatric and neurological conditions represent a significant opportunity to study the relationships between these data sources as novel, complex endpoints for clinical trial research.

Convergences and Intersections Between NeuroTech and SpaceTech in Safe, Human-Centered Longevity-Related Therapeutic Validation

In the case of NeuroTech, the transition to human validation may be an even smaller leap than SpaceTech, which is leading the way. However, NeuroTech is similar to SpaceTech insofar as it is closer to human validation. The human brain evolved very rapidly after our ancestors began to stand up and walk. This standing position favored the use of hands and tools, a key aspect in the ongoing human evolutionary process. Our brain can continuously adapt to new situations and demands from our environment. Like children progressing through their developmental stages, astronauts have to adapt step-by-step to a completely new and weightless environment. Balance, movement, and other brain functions are affected under zero-gravity or microgravity conditions.

Space cognitive neuroscience seeks to understand how the brain and mind react to the special environmental conditions of space. Among the factors inherent to the space environment that may affect the human brain and mind are microgravity, radiation, weightlessness, acceleration, noise, and stress. The study of the human brain in space has become an essential subject in recent years. The first studies on space neuroscience go back to 1962 during the Russian Vostok-3 mission, when some sensory-motor studies were carried out. On Earth, new brain imaging techniques, neuropsychological assessment tools, and other physiological measures have been developed to enable very detailed studies of brain activity and cognitive functioning. For neuroscientists and psychologists, it is of high relevance to understanding the underlying neurocognitive and neuropsychological parameters of space flight. Unfortunately, standard brain imaging techniques – for example, functional magnetic resonance imaging (fMRI) – are not applicable in space, due to the payload restrictions of space missions and costs. The European module "Columbus," part of the International Space Station (ISS) was equipped with an electroencephalography (EEG) system as a tool to research the link between weightlessness and central nervous system (CNS) activity.

Neurocognitive tests, electrophysiological measurements, and other related methods are commonly used in space to assess brain activity, neurocognitive and behavioral status, and the mental health of astronauts.

Earth-based research may also help to further our understanding of how the brain functions in space. This is achieved using so-called space analog environments and space simulations, both of which are defined by their extreme environmental characteristics. The most well-known space analogs are the Antarctic and Arctic stations, desert and submarine installations. At these facilities, crewmembers spend months in isolation and harsh weather conditions performing various tasks and procedures similar to those carried out in space missions. Other research paradigms that help in the field of space neuroscience include bed-rest experiments. When astronauts return from a long flight, their bodies need days to recuperate from the effects of living in weightlessness. In bed-rest experiments, the subjects remain in supine position on a bed for long periods, with the head tilted slightly downwards while performing different tasks. Bed-rest studies resemble some aspects of weightlessness, allowing scientists to probe how the body reacts to microgravity and to test methods for keeping future astronauts fit and healthy.

A well-known threat to the success of space missions is the inadequate and ineffective performance of the crew. Results from Earth-based research highlight the importance of studying the effects of stress on cognitive performance. Cognitive and prone performances deteriorate under stress.

We can, therefore, expect similar effects in the stressful environment of a space mission and in extreme environments and simulations. Various psychomotor functions are degraded during space flight, among them central postural functions involving hierarchically organized brain areas, including the motor cortex in frontal lobes, basal ganglia, the vestibular system in the midbrain and cerebellum, and the speed and accuracy of aimed movements associated. The primary motor cortex, cerebellum, visual cortex, and internal timekeeping are related to the prefrontal cortex and striatum, while attentional processes are distributed in different brain areas, such as frontal and parietal cortex, superior colliculi subcortical region, frontal eye field and anterior cingulate cortex, limb position sense including the primary somatosensory cortex and cerebellum, and the central management of concurrent tasks involving mainly the prefrontal, temporal and parietal cortex and basal ganglia.

Such psychomotor deficits have been implicated as the causes of accidents in space. It remains unknown to what extent these observed deficits might change during long-term space missions, as on a flight to Mars, for example. Results from Mars-500, one of the longest-ever space mission simulations revealed that a variety of neurological and psychological factors, such as circadian rhythms and social behavior were affected by the characteristics and special environment of the mission.

Summary

- No discussion on the topic of Practical Human Longevity would be complete without touching upon cognitive health, psychological aging, and neurological Longevity.
- Without the Healthy Longevity of the brain, the Healthy Longevity of the body is inconsequential, as a state of optimal bodily health would be meaningless to an aging and functionally-declining mind.
- Furthermore, bodily rejuvenation without cognitive rejuvenation would be in many senses impossible given that the health of the body affects the health and performance of the brain and vice versa.
- The neurobiological, cognitive, and psychological Biomarkers of Human Longevity Industry is steadily growing. Many biomarkers and biomarker panels have high levels of technology-readiness and market-readiness and have reached the point of consumer applications.
- There is a clear need for additional technological advances that can overcome the expensiveness, complexity, and one dimensionality of existing biomedical imaging tools such as CT or MRI. There is also a clear need for systems that enable the simultaneous and integrated analysis of different types of neurobiological data.
- Nonetheless, the field has yet to be assembled into an industry-standard consensus framework of biomarkers for the purpose of industry stabilization and the translation of theory into practice. This continues to be perhaps its most pressing need.
- Fortunately, however, the current state of NeuroTech, much like Space Medicine, already uses a more human-centered approach to R&D and clinical translation than most other sectors of biomedicine, and serves as a modern precedent for safe, tangible, human-focused therapeutic validation.
- As such, the practically oriented engineering mindset behind the creation of consensus frameworks of cognitive and psychological Biomarkers of Human Longevity should already be much more palatable to the NeuroTech industry professionals than in other sectors.

Chapter IX: Longevity Economy 2.0: Biomarkers of Human Longevity as the Major Catalyst to Increase Investments and Ensure the Sustainable Growth of the Longevity Financial Sector

Having established the risks currently inherent in an investment in healthy life extension and the immediate benefits of the biomarker revolution in expediting clinical validation, we now turn to the topic of how these developments can be used to get the wheels of Longevity Finance turning and get the capital flowing. Because without financial liquidity, there is no future Longevity Industry.

We begin by exploring a specific sector of the financial industry – insurance and InsurTech – that is already onboarding Biomarkers of Health, Aging and, Longevity into their business models and practices in order to transform a source of risk (aging) into a source of profit, growth and stability (Longevity), which serves as an ideal case study of how other sectors (such as the pensions industry) can do the same.

We will then return to the major thesis of Chapter V – that Biomarkers of Human Longevity, in combination with related approaches to the tangible, safe human validation of Longevity therapies and technologies is the major mechanism capable of de-risking Longevity investing. This chapter will illustrate in a deeper manner how this thesis applies equally to the broader financial industry, and in particular to retail banks, investment banks, exchange-traded funds, and stock exchanges.

This broader exploration into the use of Biomarkers of Human Longevity as a mechanism for de-risking business models and optimizing strategic decision-making within the global financial industry will also serve to show how the Longevity Industry's continued growth and stability will soon depend upon the involvement of larger financial institutions, markets, instruments and systems, and how the use of tangible, realistic and safe human-centered approaches to the validation of Longevity therapies and technologies by these financial institutions is the only mechanism capable of bridging the Big Liquidity Gap between Longevity start-ups and the conservative institutional investment and finance communities.

Once a tangible framework of metrics is in place to perform due diligence and valuation of Longevity companies in a realistic and relevant way, to forecast clinical outcomes, and to protect investors from the risk of failed translation of results from mice to humans, which can be accepted in a consensus manner by investors and the broader financial industry, it will unlock untold trillions of investment capital to fuel the further growth of the Longevity Industry, all the way to Longevity Escape Velocity.

By illustrating the above, the present chapter will ultimately chart the specific scope of the investment and financial approaches, systems and instruments, and infrastructures required to transform these practices and frameworks from tools being used by a small minority of Longevity investors and financial entities into industry-standard practices for the mutual benefit of investors, companies, the general public, and the industry as a whole.

Biomarkers of Human Longevity and InsurTech 2.0

In " **Longevity Industry 1.0: Defining the Biggest and Most Complex Industry in Human History**[1]," we predicted the growing trend within the health and life insurance and InsurTech sector of "gamifying" Longevity by offering financial incentives (such as discounted rates on life and health insurance) in exchange for clients maintaining a healthy lifestyle and meeting specific health-related goals.

This trend continues, and the most sophisticated and prospective approaches within this domain utilize Longevity Biomarkers as the basis for tracking whether clients are meeting their healthy lifestyle goals. Epigenetic clock tests that indicate an individual's biological

[1] www.longevity-book.com

age, are currently being marketed to life insurance companies to help them calculate an individual's life expectancy. The first epigenetic test marketed to life insurance companies and reinsurance companies was developed at the University of California, Los Angeles, and licensed to Life Epigenetics, a subsidiary of GWG Holdings.

Since 2017, Life Epigenetics has been collecting biological samples and analyzing data from life insurance policyholders. It has announced that it was pilot testing its epigenetic technology with two life insurance companies. In 2018, GWG's new life insurance company YouSurance noted on the homepage of its website that "epigenetics is the future of life insurance." Accurate age and rate of aging assessments using epigenetic data could prove valuable to life insurance companies in underwriting by allowing them to predict policyholders' life expectancy more precisely.

Health insurers have started to reward their members for better health habits. For instance, Oscar Health has begun to collect movement data through wearables and offers a $1 Amazon gift card every day to its members for hitting their step goals, up to $240 a year. However, the relationship between movement data and health has not been clinically established, and Oscar Health's attempts to control or assess health risk with wearable data fell short of expectations.

A similar reward approach was proposed by VitalityHealth insurance company. In addition to activity tracking, VitalityHealth also introduced a concept of Vitality Age, providing rewards and benefits to individuals over 70 but predicted younger by a simple survey-based test. While different wellness programs are gaining popularity, it remains a challenge to identify healthy behavior with clinically established health risk mitigation or prevention. More fundamentally, insurers need a rigorous overall measure of human health risk, and Longevity biomarkers can significantly benefit and strengthen current approaches.

As just one example, YouSurance, a digital managing general agency (MGA), is the first company to use epigenetic biomarkers to assess life insurance applicants' health and lifespan. YouSurance's patent pending process of predicting biological aging and all-cause mortality is step-change technology for the life insurance industry.

While the InsurTech sector is the leading example today of financial companies onboarding the practical application of Biomarkers of

Longevity, we believe that it is just the start of a much larger trend that will involve the Financial Industry (across many sectors) incorporating the use of Biomarkers of Human Longevity – and potentially other market-ready, tangible forms of human validation described in previous chapters – into their products, services, and strategies in a several different ways.

While this trend is interesting in its own right, its importance to the health and stability of the entire Longevity ecosystem extends far beyond this as a novel form of growth and diversification within the Global Longevity Financial Industry. ***Biomarkers of Human Longevity will come to drive, support, and serve as the foundation for several significant developments and forms of industry infrastructure that are required to sustain (and even accelerate) the Longevity Industry's current trajectory of growth – while simultaneously neutralizing some of its largest and most dangerous sources of potential destabilization.***

The Need for New Investment and Financial Infrastructure to Support the Shift from Experiments with Mice to Human Validation

While these approaches are being on-boarded by some specific segments of the financial industry, and by a minority of Longevity investors, it is necessary to describe the need for such approaches to be embraced by the wider majority of investment firms – as well as the market regulatory systems and infrastructure governing industry financing – that will shape the Longevity Industry investment landscape.

Smart investors should prioritize funding for Longevity companies that possess at least some level of human validation, considering the large translational gap in successful results from model organisms to humans, starting first of all with Biomarkers of Human Aging and Longevity, and possibly encompassing other approaches to safe, human-centered validation described elsewhere in this book. By demanding that these practices become a standard part of scientific and technological due diligence, they will be hedging their own risk while also helping to neutralize an already-prevalent and dangerous trend that has the potential to weaken investor sentiment and topple the entire industry – to the extreme detriment of humanity.

Beyond this, however, given the increasing number of Longevity companies that are holding IPOs, there is also an ever-increasing danger of additional public examples of declining market capitalization due to clinical trial failures by public Longevity companies. If this trend continues, and more tangible human-centered approaches for Longevity company valuation and clinical outcome forecasting fail to become a standard industry-wide toolset for Longevity investing and pre-IPO due diligence, it could destabilize investor sentiment and prove fatal for the industry's continued growth, diversification, and long-term stability.

The Challenge and Opportunity of Longevity Industry Liquidity GAP

Of all the factors and industry components that limit the ongoing growth of the Longevity Industry, liquidity is the most important. It can simultaneously enable the acceleration of its developmental trajectory and the stabilization of its prospects as an industry. As the size and scope of the industry continues to grow, the time has come for much larger, and more conservative, investment and financial institutions to become actively involved.

While Longevity IPOs enable exposure to an increasing number of retail investors, they carry certain risks inherent in the fact that the majority of such IPOs were held on the basis of preliminary positive results in model organisms rather than humans. Furthermore, there are many other mechanisms, systems, and approaches to increasing the amount of institutional and retail investor exposure to the Longevity Industry that have not yet been leveraged to their full potential.

There are enormous near-term opportunities for liquid Longevity indices and derivatives designed for individual investors (and HNWIs in particular); these can enable exposure to the rising tide of the Global Longevity Industry. This is a key mechanism for matching Wealthspans to rising Healthspans. Existing InvestTech solutions have the potential to give investors exposure to relevant Longevity companies in a highly liquid and tradable manner, but these are not yet being utilized sufficiently.

Just as more modern, technologically sophisticated approaches to investment target analysis, due diligence, and valuation (based, first and foremost, on Biomarkers of Human Longevity) are required, so too are more modern approaches to dealflow, investment strategy, financial structuring and execution required.

During the past several years, a clear disparity has emerged between various DeepTech sectors, particularly between the Longevity start-ups on the one hand and the entire business model of venture capital funds on the other. There have also been big gaps and disruptions in the global financial system. There is a large discrepancy between the high amount of progress in science and technology and the outdated bureaucracy in which most financial institutions remain mired, as though both exist in parallel universes.

This is happening for multiple reasons: at one end, DeepTech start-ups (including Longevity start-ups) have been suffering from a lack of access to relevant levels of seed investment. Although venture funds are by definition supposed to prioritize investments into the most disruptive technologies and start-ups, in reality, most of them prefer to avoid DeepTech sectors or to enter investment rounds at a much later stage. As a consequence, the start-ups are forced to deal with angel investors. This creates a growth gap phenomenon known as the "Death Valley": 99% of DeepTech start-ups do not survive the stage of growth between seed financing and the beginning of revenue generation or even "A" rounds.

At the other end, there is an extreme abundance of significant assets being held and preserved in bank accounts or in comparatively stable derivatives on a scale of tens of trillions. For instance, there is a tremendous volume of money on the scale of at least €1 trillion currently being stored in Swiss bank accounts with ***negative interest rates***, which is itself an ***illogical phenomenon*** in the modern financial world. Yet, the owners of these financial assets would prefer to avoid investing them in venture funds, which are equally reluctant to invest in Longevity start-ups. **Thus, the major source of these disparities comes down to the issue of illiquidity: a very high number of financial assets end up being preserved rather than invested.**

There are enormous amounts of financial assets being conserved within the umbrella of family offices (and we can estimate that there are at least 10,000 of these) and hundreds of very conservative financial institutions, which hoard this capital simply due to a lack of safe, stably growing, and predictable financial derivatives in which to invest. There is no relevant methodology to assess the number of these assets as they tend to keep this a secret. But they amount to at least several tens of trillions of euros.

Many owners of these vast sums of wealth are, nonetheless, personally interested in Longevity, both as a prospective market with the capacity for unprecedented growth and as a means to their own personal life extension. Nowadays, already many of even the most conservative investors, and the owners of the largest financial assets, now very clearly understand that the industries of AI and Longevity are two of the most promising sectors to invest in, with full confidence that such investments will lead to relatively low-risk and stable profitability in the long term.

However, due to the lack of liquid tradable and bankable financial instruments tied to the Longevity Industry, the owners and managers of these assets still prefer to avoid any significant investments in this sector. Moreover, in many cases, angel venture investors are operating as sharks, exploiting this gross illiquidity to their financial advantage, to the detriment of Longevity and other DeepTech start-ups. Therefore, financial innovations that can provide liquidity to Longevity companies and technologies – and form a bridge between the Longevity Industry and conservative financial markets – would inevitably allow for the injection of at least 1% of the tens of trillions of euros currently lying inert as "lazy money" within the Global Longevity Industry. In practice it will mean investments of hundreds of billions of euros into Longevity Industry.

And the Longevity Biomarkers revolution may indeed serve as one such innovation necessary to unlock this untapped ocean of wealth and liquidity. In order for the large, conservative investment and financial communities to begin onboarding Longevity as a strategic priority, they will require significant innovations in both the ways that investment targets are identified, valuated and due diligence (in a manner that de-risks the potential for failed clinical translation in humans, and utilizes more tangible, sophisticated mechanisms of clinical outcome forecasting), to feel as safe in their investments as they do for more predictable and forecastable sectors (e.g., real-estate and commodities).

However, equally fundamental adjustments and innovations will be required in terms of the technologies, instruments, investment vehicles, and markets used to structure dealflow and investments.

The specific financial institutions, entities, and technologies utilizing these frameworks for investment target due diligence and valuation will need to be readjusted as well, either via the establishment of dedicated Longevity-specific subsidiaries, departments, or affiliates of existing financial institutions, markets, and systems or via the creation of entirely new ones.

Deep Knowledge Group's Plans to Launch the World's First Longevity Investment Bank

Deep Knowledge Group's plans to launch the world's first Longevity Investment Bank by 2023 were profiled in a recent article published by fDI intelligence, a central part of the Intelligence portfolio on the investment products and services from The Financial Times. This indicates, together with a number of other activities, that the Longevity Industry is quickly evolving from a private equities-dominated space into the more mature realm of public markets, IPOs, and institutional investors.

The article "***Forever young? Biotech's next frontier: Investments into the Longevity sector is stepping up [sic], bolstered by the pandemic***[1]," also described the ongoing work of Deep Knowledge Group to de-risk Longevity investments, bridge the industry's liquidity gap, and pave the way toward Longevity Stock Exchanges and financial derivatives:

"***Dmitry Kaminskiy***, general partner at ***Deep Knowledge Group***, aims to create Longevity exchange-traded funds and issue Longevity derivative products in a bid to provide institutional investors with exposure to this growing part of the financial market. This can then be redeployed into additional research, creating a self-inducing cycle, he hopes... Mr. Kaminskiy's ultimate goal is to apply for a banking license for a Longevity bank, followed by the establishment of a ***Longevity investment bank*** – and further down the line, a ***Longevity stock exchange***."

As the appetite for exposure to the Longevity Industry among conservative, institutional, and retail investors continues to grow, the natural relevance and logical evolution of such type of specialized financial institution will become increasingly clear and necessary.

[1] www.fdiintelligence.com/article/79406

The closest analogy to this concept is the Silicon Valley Bank, a specialized investment bank focusing on comparatively matured start-ups in Silicon Valley (California, USA), but which also has branches in London, Hong Kong, Beijing, Israel, Germany, and Canada.

The fundamental purpose set for the Longevity Investment Bank is the creation of a platform to service corporate entities intending to trade Longevity financial assets and instruments. The Bank will enable involved parties to manage their investments, thereby creating innovative environments with advanced FinTech and InvestTech solutions.

Much like Deep Knowledge Group's Longevity-focused investment arm, Longevity.Capital, this Longevity Investment Bank will utilize a data-driven approach to investment target identification, due diligence, and valuation that utilizes tangible frameworks of metrics based upon human-centered validation approaches like Biomarkers of Human Longevity to more clearly forecast clinical outcomes and to protect investors from the risk of failed translation of results from mice to humans.

Longevity Retail Banks

Although investments in the industry are skewed toward the biomedical industry, we expect to see rapid growth in tech-oriented sectors that emphasize high levels of market readiness as the Global Longevity Industry continues to mature and diversify. One specific sector we expect to see major growth in over the next few years is Longevity FinTech and the realm of Longevity-focused retail banking.

One vanguard project that Deep Knowledge Group has supported within this domain is the Longevity Banking Card[1], a specialized FinTech project offering the first Longevity-focused banking products and services to a growing market of individuals and businesses embracing the notion of Health as the New Wealth and seeking access to integrated AgeTech-HealthTech-WealthTech solutions.

The company aims to make the mobile banking experience easier and safer for people by developing new financial products specifically designed for customers planning to live extra-long, healthy lives

[1] www.longevity.cards

while remaining high-functioning and financially stable. It is creating an expertly-curated marketplace of validated products and services capable of optimizing Healthspan, Wealthspan, and personal Healthy Longevity among their clients.

One more distinctive feature will be focus on specialized financial products for seniors, which provide unique banking services for people aged 65+ years old with a full scope of features that will help older people get easy access to banking products and Longevity-related services.

In the longer-term, Longevity Biomarkers will serve a fundamental role in the company's activities. Much like the InsurTech is beginning to gamify Longevity by offering discounts when clients meet their Healthspan, users of Longevity Banking Card will receive Longevity Incentive Points, redeemable for products and services in the Longevity Marketplace, for maintaining and surpassing their personal Healthspan goals, and the use of Longevity Biomarkers for Healthspan tracking and maintenance will play an increasingly important role as the scope of obtainable biomarkers continues to grow.

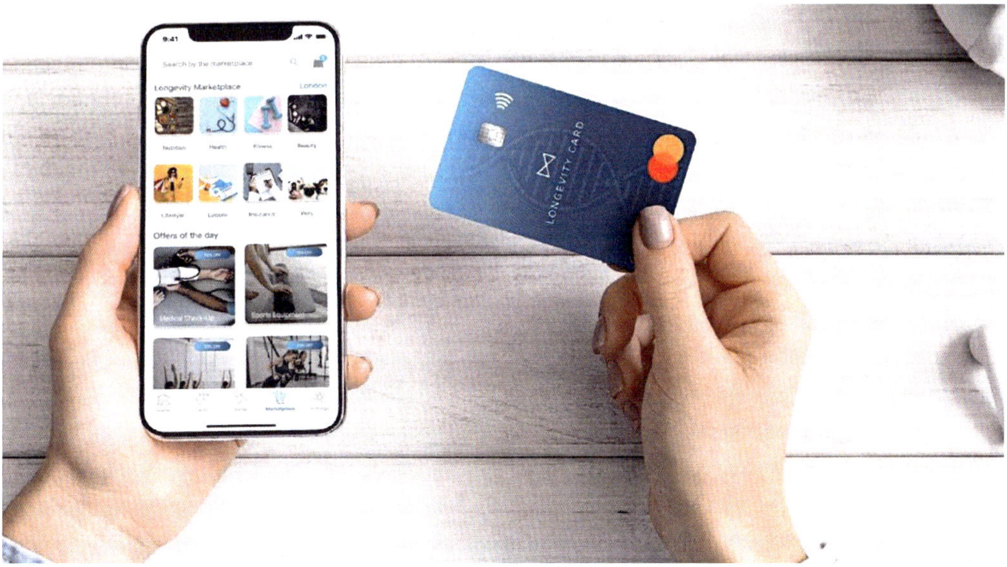

Source: Longevity Card[1]

Deep Knowledge Ventures and Longevity.Capital provided the seed funding for Longevity Card, which they are treating as a vanguard project to explore the Longevity banking and financial services mar-

[1] www.longevity.cards

ket. They also intend to use it to formulate and adjust region-specific strategies toward the goal of establishing a formal Longevity Bank in the UK and EU in 2023, supported by the sophisticated Longevity Investment Analytics of Aging Analytics Agency.

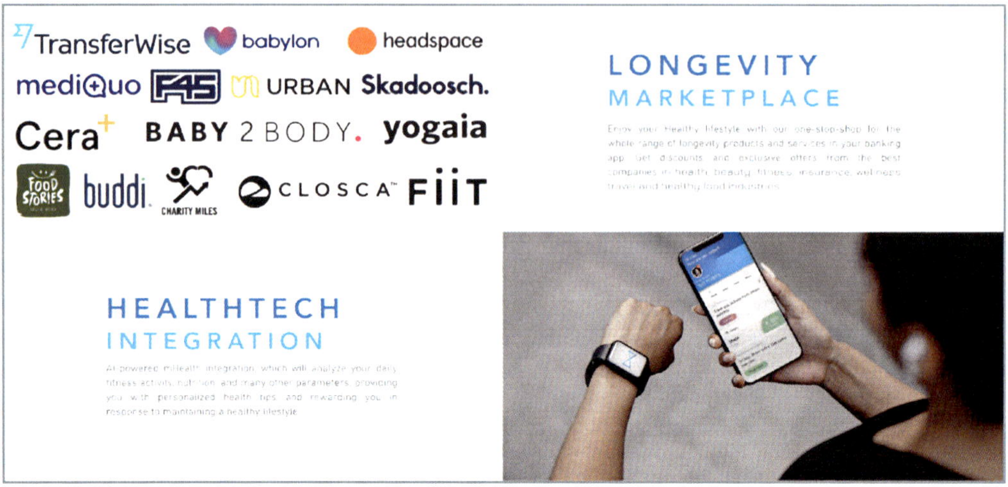

Source: Longevity Card[1]

Longevity Novel Financial Instruments

Longevity-based derivatives are going to be a preferred option for investors, such as survivor bonds, that pay a coupon based on the "survivorship" of a stated population group along with Longevity notes which reference a pool of pre-defined lives. Due to the rise in such Longevity-themed financial instruments and investment options, the government mortality indices will be closely watched in the coming years. There will also be a significant increase in investment to adapt existing technologies and infrastructure to the aging population.

Today, the majority of Longevity indices and derivatives are packaged as risk management solutions. The current synthetic Longevity products are emerging Longevity indices, Longevity notes, Longevity swaps, and other Longevity derivatives such as survivor bonds. The attraction of Longevity-based financial products that take on Longevity Risk is that they are standardized and tradable, which allows massive liquidity to larger institutions that buy them,

[1] www.longevity.cards

such as pension funds and insurance companies. Since the Longevity Risk on insurance and annuity contracts is swappable, the secondary insurance market may see minimal increases in liquidity, although life insurance products and annuities may not be very liquid in the future. These kinds of practices are the simplest and least modern form of this general trend.

We can also envision the emergence of financial instruments and analytics tied to Biomarkers of Aging (which serve as proxies for measuring the current state of individuals' and populations' biological aging) and to the technological and scientific validation of Longevity companies' therapeutic pipelines and the more modern and sophisticated approaches for validation described in Chapter VI: A Proposed Solution to The Longevity Industry's Biggest Risk: Modern Tools for Safe Human Validation to De-Risk Investment Decision-making and IPOs.

More specifically, we envision the development of at least three types of financial analytical products. All of these products will be derived from sets of Biomarkers of Aging and Longevity:

- An analytical panel designed for the use of hedge funds and investment banks to predict the success or failure of particular molecules in the later stages of clinical trials and provide investors with the signals to form long or short positions;
- Advanced technological due diligence (enhanced SWOT and comparative analysis) for venture investors to evaluate the claims of emerging companies on whether their technologies can deliver actual results on humans ;
- An analytical panel for Healthcare InsurTech companies focused on retail clients. Currently, we are aware of at least six companies working on similar types of solutions, and we can envision the emergence of another 10-20 such companies in the next 2-3 years.

The market of the Longevity-derived financial instruments shows a positive trend in terms of both number of deals and the allocation amount. The graph below depicts the amount of the allocations made with the Lonfevity-derived Swaps. As can be seen, despite the decline during the 2016-2018 period, the importance of such instruments was "rediscovered" again in 2019 and the following year.

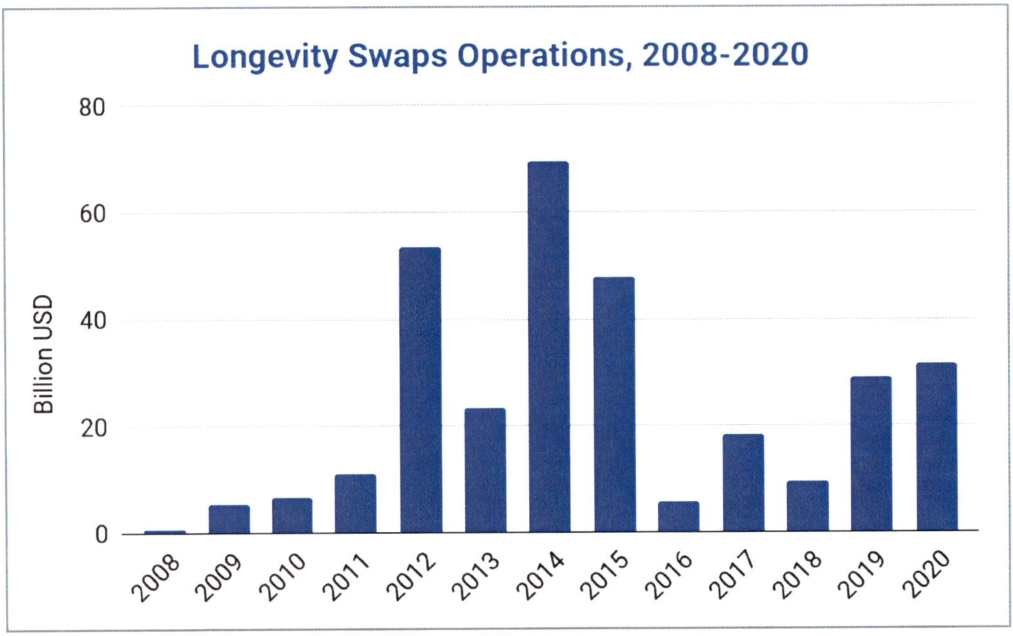

Source: Longevity Derivatives and Financial Instruments Report[1]

The last deal using Longevity swap was signed in May 2021 between Swiss Re as an arbitrageur and The Trustee of the ICL Group Pension Plan, a Fujitsu pension scheme, as a hedger. This Longevity risk transfer insured **£3.7** billion of hedgers' liabilities and covered pensions in payment for approximately 9,000 members.

Other institutions of the swap deal infrastructure include:

- Willis Towers Watson as actuarial and transaction adviser to the hedger
- Gowling WLG LLP and Momentum Investment Solutions and Consulting as legal and investment adviser to the hedger
- Pinsent Masons LLP as legal adviser to the hedger
- Insight Investment as calculation agent, collateral manager, and collateral valuation agent.

Daniel Harrison, Global Head of Longevity Solutions at Swiss Re stated that: "There is a compelling rationale for pension plans and insurers to transfer their Longevity risk to reinsurers. We have a natural offset with

[1] www.invest-solutions.tech/longevity-derivatives

our mortality business, the capacity to write the business onto our balance sheet, and the expertise to tailor the transaction to meet our client's needs."

As the predictability of the standard approaches has experienced a significant decline during the COVID-19 pandemic, we foresee the growing interest in new, advanced approaches in the Longevity-related financial and insurance subsectors. This interest will accelerate the development of the Longevity companies operating in the aforementioned areas, attracting additional investments in the industry.

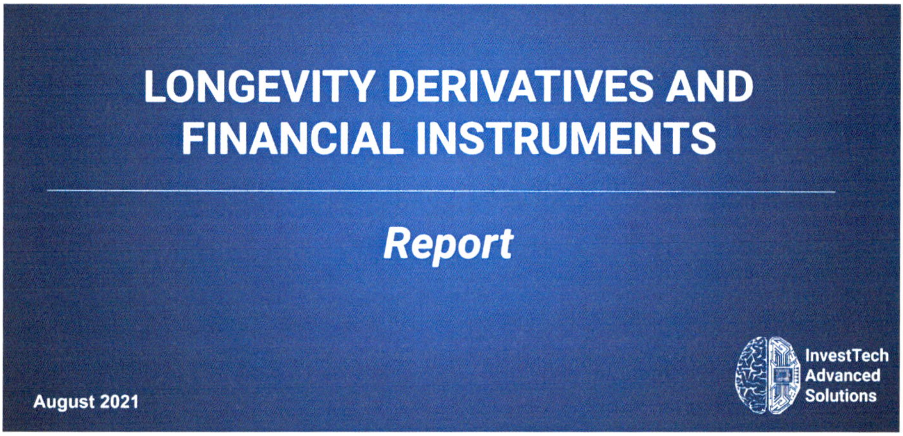

Source: Longevity Derivatives and Financial Instruments Report[1]

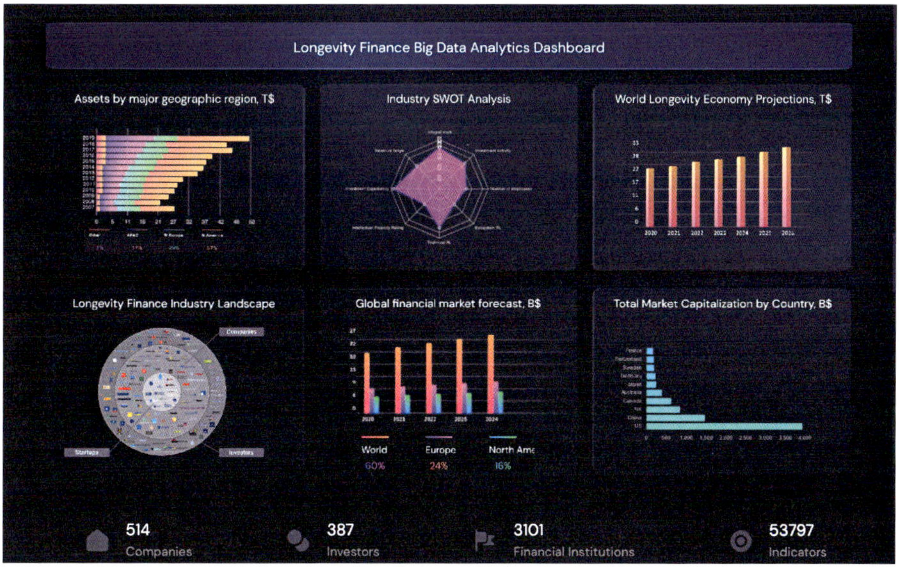

Source: Longevity Finance Big Data Analytics Dashboard[2]

[1] www.invest-solutions.tech/longevity-derivatives
[2] www.deep-innovation.tech/longevity-finance-dashboard

The additional information about the Longevity-tied financial instruments can be found in the report, published by InvestTech Advanced Solution, where profound analysis of the market, as well as uniquely developed assessment models are presented. For the interested readers we also suggest visiting our Longevity Finance Big Data Analytics Dashboard, where one can obtain all relevant information about the financial institutions participating in the Longevity Industry.

Longevity Stock Exchange

The next, more innovative method of neutralizing the disparities and inefficiencies behind the Big Longevity Liquidity Gap is the establishment of a specialized stock exchange for Longevity companies. Such an entity would be the first of its kind in the world and if being integrated with the Global Longevity start-ups ecosystem – could easily secure the position of a leading, progressive Longevity financial hub. The ultimate goal of this project would be the deployment of a Longevity Industry index (similar to the NASDAQ-Composite index, which serves as an indicator of expectations on the growth of the US tech industry).

The first nation to establish a specialized Longevity Stock exchange will have effectively created something resembling a perpetual motion machine for the further growth of its national Longevity Industry, having built an engine for providing its companies with sufficient investment and accelerating the market-readiness of their technologies, products, and services. The nation that establishes market places for shares in publicly traded Longevity companies – as well as financial instruments and derivatives built as a second layer upon its Longevity Industry – would be capable of attracting several trillions of euros of financial liquidity on such exchange.

Establishing a Longevity Stock Exchange would require the public listing of at least 100 Longevity-focused companies to create good enough diversity and potential volume for trading. Thereafter, the more advances there are in Longevity, the more even the most conservative investors will want to invest in this industry. By that point, they will have been well advised that Longevity is an industry like no other. It is by such means that increased global aging population will be transformed from a threat into an unprecedented opportunity.

However, the establishment of such a stock exchange should involve the use of modern, sophisticated approaches to the human-centered validation of Longevity therapeutics and technologies (including Biomarkers of human Longevity and the other platforms for human experimentation and validation discussed previously) as a necessary prerequisite for companies to be listed on such exchanges.

As larger, more conservative investment and financial institutions, instruments, and markets begin entering this space, their expectations in terms of the stability and reliability of the markets they are gaining exposure to, and of the investment decision-making and capacity for reliable predictive forecasting within those markets, will need to be matched. In other words, for Longevity to be truly embraced by these entities and markets, framework for de-risking, reliable predictive forecasting, and stabilization of strategic investment decision-making will need to become industry-standard components utilized by the majority of key institutions, instruments, and markets.

Longevity Composite Indexes

To take this concept even further, suppose that nations were then to build a Longevity Index as a second layer on top of this Longevity Stock Exchange in order to create the analog of a NASDAQ Composite Index[1] for HealthTech, Preventive Medicine, Precision Health, and Longevity industries. This could create one of the most predictable, stable, and rapidly growing investment opportunities, capable of attracting trillions of euros worth of "lazy money" from conservative investors. Furthermore, because Longevity is very much a science- and technology-driven industry, it is far more easily predictable than, for example, the larger tech or real-estate market – but only if modern, relevant, human-centered approaches are used to de-risk the translational gap from model organisms to humans and enable clear, dependable forecasting of clinical outcomes.

The threats and opportunities of Global Longevity are of increasing interest to global finance and an increasingly frequent topic of discussion at international conferences, where there is more and more consideration among attendees of the future shape of global finance. The increased liquidity that these would provide would set in motion

[1] www.investopedia.com/terms/n/nasdaqcompositeindex.asp

a self-perpetuating cycle of Longevity Finance: with greater progress in achieving healthy Longevity, more owners of wealth will want to invest in the repeatedly reinvigorated labor force endowed with a greater Healthspan, leading to further growth and greater Healthy Longevity.

Major Benefits of Longevity Derivatives and Instruments Over Traditional Longevity VC

Hedge Fund-style Liquidity with VC Objectives	Bets on the Success of the Entire Longevity Industry	Significantly De-Risks Exposure to Longevity

Source: Longevity Investment Digest[1]

Financial Instruments and Futures Markets Tied to National Longevity Economies

Another type of Longevity-focused financial products and market that we can expect to emerge in the next several years are financial instruments and Futures Indexes tied to either the current or predicted future state and performance of entire National Longevity Economies with respect to Healthy Longevity and Aging Population. And Biomarkers of Human Longevity will come to play a very necessary role in measuring the current state (and real-time changing trajectory) of population-level National Healthy Longevity.

Such instruments may become an integral part of general national Longevity and healthcare infrastructure, contributing to Longevity governance activities. However, to efficiently reflect the state and the prospects of entire Longevity economies, they should be backed by cutting-edge Longevity governance analytics, benchmarking, and forecasting. A prototype of a big data analytics system of this kind has been developed by the Aging Analytics Agency in the form of the Longevity Governance Dashboard.

The ability to invest in individual countries' Longevity-based economic conditions (e.g., social security) is an opportunity for institutional investors to try and reach untapped markets and produce significant alpha while taking on minimal risk. The emergence of governmental Longevity-based indices would add another avenue for investors to

[1] www.aginganalytics.com/investment-digest

safely park large amounts of money with large investments on the long and short bias on systematic Longevity-based risks taken on by governments around the world.

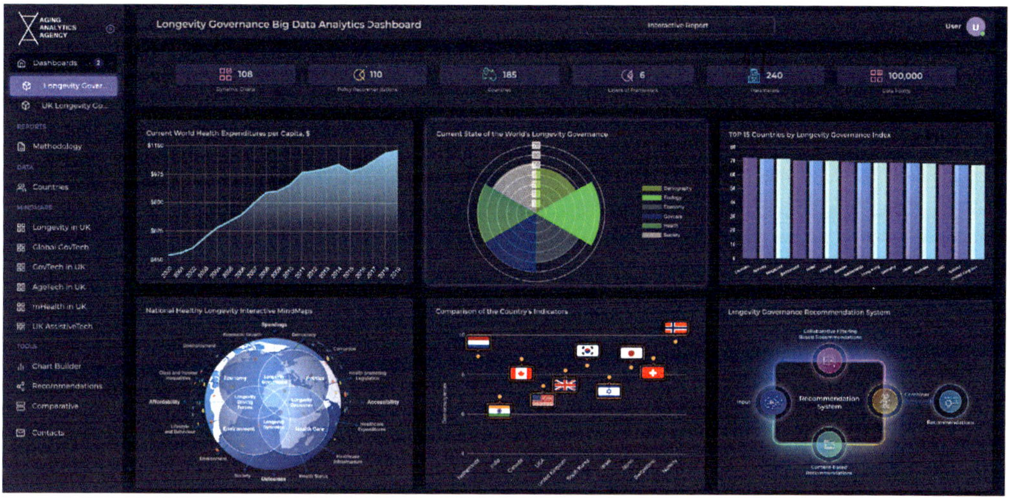

Source: Longevity Governance Big Data Analytics Dashboard[1]

These types of instruments can also be broken down in a more granular fashion, tied to specific global regions, and to specific individual components of entire national economies, such as the current tate of Longevity success and progressiveness of different regions' social security systems, national healthcare systems, government-funded pension funds, level of HALE, the gap between HALE and unadjusted life expectancy (and the specific trajectory of certain nations growth in HALE as measured by Biomarkers of Human Longevity), and others.

Such instruments and markets could be either structured and traded individually or in the form of segregated baskets of financial products. This futures market would create a framework for forecasting positive and negative scenarios surrounding the intersection of what we refer to as the Opposed Longevity MegaTrends of Aging Population and Advanced Biomedicine.

Meanwhile, in addition to creating a new type of Longevity-focused investment (taking long or short positions depending on how well specific regions are facing the challenge and opportunity of National Longevity and Aging Population), this type of system would also provide tangible inputs of investors and financial corporations

[1] www.deep-innovation.tech/longevity-governance

on the most prospective regions to actually invest in from a Longevity Industry standpoint.

For the readers interested in the development of the financial sector of the Longevity Industry, we suggest to keep an eye on the new releases concerning the "Longevity Financial Industry" book, which will include a profound analysis of the major topics related to the aforementioned sector and is going to provide additional insights relevant for all parties involved in the Industry: investors, companies and financial institutions representatives, as well as general public.

"Longevity Financial Industry" book[1]

Possible Forms of Futures Markets Based on National Longevity Economies

Instruments Tied to Current State of Health-Adjusted Life Expectancy (HALE)	Futures Betting on Future HALE and Gap Between HALE and Life Expectancy	Instruments & Futures Tied to National Healthcare Economies
Instruments & Futures Tied to National Pension Systems	Instruments & Futures Tied to National Social Security Systems	Instruments & Futures Tied to National Longevity Financial Systems

Source: Longevity Investment Digest[12]

Biomarkers of Human Longevity as the Key for Longevity Financial and Investment Prospects

There is one thing at the heart and foundation of all these potential approaches, technologies, products, instruments, and infrastructures at the intersection of the Longevity Industry and the global financial

[1] www.longevity-book.com
[2] www.aginganalytics.com/investment-digest

market. This is the array of market-ready solutions to the safe, human-centered validation of Longevity therapeutics. Foremost among them are Biomarkers of Human Longevity, as well as in silico human modeling and other modern techniques for validation, due-diligence, forecasting and de-risking.

In some cases, the use of these technologies is fundamental to the product, service, or approach itself. This is the case with, for example, the use of Biomarkers of Human Longevity in InsurTech or for tracking population-level Healthy Longevity in financial instruments tied to the health (or Healthy Longevity) of specific populations. In other cases, these modern approaches to human validation serve as the major safeguarding factor for the de-risking of institutional, venture, or retail investments and preserving investor sentiment. Either way, it should be clear, at the close of this book, that the paradigm shift from model organism studies to safe and effective human validation is the foremost catalyst for protecting and ensuring the stability of the Longevity Industry as it stands today, as well as its future developmental trajectory.

Summary

In this chapter, we provided an overview of the ways in which the Financial Industry can take on board some of the topics related both to Biomarkers of Human Longevity and to modern, sophisticated approaches and solutions for validation of Longevity therapies. We also charted the specific scope of the investment and financial approaches, entities, and infrastructures required to truly transform these practices from tools being used by a small minority of Longevity investors (such as Deep Knowledge Ventures and Longevity. Capital) into industry-standard practices – to the mutual benefit of investors, companies, the general public, and the entire Longevity Industry itself.

The major take-aways from this chapter include:

- Human Biomarkers of Longevity are already being on-boarded by specific sectors of the global financial industry – including in particular life and health insurance, and InsurTech, which are using biomarkers to gamify health and offer incentive benefits to clients who maintain their health goals.

- Smart investors need to prioritize funding for Longevity companies that employ some combination of human-centered validation approaches in order to demonstrate the safety and efficacy of their therapies – starting first and foremost with Biomarkers of Human Longevity – both for their own benefit and the benefit of the industry as a whole.
- While a small minority of investors are utilizing such human-centered frameworks and re-tuning their investment target identification, valuation, and due diligence processes to make strategic decision-making more modern, relevant and reliable, and to de-risk the translational gap from model organisms to humans, such frameworks need to become industry-standard tools in order to provide the levels of stability and predictive reliability expected of large, conservative investors and financial institutions.
- Once such frameworks and approaches are accepted as authoritative by investors, this will substantially de-risk the practice of Longevity investing, making it as safe, stable, predictable, and lucrative as for example real estate sphere and commodities. This will unleash an unlimited source of investment capital capable of bridging the Big Longevity Liquidity Gap, and enable the establishment of the kinds of financial entities, markets, and systems required to drive the evolution from Longevity Industry 1.0 to 2.0, including Longevity Investment Banks, Stock Exchanges, ETFs and Index Funds, and novel financial instruments – which will need high degrees of industry stability and reliable predictive forecasting to operate in a relevant way.
- This is critical to the Longevity Industry's future, which has reached the point in its ongoing growth where the active involvement of large financial institutions is required to fuel its ongoing growth and the real-world implementation of benefits for humanity and socioeconomic dividends.
- By that time there will need to have novel financial institutions in place capable of utilizing these frameworks and approaches in practice, either established from the ground-up or adapted from existing institutions, markets, instruments, and infrastructures.

Deep Knowledge Group

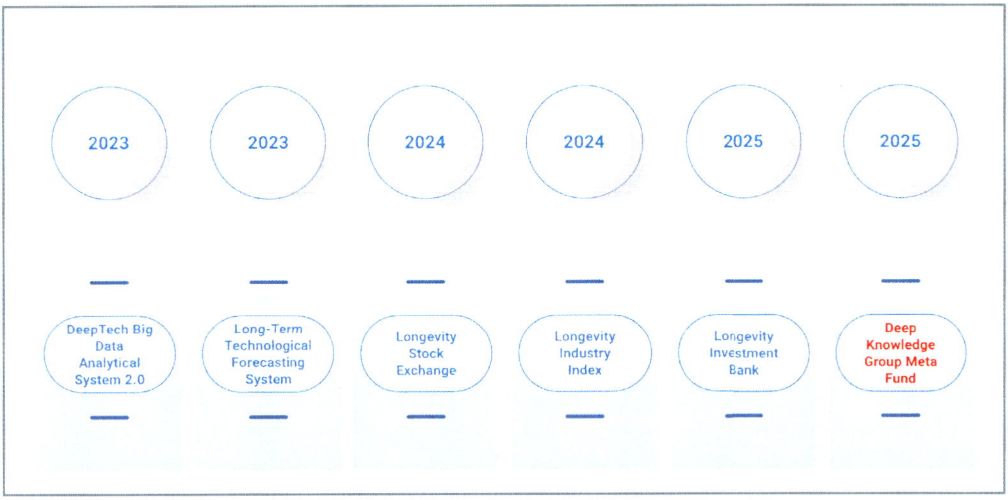

Source: Deep Knowledge Group[1]

Deep Knowledge Group's Longevity Finance Activities

Data Science-Driven Financial and Investment Projects

DeepTech and Longevity Industry Financial Advisors

DEEP KNOWLEDGE VENTURES

LONGEVITY CAPITAL

Built by Deep Knowledge Group:

- Longevity.Capital
- Longevity Banking Card
- DeepTech and Longevity Industry Financial Advisors
- AMCs and Performance Trackers
- Longevity Investment IT Platform
- Big Data Investment Analytics Dashboard
- Longevity-Focused Structured Financial Products
- Automated Due Diligence, SWOT Analysis, De-risking

Deep Knowledge Group's Future Plans:

- Longevity InvestTech Platform
- Longevity Investment Bank
- Longevity and DeepTech Stock Exchanges
- Longevity Index
- Longevity MTF
- Longevity IP Marketplaces

[1] www.dkv.global

Chapter X: DeepTech Engineering Toward the Concept of a Comprehensive Digital Human Avatar 1.0, 2.0, 3.0 and 4.0

As the number and range of applications of Longevity Biomarkers continues to grow, the evolution toward increasingly comprehensive and sophisticated virtual models of the human organism, which can be summarily referred to as the concept of a Digital Human Avatar – encompassing and integrating an increasing number of individual biomarkers and an increasing diversity of markers (both biological and non-biological) – becomes not just an evident prospect, but the likely endpoint of these converging trends.

Much of this is tied in with the increasingly necessary (as opposed to complementary) role the AI will play in the development, validation, and personalized application of biomarkers generally – and of Biomarkers of Human Longevity in particular – combined with the increasing volume of individual and population-level biomarker data that is readily obtainable through consumer-grade devices, products, and services, as well as the rise of the field of digital biomarkers.

Digital Biomarkers and Digital Panel Platforms are necessary for the widespread usage of Digital Avatars and the concept of storing all relevant persons' health and non-health data inside one specific data storage. This is a sector currently being spearheaded by the companies Deep Longevity, Biocanic, LifeOmic, Onegevity, 122 Labs, and Seqster, shown in their industry context below:

At the same time, the intensive application of AI to all stages of Longevity and Preventive Medicine R&D and healthcare has the potential to accelerate the clinical translation of already validated and also yet experimental biomarkers toward diagnostics, prognostics, and therapeutics. It can empower patients to become the CEOs of their health through continuous AI-driven monitoring of minor fluctuations in personal biomarkers.

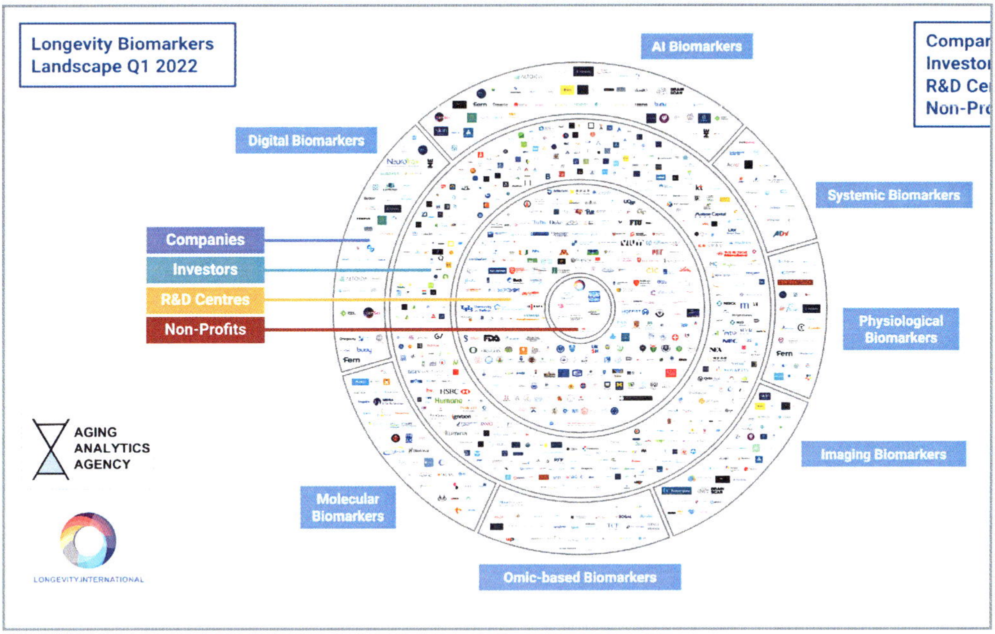

Source: Longevity Biomarkers Ecosystem[1]

In the coming years, we can expect the convergence of AI, novel aging biomarkers R&D, and digital biomarker applications to accelerate the further development of AI-based Biomarkers of Aging and Longevity. This synergy of AI, Big Data usage, and Biomarkers can lead to the evolution of a Digital Avatar concept that will track the health status of each individual; measure in near to real-time the effectiveness of therapeutic and lifestyle interventions on health, Longevity, and performance; enable personalized diagnostic and prognostic assessments and predictions; and offer personalized data-driven and evidence-based recommendations.

I have already introduced in Chapter VI what I consider to be the characteristics of a Digital Human Avatar and the current, real-world state of its ongoing development. Here, I will provide my vision of its future and outline the evolution of increasingly comprehensive, diverse, and powerful Digital Human Avatars – and what could be considered to qualify as Digital Avatar 2.0 and the next-generation technologies Digital Avatars 3.0 and 4.0.

[1] www.longevity.international/biomarkers

The Evolution of Digital Human Avatar 1.0 to 2.0, 3.0 and 4.0

We foresee a five-stage evolution of Digital Human Avatars from 0.5 (the current state of Avatars available on the market) to 1.0, 2.0, 3.0, and 4.0. This classification framework is based on a number of specific factors including:

- their technological sophistication;
- their use of AI for assessment, predictive analytics, and personalized recommendations;
- the frequency with which they measure the biomarkers that they encompass;
- the overall scope of markers impacting health, Longevity, performance, and QALY across the full human lifecycle, from biology to behavior, technology, and environment that they are capable of measuring, integrating and predictively analyzing.

We have taxonomized them in greater detail below, describing the distinguishing characteristics that qualify them as an Avatar 1.0, 2.0, 3.0 or 4.0, as well as the timelines in which we can expect them to emerge.

Digital Human Avatar 0.5

This category represents Digital Human Avatars that are available on the market today, consisting of a robust panel of biomarkers adjusted to the individual. This category encompasses all known and validated biomarkers of disease and health and some biomarkers of Longevity. It is used mainly for the purpose of early diagnostics within the scope of Preventive Medicine (biomarkers of disease) and Precision Health (biomarkers of health).

This includes "classic" disease-focused prognostic and diagnostic biomarkers (laboratory parameters that doctors can utilize to help in decision-making in terms of diagnosis and selecting a course of treatment), as well as biomarkers of general health and fitness (such as those obtainable via wearables).

The main distinguishing feature of this avatar is the attempt to measure, synthesize and integrate a comprehensive set of biomarkers into a single system for the purpose of individualized health assessment, early diagnosis and prognosis of disease, predictive analytics, and health-status forecasting.

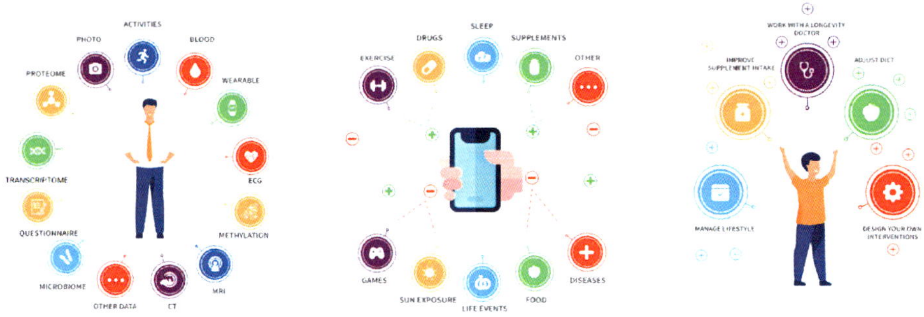

Source: Young.ai[1]

We know of at least six companies that could be said to have developed at least a prototypical Digital Human Avatar, thus placing themselves at the cutting edge of biomarkers development. These include Deep Longevity, Seqster, 122 labs, Biocanic, LifeOmic, and Onegevity.

We can estimate that these companies will be capable of further developing their systems to a level of development characteristic of a Digital Avatar 1.0 (described below) already before the end of year 2022. We also predict that some of them may be able to attain this state of sophistication, comprehensiveness, and real-world utility less than a year from now.

AI and Digital Platforms: Toward a "Digital Avatar"

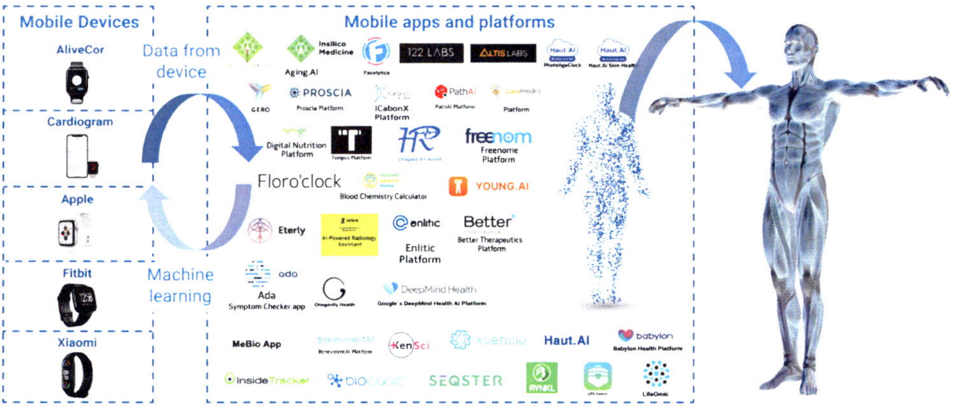

Source: Longevity Biomarkers Ecosystem[2]

[1] www.young.ai
[2] www.longevity.international/biomarkers

Digital Human Avatar 1.0

These avatars would constitute the first fully fledged Digital Human Avatars and would encompass a level of comprehensiveness, real-world usefulness and technological sophistication capable of enabling in practice the core functionalities first envisioned by the concept of a Digital Human Avatar.

Its defining features would include measuring and integrating a truly comprehensive set of biomarkers of health, disease, and Longevity in order to create a virtual profile of individual humans by aggregating, storing, and dynamically integrating as much health data as possible to a given individual, encompassing multiple types of data through a variety of devices. Instead of representing a profile of an individual limited to unidimensional, disease-linked biomarkers, it will represent the whole organism from an overarching health perspective and focus mainly on disease prevention and extending Healthspan.

It is very likely that these systems will also store health and life data in a more technologically sophisticated manner than is common today, such as through cloud storage facilities or blockchain-based storage with advanced cybersecurity, to ensure that individuals have full control of their own data. It will be common for users to selectively disclose their data to select third parties including doctors, experts, consultants or even analytical companies capable of performing deeper analyses using their data. In most cases the companies developing these systems will have some capacity to provide personalized analytics, but I would suspect that many individuals will desire second opinions from other sources, including companies developing other Digital Avatars with different technical compositions of specific biomarkers.

Once again, given the current state of AI and digital biomarker platforms, we can reasonably estimate that Digital Human Avatars capable of being classified as 1.0 will be market-ready by the end of 2022.

Digital Human Avatar 2.0

Digital Human Avatars of the 2.0 level will feature all characteristics encompassed by 1.0 avatars, but with a larger scope of measurable markers, and with more technologically sophisticated analytical capabilities, and a higher frequency of biomarker measurement, health assessment, and personalized recommendations.

They will likely encompass a wider range of Biomarkers of Human Longevity than 1.0-level systems, including biomarkers that allow us to understand the biological ages of tissues and cells and compare them with chronological age: DNA methylation or other epigenetic clocks, mitochondrial health, genomic instability, frailty index, protein expression, telomere length, stem cell exhaustion, grip strength, circulating factors, autophagy, etc.

This type of avatar would enable individuals to track their own biological (as opposed to chronological) age, predict the likely effectiveness of therapeutic and behavioral interventions on their Healthspan by measuring the effect of subtherapeutic doses and track the real effectiveness of such interventions in a more dynamic manner.

One of the key characteristics distinguishing it from 1.0 avatars will be its capacity to provide current health assessment and future health (predictive) analytics at a higher frequency (e.g., daily measurement and adjustment), in such a way as to actually constitute a rudimentary platform for safe self-experimentation.

It will be capable of measuring (and to some extent extrapolating) the effect of various behavioural, lifestyle and therapeutic interventions on biomarkers of health, disease and Longevity, and of recommending specific adjustments to individuals' health and Longevity regimen (e.g., exercise, diet, supplements, lifestyle, etc.).

These recommendations will include the incorporation of new interventions and factors (e.g., new supplements, foods, or exercises); the removal of certain interventions from their existing regimen; and time-intensity of dose-specific adjustments to what could be referred to as the "amplitude" or "intensity" of existing interventions.

In other words, it will provide an AI-mediated system capable of enabling individuals to fine-tune their own health, disease prevention and Healthspan-extending lifestyle, behavioral, and therapeutic regimen for maximum personalization, precision and real-world effectiveness.

Besides enabling assessment, analysis and medium-frequency feedback-mediated intervention adjustment and personalized fine-tuning, these systems will also be capable of reliable forecasting of health and Longevity (including the predicted effects of lifestyle, behavioural and therapeutic regimens) across timelines of 1-5 years ahead.

Based on the current rate of development in AI and digital biomarker platforms, we would assume that such technologies and platforms may be available in prototype by as early as 2023, and available in a more robust, validated, and developed form by 2024-2025.

AI-Driven Biomarkers and P4 Medicine

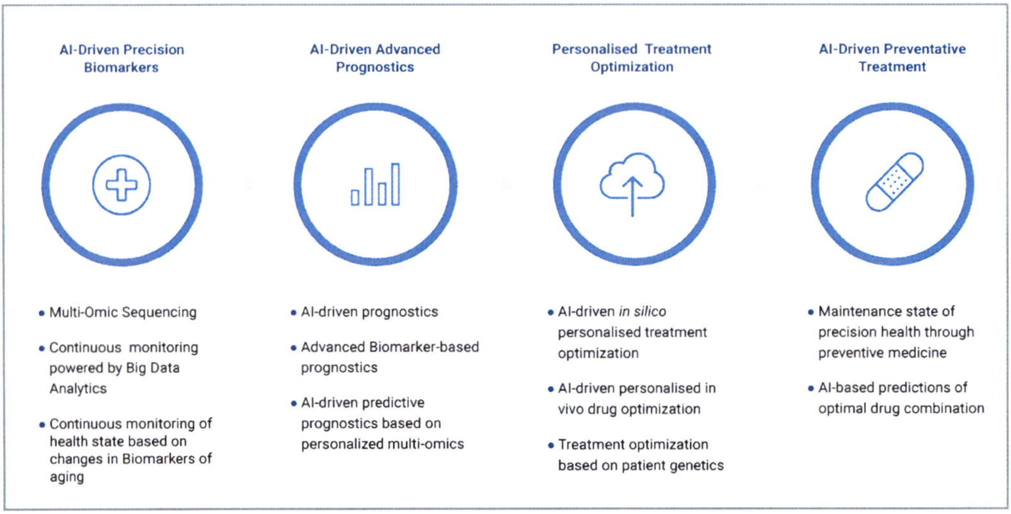

Source: Longevity Biomarkers Ecosystem[1]

Digital Human Avatar 3.0

Systems that qualify as Digital Human Avatars at the 3.0 level will feature a greater degree of comprehensiveness (in terms of the scope of biomarkers of health, disease, and Longevity they are capable of measuring and synthesizing), analytical capability, and technological sophistication than 2.0-level systems.

Importantly, a major distinguishing feature between 2.0 and 3.0 level systems will be their capacity to encompass a range of personalized non-biological markers of health, wellness and Longevity, including such domains as financial wellness and Wealthspan, psychological health, social activity, and social wellbeing, as well as the individual's specific technological ecosystem (the scope of external devices and technologies that impact their health, wealth, Longevity, QALY, and performance).

These systems will also measure external non-biological factors, which collectively can be classified as environmental markers, includ-

[1] www.longevity.international/biomarkers

ing factors of their local environment (e.g., their city, municipality, or country) impacting health, performance, and QALY, such as environmental hazards (air and water pollution, dietary impurities and contaminants, electromagnetic pollution, occupational hazards, etc.).

They will feature higher-frequency feedback (measurement of the effects of interventions on biomarkers, and adjustment of intervention composition, dosing, intensity, duration and amplitude to fine-tune their safety and effectiveness) that is closer to real-time, and will feature some degree of embedded sensors capable of measuring biomarkers internally (beyond the scope of markers that can be measured via traditional wearables).

We can also expect the degree of technological sophistication featured by embedded sensors to have progressed to the point of incorporating some degree of computational hardware and internal predictive analytics themselves. The overall level of integration and synergy between embedded sensors and external AI systems will be greater, and this will enable a higher degree of accuracy, personalization and precision of forecasting, recommendations, intervention adjustment, and fine-tuning. This general trend will be further supported by the exponential advancement of AI, machine learning, deep learning, and big data analytics systems. These synergies will be even further multiplied by the growing sophistication of the companies developing Digital Human Avatar systems and by the ever-growing number of individuals using these systems (which will increase the amount, accuracy and real-world relevance of data that can be consensually and anonymously extracted from users by the companies developing these systems).

All of these factors will create a self-inducing cycle of enhancement, enabling the Avatars themselves, the Avatar companies, analytical companies, individuals, and doctors to provide increasingly accurate, relevant, precise, and personalized assessments, forecasts and recommendations. These assessments and recommendations will be focused not just on health and lifestyle but on tangible therapeutic interventions and treatment via AI analysis of very exact data points guided and supervised by specialized Longevity Medicine doctors and physicians – collectively enabling a system that can reliably maintain a state of Precision Health.

By encompassing a more comprehensive set of factors impacting individual health, Longevity, performance and QALY, these systems will not only be capable of providing recommendations and predictions on shorter timescales than 2.0 (allowing for near real-time adjustment of interventions), but will also be capable of providing predictive analytics and recommendations on longer time-scales as well, on the order of 5-10 years (for example, recommendations on changing one's social environment or occupation, relocating to a different city or municipality, eliminating certain systemic stress factors that cause long-term negative impacts.

While the feasibility of obtaining, analyzing, and integrating these types of data might at first glance seem highly challenging, this is the inevitable outcome of the full-scope digitization of personal and societal activity. Therefore, it is a question not of whether it will happen, but of when. As the types of data that can be obtained on individuals via personal devices (including smartphones and wearables) and through various types of social infrastructure (such as digitized government databases, social media platforms, and the digital trail of activity left by people through their interactions in the digital economy) continue to increase, the real-world capacity for consumer-grade software and AI to track, measure, integrate, and analyze this data – and offer various types of assessment and predictive forecasting based upon this data – is increasing proportionally.

Individuals' personal financial and behavioral data is already digitized, tracked, analyzed, and predictively modeled by the financial institutions and other types of commercial corporations that are using multiple sources of digital data, including analysis of bank transactions, communications, and internet searches. Nearly every aspect of daily activities leaves a digital trail that is being monitored, stored, and analyzed.

Meanwhile, governments have long been tracking, integrating, and digitizing a wide variety of environmental parameters simply to meet the logistical challenge of maintaining the environmental, ecological, and economic integrity of their populace at local, municipal, and national levels. They have also been utilizing digital technologies to integrate these different datasets together to an increasing extent in order to conduct efficient predictive analyses, policy formulation and execution, and other types of governmental planning. In many ways, the breadth, depth, and sophistication of external (e.g., environmental and societal) data tracking, monitoring, and integration is decades ahead of individual data tracking, which makes the integration of such environmental data with personal data a much easier task than it might first appear.

Based on current rates of development, we can expect Digital Human Avatar 3.0-level systems to be ready in prototype by 2024, and to be market-ready by the year 2025-2026.

Digital Avatar 4.0

Digital Human Avatar 4.0-level systems will encompass at least the same degree of technological sophistication and analytical capabilities, and the same scope of biological and non-biological markers as 3.0-level systems, with a key distinguishing feature: they will also incorporate a comprehensive panel of societal and infrastructural factors that influence an individual's health, QALY, Healthspan, Wealthspan, and performance, which collectively will serve to measure the level of infrastructural support provided to them by their society and government to optimize their health, wealth, and QALY.

This ecosystem of markers would encompass such factors as the general state and efficiency of their national and municipal healthcare systems (e.g., to what extent they support Preventive Medicine techniques); the technological eco-system accessible to them; the state of their national and municipal economy; their material and immaterial supportive infrastructure (e.g., education, access to Precision Health therapies and lifestyle interventions, etc.); the presence or absence of government-led National Longevity Development Plans and Longevity Industrial Strategies; and the extent to which their municipal and national governments are prioritizing the health and wealth of their populace.

In other words, the key innovation underlying the concept of the Digital Avatar 4.0 is the capacity to integrate personal (internal) and intra-personal data with a much larger scope of data pertaining to an individual's immediate (e.g., municipal) and long-term (e.g., national) external environment, particularly with respect to the state of their region's technological, economic, geopolitical, and healthcare-related strength, stability, and comparative performance.

Due to the comprehensiveness of the scope of markets that 4.0-level systems will encompass, they will be capable of providing predictive analytics, forecasts, and long-term recommendations on the order of 10-25 years ahead, while also being able to provide more accurate, personalized, and reliable predictions and recommendations on 3.0-level timescales (5-10 years ahead) and 2.0-level timescales (weekly and daily recommendations, and high-frequency, near real-time biomarker measurement and intervention adjustments).

It is likely that these avatars will utilize a wide variety of embedded and non-embedded sensors to measure the markers that they encompass, that the majority of their back-end AI and computational infrastructure (for predictive analytics) will be mediated via cloud computing systems, and that the Avatar interface itself will be stored on individuals smartphones.

Based on current rates of technological development, we can expect Digital Human Avatar 4.0-level systems and platforms to be market-ready between 2025 and 2027. Meanwhile, we can expect that these systems will be validated and refined to such an extent by the year 2030 that the majority of individuals living in developed countries will be using them in one way or another. While they will initially be comparatively expensive, we can expect them to decline in price fairly quickly, in a manner similar to smartphones. We can also expect these systems to become standard components of national healthcare systems, covered by the majority of modern health insurance providers.

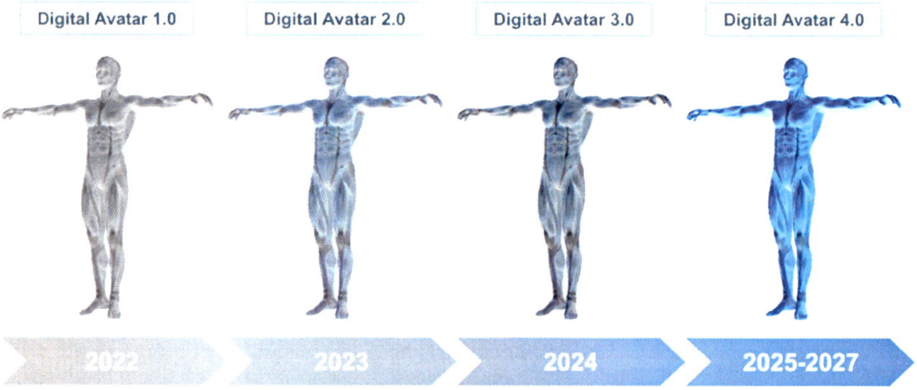

Source: Longevity Biomarkers Ecosystem[1]

Lastly, we can expect that by 2030 the degree of technological sophistication achieved by 4.0-level systems, in combination with general advances in Longevity Medicine and Precision Health, will be enough to provide assurance to the the proportion of Digital Avatar 4.0 users below the age of 60 that they have achieved ***Longevity Escape Velocity*** (LEV). This means that their biological age has been ***frozen***, and they can expect to systematically remain biologically at the same age or even younger than each year of chronological age that accumulates past the point of having reached LEV.

[1] www.longevity.international/biomarkers

4.0-level systems marks the final step toward a truly full-scope ***Virtual Twin*** – virtual emulation digital model of the individual person, encompassing personal biological and non-biological (e.g., psychological, social, financial, and technological) markers, external (e.g., environmental) markers, and societal (economic, geopolitical, and infrastructural) markers to track, assess, and predictively forecast the HALE, QALY, and performance of individuals in the context of the physical and societal environments in which they live.

Digital Human Avatar Framework: 1.0, 2.0, 3.0 and 4.0

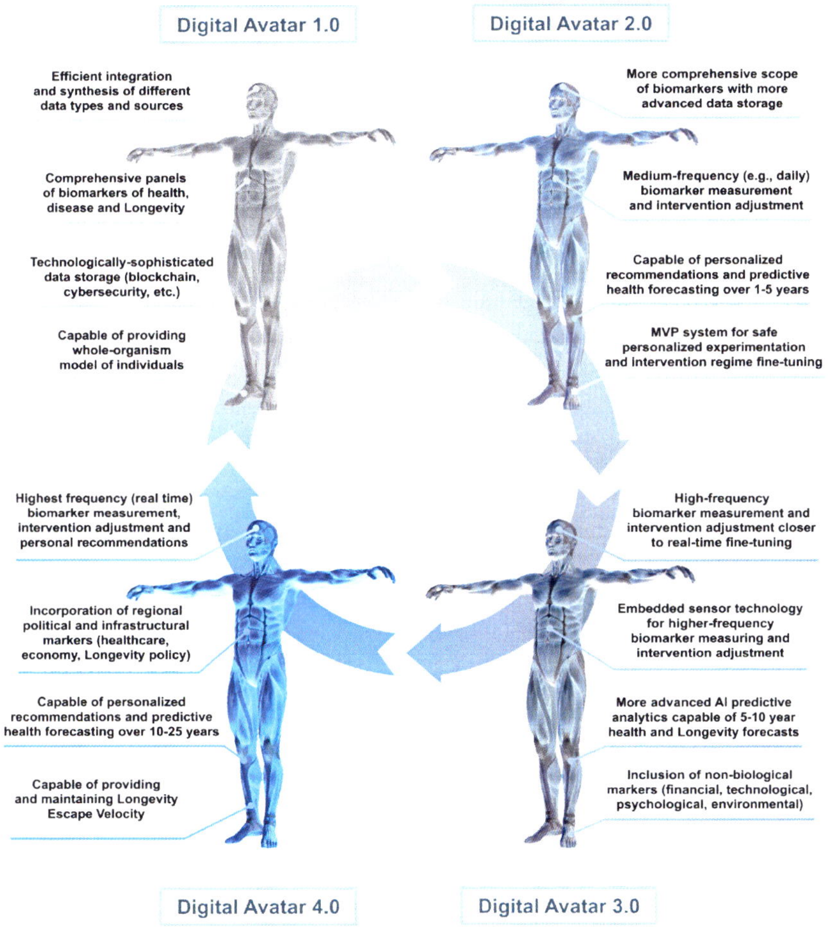

Source: Longevity Biomarkers Ecosystem[1]

The systematic evolution of the power, sophistication, and comprehensiveness of the concept of a Digital Human Avatar, from 1.0 (minimal viable stage) to 2.0 (significantly comprehensive with embedded strong AI

[1] www.longevity.international/biomarkers

predictive analytics engine), 3.0 (maximally comprehensive and robust), is inevitable. However, the speed of this process will be determined by the extent to which individuals, corporations, and municipal and national governments recognize the domain of Biomarkers of Human Longevity and the wider trend and trajectory that it represents (the use of AI and high volumes of biological and non-biological data points as tools for the optimization of HALE and QALY). This will be the critical catalyst to optimizing strategic decision-making; achieving rapid real-world impact on Practical Human Longevity; and improving, maintaining, and extending the health of individuals, entire nations, and, eventually, global humanity.

Summary

- Deep Knowledge Group has a broad conception of Biomarkers of Human Longevity, encompassing non-biological markets such as lifestyle, behavior, psychology, social activity, finance, and environment.
- The increasing capacity for inexpensive, consumer-grade devices to be used as tools for the aggregation, storage, and analysis of high volumes of individual biological and non-biological data, is naturally leading toward the creation and adoption of virtual models of individual humans, or what could be considered as Digital Human Avatars.
- The increasing digitization of all aspects of life, on both an individual level and a nation-wide governmental, environmental and economical level, means that the abundance of data – and the technological capacity to monitor, synthesize, integrate, and predictively analyze that data – is more advanced than it ever has been before.
- These trends are naturally converging toward the evolution of increasingly comprehensive Digital Human Avatar (Virtual Twin) that will encompass literally all aspects of individual, intra-personal, and external factors (technology, environment, economy, policy and geopolitics, healthcare, infrastructure, and more). This will enable the development and consumer-level, widespread adoption of highly precise and powerful models to assess, measure, predictively forecast, and optimize (via personalized recommendations and high-frequency monitoring and adjustment of internal and external factors) human health, Longevity, QALY, performance, safety, and wellbeing.

Afterword by Dmitry Kaminskiy

No specialist or expert in any industry can forecast the sustainable growth of their industry without tangible metrics and tools to measure the actual result of the products.

Based on my personal experience, I am witnessing that (unfortunately) in the Longevity Industry, numerous intelligent people and many scientists are working blind due to the lack of such metrics. These committed and dedicated people devote their lives to the mission of extending lifespan and human Longevity by eradicating diseases, but they operate blindfolded in a new and uncharted field, as they lack a good set of metrics or tools to measure results.

In short, the industry lacks a relevant framework for the transparent assessment of aging research for practical application in humans, and it lacks consensus on industrial standards for Biomarkers of Aging and Longevity.

There is, quite simply, almost no sector or aspect of the Global Longevity Industry which would not be positively impacted, de-risked, and stabilized by the creation and industry-wide onboarding of a consensus on the framework for Biomarkers of Human Longevity.

Chapters II and III of this book summarized the various ways that Longevity clinical trials have not yet resulted in regulatory approval for any one fully concern-free Longevity drug or therapeutic, and the ways in which biomarkers and other related technologies for safe human validation can be used to stabilize and modernize the process of Longevity clinical testing. The lack of a tangible framework based on human validation for Longevity company valuation and due diligence, especially in the case of companies seeking to go public, creates significant risks and issues for the continued growth and stability of the industry.

Chapter IV substantiates the concerns presented in Chapters II and III by analyzing 12 notable cases of publicly traded Longevity companies experiencing changes in their market capitalization and stock price dynamics following the release of clinical trial results, including market capitalization declines as a result of clinical trial failures (i.e., an inability to replicate previous positive model organism results in humans), demonstrating that these issues are already present in the industry and that they will continue to grow until a more tangible and reliable set of metrics for company due diligence and valuation – based on some level of human validation – will be established and widely adopted.

Chapter V detailed the way in which the adoption of a true consensus framework of biomarker panels in other more traditional "one-disease" biomedical industries, such as oncology, has historically resulted in rapid industry maturation, investment de-risking, acceleration of R&D and clinical translation efforts, and the implementation of effective real-world impact on actual patients – which collectively stand as a historical precedent for many of the positive impacts that biomarkers will have on the Longevity Industry forecast.

Chapter VI outlined a preliminary framework of metrics and technologies that can serve both as a basis for safe self-experimentation and for company valuation and due diligence based on human validation. It also described how the widespread adoption of these and similar approaches has the power to substantially de-risk investments in order to both increase the total volume of capital available for funding further industry progress; and to neutralize the risk of widespread declines in investor sentiment as a result of too many public Longevity companies failing to replicate in human trials their previous preliminary positive results in model organisms.

Chapter VII charted the ways in which Space Medicine stands as a major historical and modern precedent for safe human therapeutic validation and the use of overwhelmingly human-centric approaches to R&D and the translation of theory into practice, as well as the many specific convergences between Longevity Medicine and Space Medicine and the ways in which they are mutually focused on the optimization of health and performance.

Chapter VIII outlined the role of NeuroTech and of neurological, cognitive and psychological Biomarkers of Human Longevity within the broader industry landscape. It illustrated how the approach to Practical Longevity could not be complete without accounting for the health of the brain and showed how NeuroTech's engineering mindset and existing emphasis on human (as opposed to theoretical mindset focused on model organisms) validation makes it likely to earlier adopt a consensus framework of Human Biomarkers of Longevity than other aging research sectors. It also highlighted additional practical convergences and complementary intersections between NeuroTech and Space Medicine.

Chapter IX detailed the ways that Biomarkers of Human Longevity are already being on-boarded by select sectors of the Longevity Financial Industry. It described how they are quite simply the only available tool for de-risking investments in an easy, effective, and market-ready way. This de-risking is necessary to allow conservative institutional investors, UHNWIs, and large financial corporations to begin actively fueling further industry growth at the exact moment when their active involvement is needed to maintain the Global Longevity Industry's upward trajectory of development.

Finally, Chapter X charted the future of Human Biomarkers of Longevity from now (2021) to 2030, outlining the ways that Longevity Biomarkers will be essential to achieving real-world results in Practical Human Longevity and how the increasing capacity for inexpensive, consumer-grade devices to be used as tools for the aggregation, storage, and analysis of high volumes of individual biological and non-biological data is naturally leading toward the creation and adoption of increasingly effective, efficient, and sophisticated virtual models of individual humans; these models can be considered as Digital Human Avatars.

The numerous developments occurring in the biomarkers research sphere are quickly converging to enable personalized predictive models of individual humans. The latter will for the first time allow people to reliably take control of their own health and Longevity in a safe manner, assessing their current state of health and testing, re-tuning, and refining in near real-time. Digital Human Avatars will eventually develop to encompass the full scope of multidimensional markers – biological, technological, financial, environmental, infrastructural, and even political – that have an impact on health, wealth, performance and Longevity.

Collectively, these chapters have made the case for Biomarkers of Human Longevity being the critical catalyst not only for maintaining the Global Longevity Industry's continued positive development and growth, but for accelerating it, for neutralizing existing risks and disproportions and for translating the potentials of existing theoretical R&D into real-world, tangible impact on Healthy Human Longevity.

Against this contextual background, it would be very clear that there are large and readily-apparent imbalances between the amount of effort and resources spent on slow-progressing sectors of the Longevity Industry and the amount of resources and effort dedicated to the development to scale industry framework for Biomarkers of Human Aging and Longevity. Why should this be the case when, as we can see, many other industries – including the biomedical industry (for example Oncology sector) – are spending an even greater amount of resources on already very well developed cancer biomarkers, with tangible results?

The onboarding of standard biomarkers framework in oncology and other biomedical sectors has resulted in more efficient, economical, and effective preventive diagnostics, prognostics, and drug development. It has substantially de-risked investments, led to rapid market maturation, and enabled reliable forecasting of clinical outcomes. As a result, sectors such as oncology are well-financed.

Without that financing, oncology would not be capable of achieving tangible results from R&D. And those efforts have clearly paid off. We could even say that a solution to cancer now exists. Whereas 10 years ago, cancer was considered close to a death sentence, nowadays, it is seen very much as a treatable disease — except for some very complicated and rare types. Most cancer deaths occur not because of limits to science and technology but because of the lack of early detection of cancer biomarkers in individual patients, as well as a lack of the correct administration and management of treatment by healthcare providers.

The Longevity Industry should take note of this. There is no reason to suspect that the development and broad adoption of a comprehensive framework of Biomarkers of Human Aging and Longevity would not result in similar levels of industry maturation, R&D acceleration, investment de-risking, improvements in the reliability of clinical outcome forecasting, and the delivery of tangible, positive clinical results in humans. Indeed, there is even more reason to expect these effects as a result of prioritizing funding and development for Longevity Biomarkers – because the complexity and challenges of human aging are even greater than cancer biology, which is why the industry is in even greater need of methods to reliably forecast clinical outcomes and validate the effects of Longevity therapies and technologies in humans.

It is clear, both generally and based on the results of research conducted by Aging Analytics Agency in the last 8 years (and especially in the last 3 years), that a lack of common standards and consensus on Biomarkers of

Aging and Longevity is the key limiting factor for the entire Longevity Industry. Relevant resources and attention should be redirected toward the development of data-driven, AI-mediated Longevity biomarkers panels.

In conclusion, we at Deep Knowledge Group, and particularly at Longevity.Capital, consider Biomarkers of Human Longevity to be among our highest strategic priorities for the next few years. To help bring forth the golden age of Longevity technologies as soon as possible, Aging Analytics Agency is working on a dedicated analysis of methods and frameworks for relevant, reliable, and modern Longevity clinical trials and safe, human-centered therapeutic validation. This analysis was previewed in the present book, which is expected to be available by Q3 2022.

Besides further funding and financing for this low-hanging high-impact sector, we should also be promoting the allocation and redistribution of human resources. Gathering the best and brightest minds to carry this domain forward is equally essential. The Longevity sphere already attracts some of the most intelligent and committed scientific and entrepreneurial talents from around the world, from the cutting edge of multiple converging fields, simply because it is itself the cutting edge of science and technology in general. It possesses the highest degree of scientific complexity, owing to the complexity of the biology of aging compared to the biology of individual diseases.

What is needed now is to promote some of the industry's brightest minds to dedicate their time and efforts to this sector in particular. I encourage as many people to enter into this field as quickly as possible if they aim to make maximum positive impact out of their work. It is already commonly accepted that many Longevity Industry professionals have devoted themselves to this field – one could say almost religiously so – for its humanitarian impact. This is certainly true for myself and for my closest colleagues and allies.

The Longevity Industry not only has the logical potential to become the wealthiest in all history but also represents:

- the most efficient method of achieving the truest and most reliable ROI imaginable (additional years of Health-Adjusted Life Expectancy [HALE] and Quality-Adjusted Life Years [QALY] for global humanity);
- the most efficient method of altruism: socially impactful and tangible way of making philanthropy;
- and it is the most ethical way of doing business.

Appendix **A**

Longevity Biomarkers Landscape Overview Q4 2022

Report Teaser

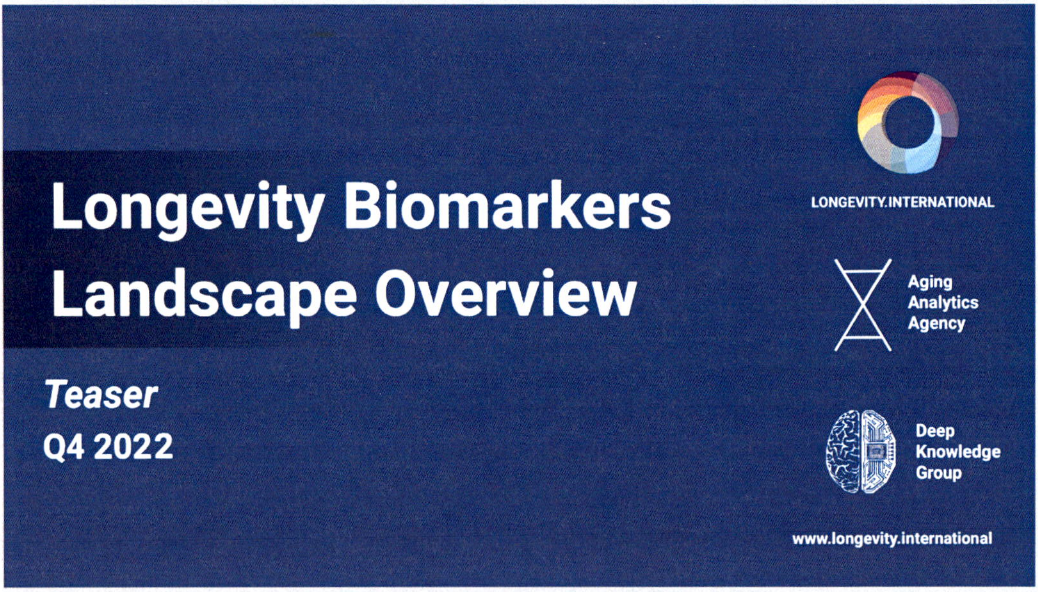

Introduction to Biomarkers

A **Biomarker,** or a **biological marker,** is a characteristic that is objectively measured and evaluated as an indicator of some biological state, condition or process. Biomarkers are used in many scientific fields and commercial activities, and subjected to several classifications that follow different criteria. Regarding health related activities:

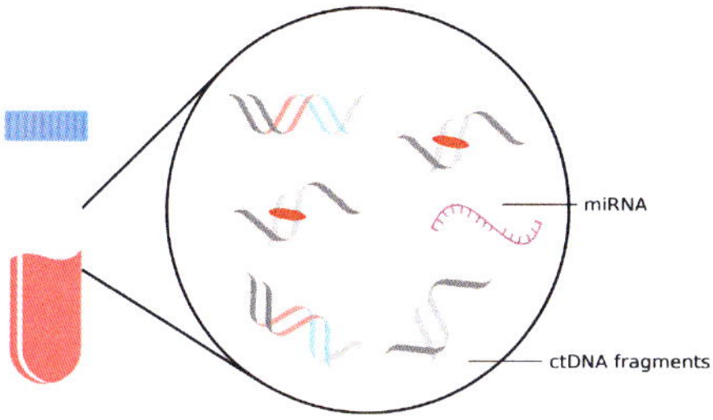

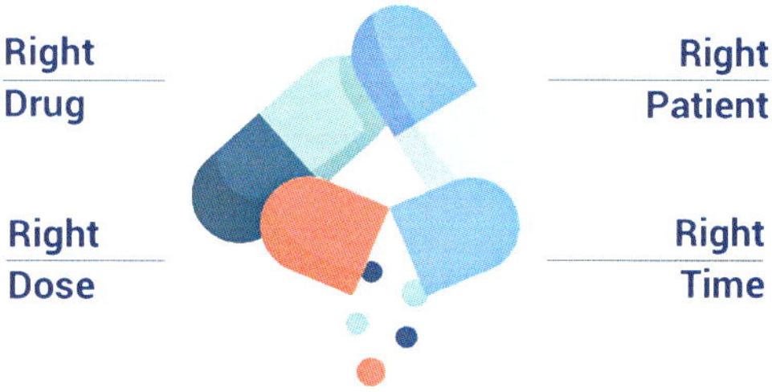

In biomedical research and clinical practice, biomarkers include measurements that suggest the etiology of, susceptibility to, activity of or progress of a disease. The use of biomarkers in basic and clinical research as well as in clinical practice has become so commonplace that their presence as primary endpoints in clinical trials is now accepted almost without question.

In drug discovery and development, biomarkers can be used to predict or identify safety problems related to a drug candidate, for patient selection for clinical study enrollment, stratification of patients during study, identification of toxic responses before they become clinically evident, or revealing an expected or unexpected pharmacological activity such as reversal, deceleration or acceleration of biological age.

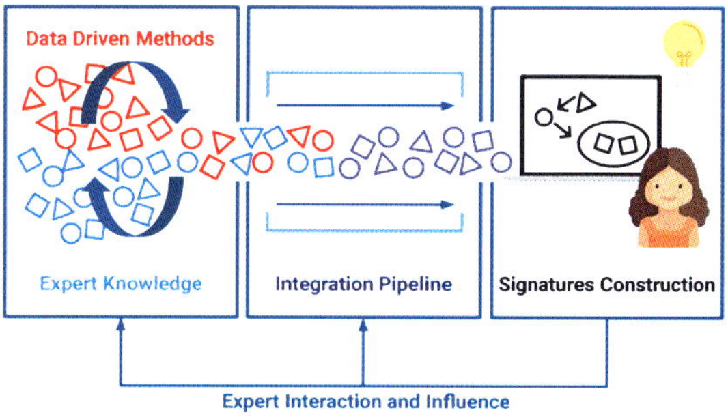

Increased knowledge around the human genome and other omics, and correlations with disease risks and progression are expected to shift the current biomarker focus. Drug companies have been increasing their investment into exploratory biomarkers that focus on areas such as infectious diseases, central nervous system (CNS) disease conditions and cardiovascular diseases to broaden their overall biomarker research and development strategies. For example, CNS disease conditions such as schizophrenia, Parkinson's and Alzheimer's are finding initial success in biomarker identification for targeted therapies.

The Role of Aging Biomarkers in Longevity Research

Lifestyle and behavioral modifications may help slow down the decline and keep the organism in the best possible state for its chronological age, a term commonly referred to as **"healthy aging."** To understand the differences between "healthy aging" and "unhealthy aging," evaluate the effects of the many lifestyle choices and a variety of emerging Longevity interventions, it is essential to be able to track the rate of aging and develop a comprehensive set of aging biomarkers.

The use of precise, actionable and available aging biomarkers with high TRL (technology readiness level) score is essential to be able to support transition from the "classical" medical paradigm (diagnosing and treating diseases when they occur) to the **prevention medicine paradigm** and more generally to **P4 Medicine** (predictive, preventative, personalized, participatory), and finally to **precision health** — creating a framework of activities and interventions for healthy people to help them stay healthy longer and never actually develop diseases in the first place, rather than treating them.

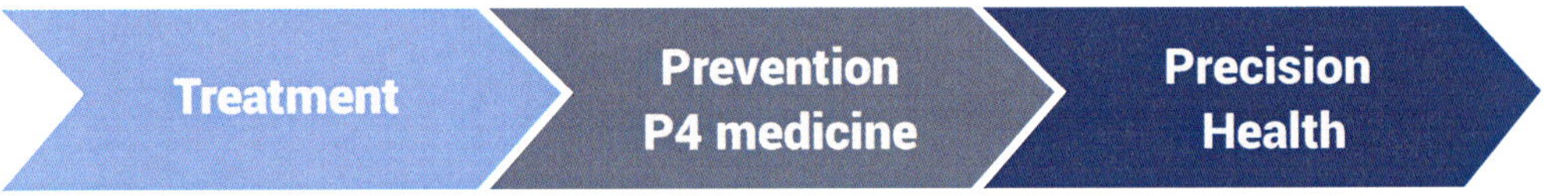

Jamie Metzl for Longevity.Technology: "First, we're increasingly understanding the biomarkers of aging. And that is giving us a language of measurement. We can assess with more precision whether certain interventions are working or not working. With the new tools of AI and machine learning we're really seeing is a super convergence of different technologies that are all pushing forward, including the science of human Longevity."

Landscape of Human Biomarkers of Longevity

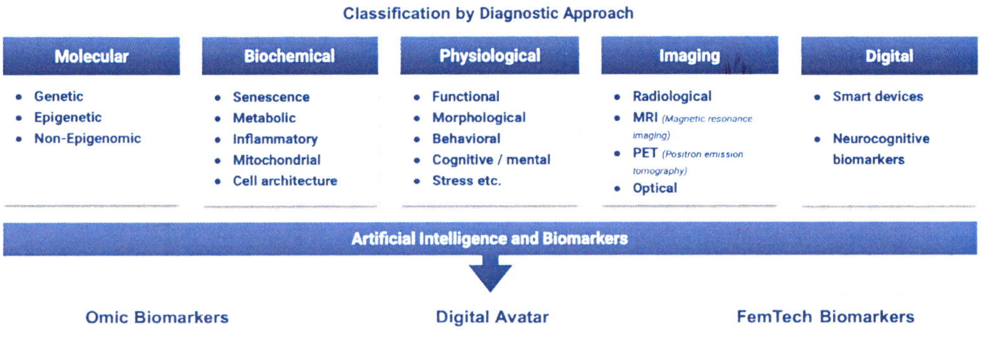

In older adults, chronological age may not be the best predictor of **residual lifespan** and **mortality**, because, with age, heterogeneity in health is increasing. **Biomarkers for biological age** and **residual lifespan** are being developed to predict disease and mortality better at an individual level than chronological age.

The **classification of Biomarkers of Ageing** can be based on different parameters, including characteristics of the individual such as age-related disease state, type of biomolecule, systematic level of diagnostics, or type of diagnostic approach.

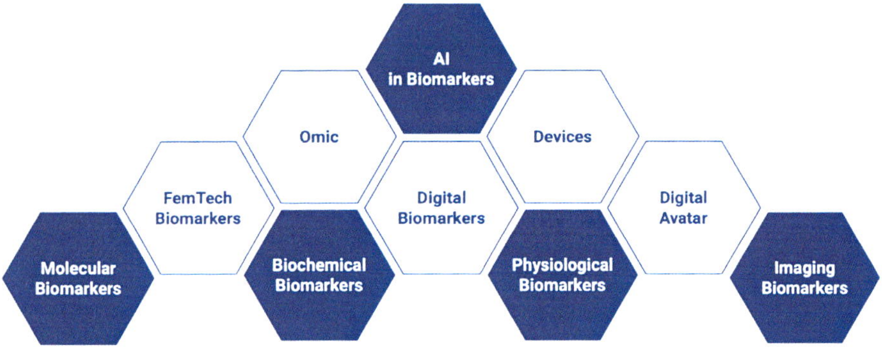

Ahead of the creation of the database, we picked the **most significant types of biomarkers of ageing**. The categorisation of companies is based on the source of biomarkers, their focus level, practical outlook, and methodology. The framework not only brings a **comprehensive view of the market,** but also **sustains relevance in advance of the development** of technologies and research approaches.

Scheme of Biological Age Prediction

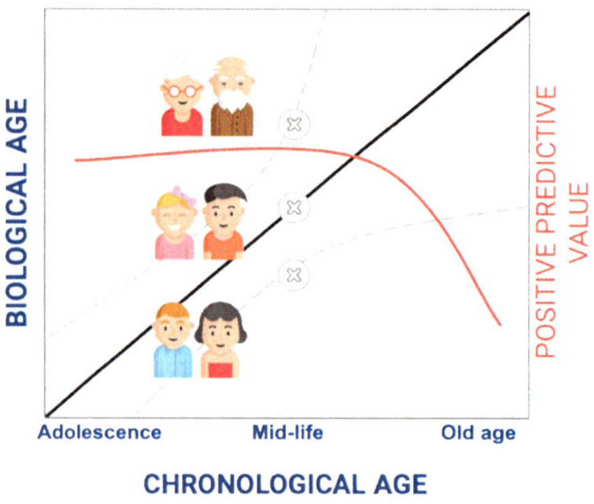

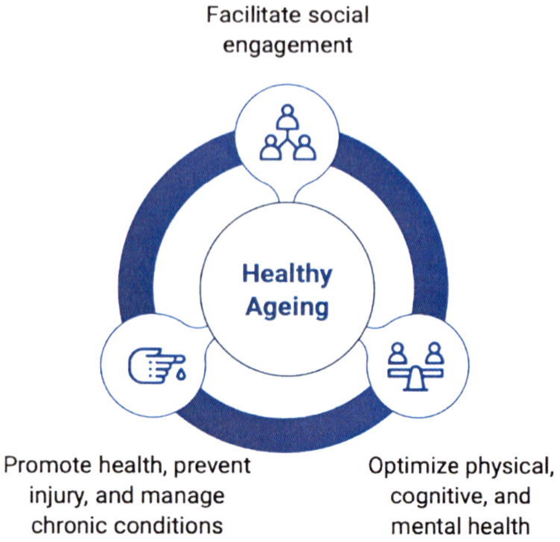

he world's **population is aging rapidly**. Between 2015 and 2050, the proportion of the world's older adults is estimated to almost double, from about 12% to 22%. In absolute terms, this is an **expected increase** from 900 million **to 2 billion people** over the age of 60. Older adults face special health challenges which need to be recognised.

Multiple studies provide dramatic statistics regarding the older cohort of patients. Approximately 80% of people develop endocrine disorders (33% of them have diabetes), 73% develop cardiovascular disease (24% have coronary heart disease), while 58% have problem with eyes and vision (18% of seniors have amblyopia).

Number of Companies by Year of Founding

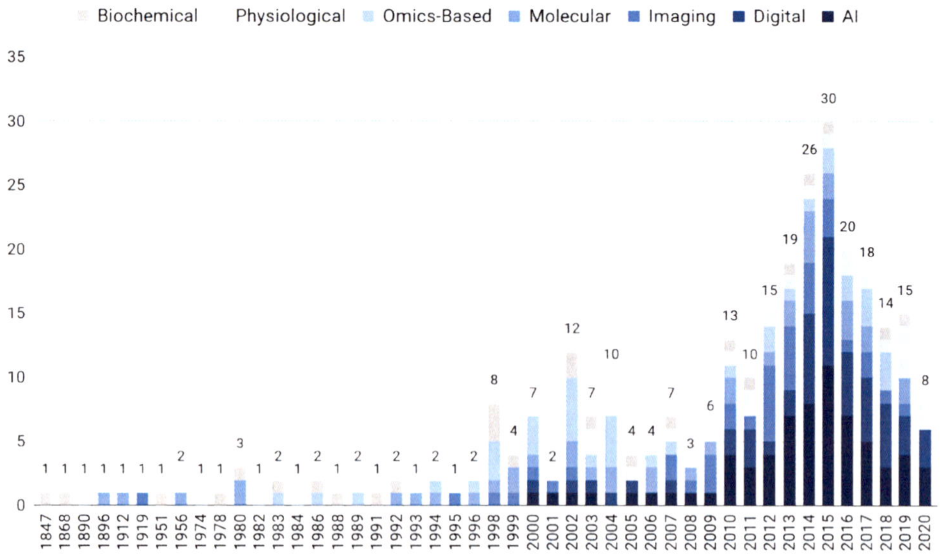

Distribution of Companies by Sector

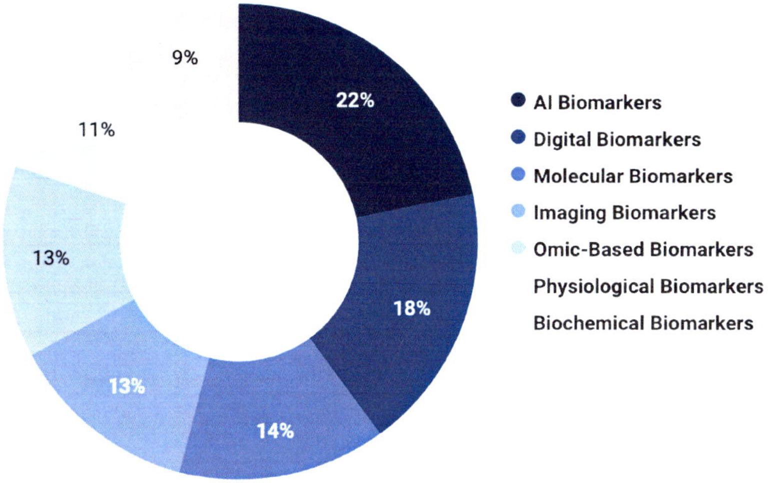

Biomarker companies are distributed relatively evenly among the different categories. **AI Biomarkers** companies constitute **22%** of all companies, while **Digital Biomarkers** companies make up **18%** of the total. Slightly fewer companies (**14%** of the total) are engaged in **Molecular Biomarkers** development, while the **Imaging Biomarkers** and **Omics-Based Biomarkers** categories each account for **13%** of the total number of companies. Despite offering a broad spectrum of diagnostic capabilities, the category of **Biochemical Biomarkers** (i.e. biomarkers of functioning of cellular pathways) includes the lowest number of companies, at only **13%** of the total.

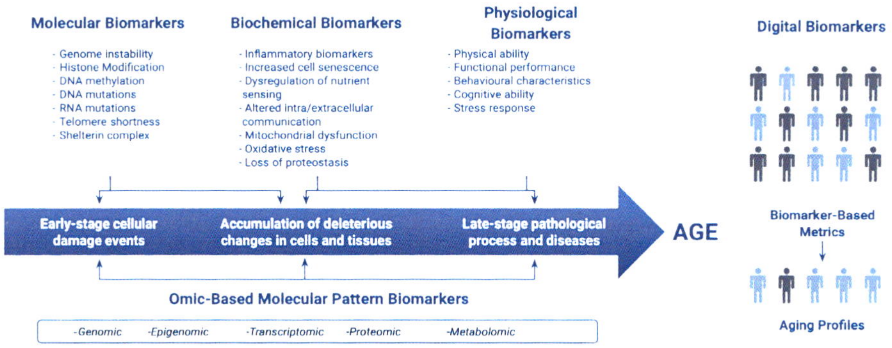

The rapidly decreasing costs of high-throughput sequencing and other massively parallel technologies, such as mass spectrometry, are enabling their use in clinical research and clinical practice. Exome and genome sequencing are already being used to aid diagnosis, particularly of rare diseases; to inform cancer treatment and progression; and, in early efforts, to create predictive models of disease in healthy individuals. These methods are now widely applied in aging evaluation and the prediction of age-related disease.

Influence of Genetics on Ageing and Lifespan

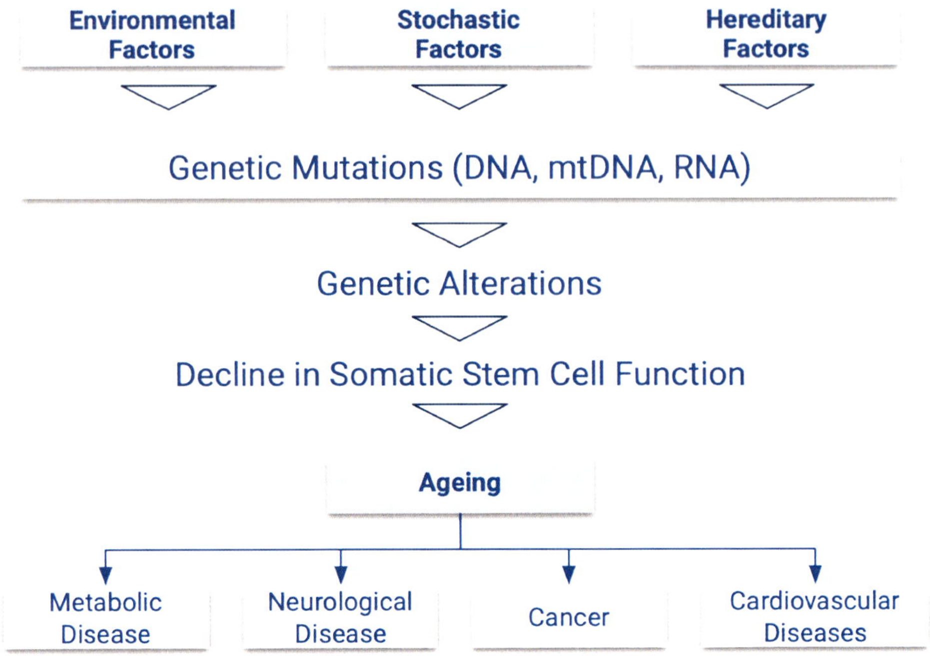

Practical Application of Age-Related Biomarkers

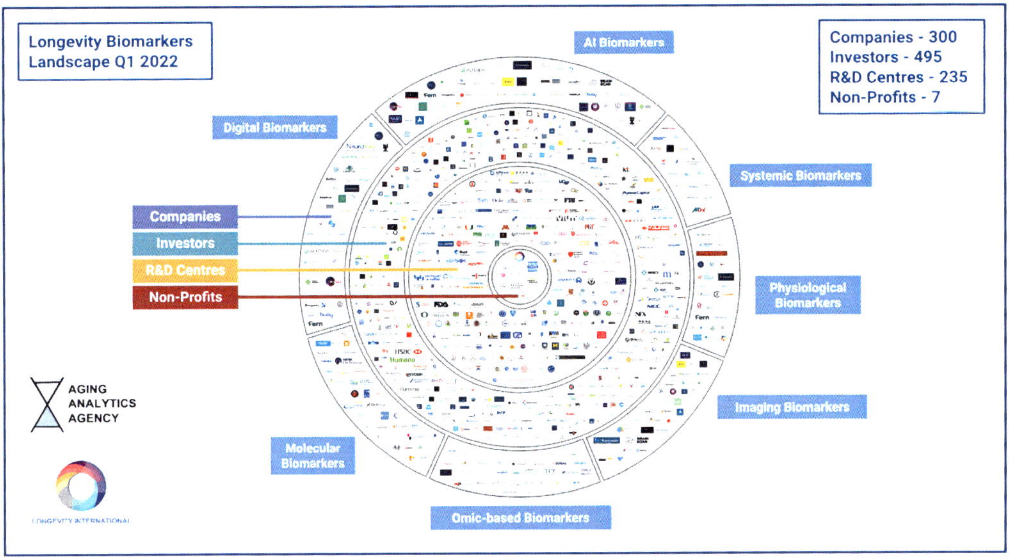

In the modern world, Healthy Ageing and well-being are widespread public and governmental objectives. The focus of research on ageing has been sharpened by recognition of the major demographic shift towards higher proportions of older adults in the population in many countries around the world, as well as the recognition that much of the cost of health and social care in economically developed countries has been concentrated in the last decade or two.

Countries with the highest share of companies in the global biomarkers market

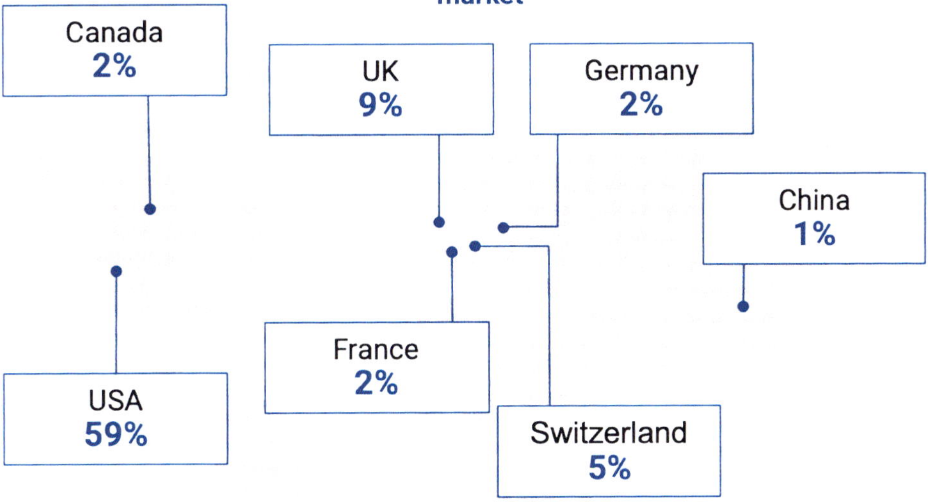

Key Companies: Molecular Biomarkers

Genetics and Ageing

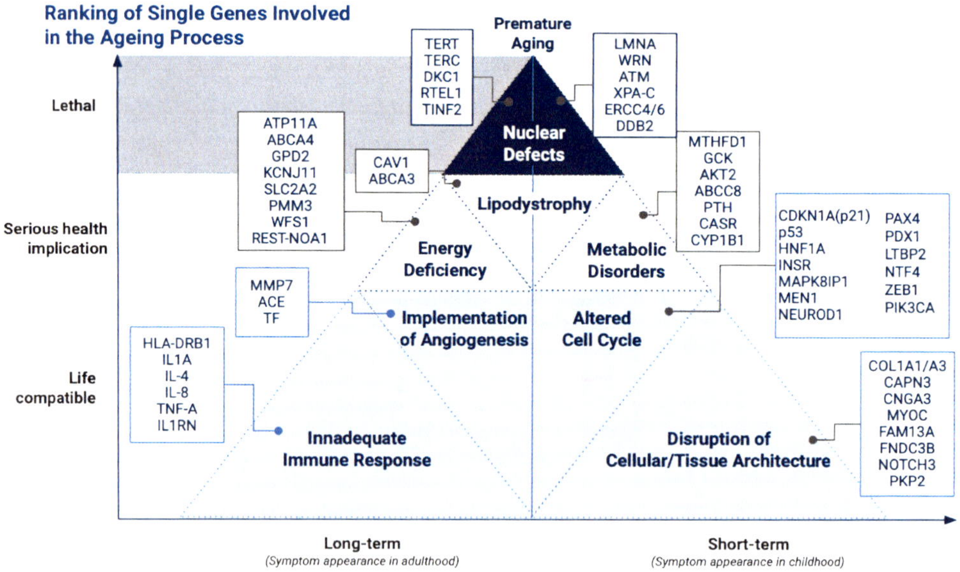

Frequency of Clinically Approved Singlepoint Gene Mutations

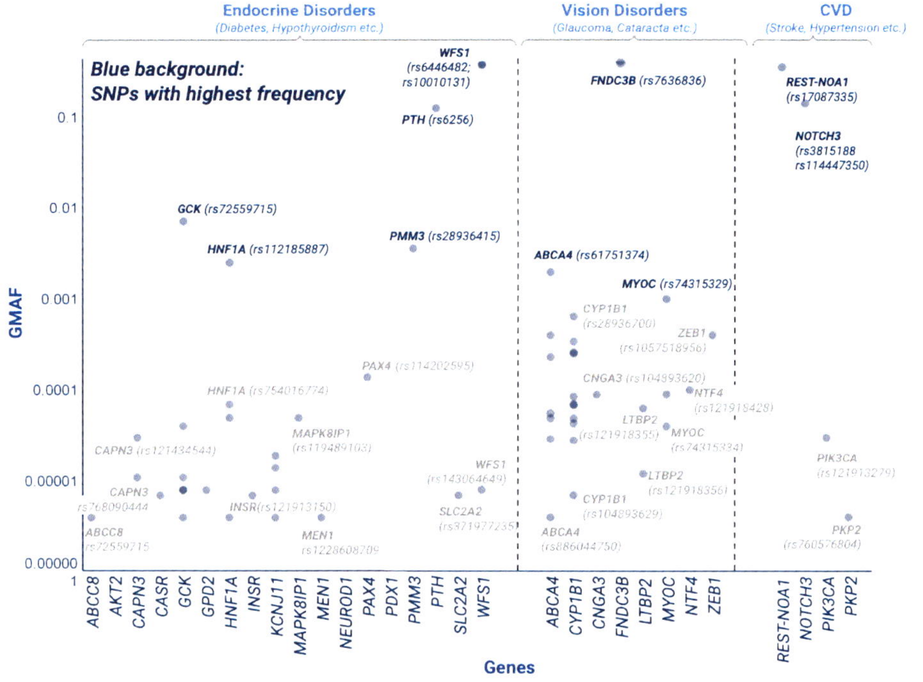

Scope of Companies with Approaches to Biochemical Biomarkers

Biomarkers of **metabolic state** are the most popular in the market as tool for assessing the **risks**, **detection**, and **monitoring** of **age-related diseases**. This can be explained by the plentiful availability of tests, low costs, reliability, and a high correlation between changes in the state of the organism and in biomarker levels. Some companies provide **senescence**, **metabolic**, **inflammation**, and **mitochondrial** markers to evaluate **ageing**.

Gradation of Companies by End-User and Biomarker Type

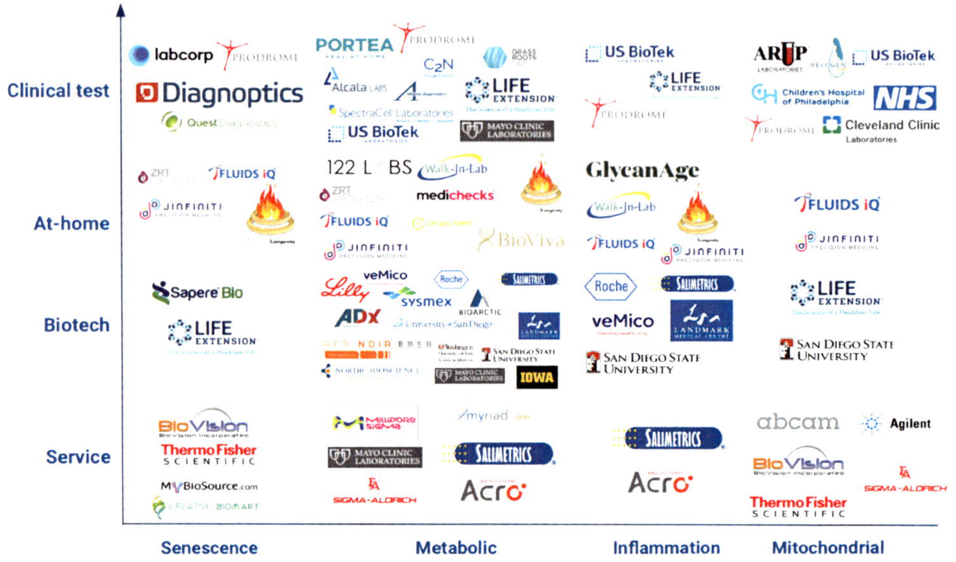

Invasivity Gradation of Biochemical Biomarkers Marketed Tests

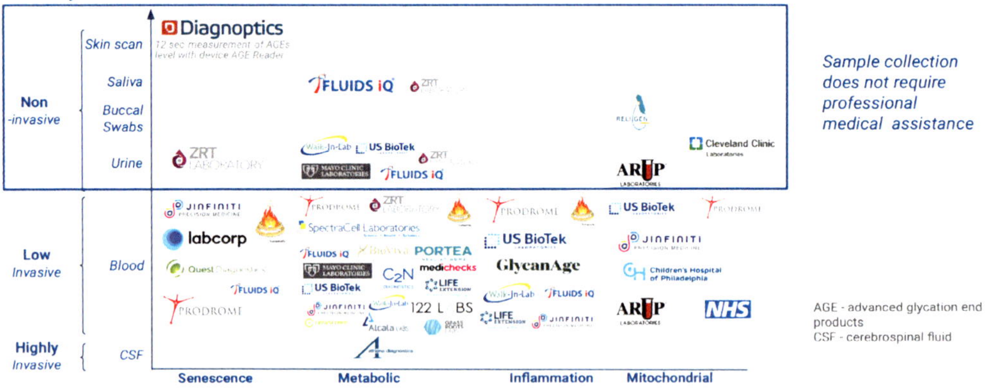

Invasivity Gradation of Innovative Diagnostic Tests

*CSF - cerebrospinal fluid

Reliability of Clinical Biomarkers

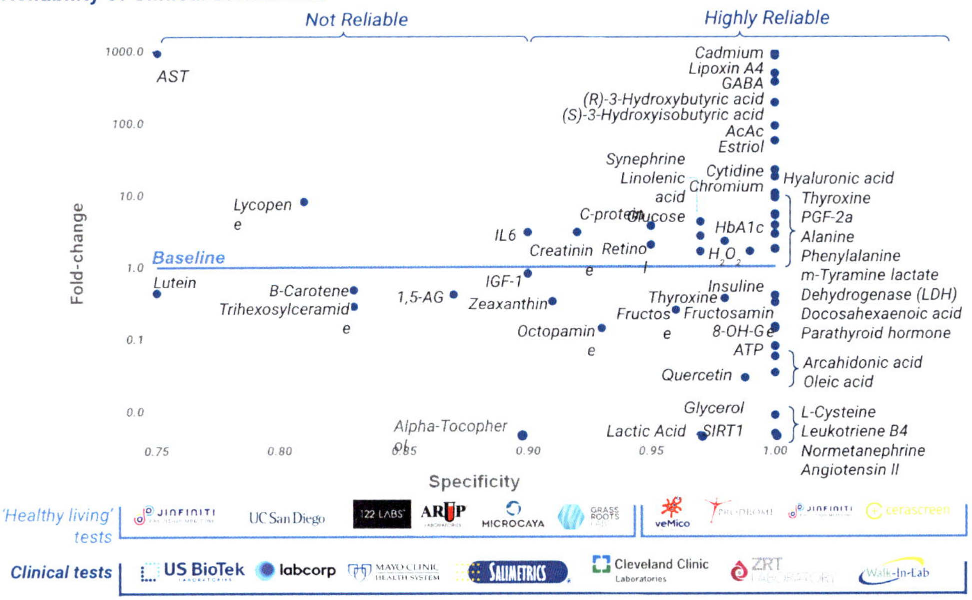

Informativity of Senescence Biomarkers

Biological Pathways	Biomarker	Biological Role	Changes of Biomarker Level in Aging	Indication	BioTech Companies	At-Home Tests	Clinical Tests
Cell cycle	**p16**	Inhibitor of CDK 4/6; **induces senescence**	⬆ **Increased**	**Heart disease Oncology**	Sapere Bio		
Hypothalamic-pituitary-adrenal axis	**DHEAS***	Sex steroids precursor	⇩ **Decreased**	**Senescence**	LIFE EXTENSION	FLUIDS iQ	
Pineal clock	**Melatonin**	Regulator of circadian rhythms, antioxidant, immunomodulator	⇩ **Decreased**	**Depression Insomnia**		ZRT	
Autophagosomal-lysosomal network	**β-gal***	Lysosomal degradation of glycoproteins and glycollipids	⬆ **Increased**	**MPS***		JINFINITI	labcorp
	NADase	NAD degrading enzyme	⬆ **Increased**	**Senescence**		JINFINITI	
DNA repair and prevention of Replicative senescence	**NAD** *(in plasma)*	NAD is a **coenzyme** for **NAD-dependent Sirtuin-1**	⇩ **Decreased**	**Senescence**			
	SIRT1*	**Prevents Aging** and **related diseases** via the deacetylation of histones; activates **autophagy**	⇩ **Decreased**	**Senescence**		JINFINITI	
DNA methylation	**DNMTs */ choline system**	Choline is a methyl donor *for* histone methylation	⬆ **Increased**	**Neuro, Stroke Heart disease**			PRODROME
Aging phenotype / Multisystem functional decline	**AGEs***	**Harmful,** may be formed via spontaneous glucose binding to the biomolecules with loss of their normal functions	⬆ **Increased**	**Heart disease Diabetes mellitus**			Diagnoptics

* *AGEs - Advanced Glycation Endproducts; DNMTs - DNA Methyltransferase; DHEAS - dehydroepiandrosterone sulfate; MPS - mucopolysaccharidoses; SIRT1 - Sirtuin-1; β-gal - beta-galactosidase.*

Informativity of Metabolic Biomarkers

System / Organ	Biomarker	Biological Role	Changes in Ageing	Indication	Biotech Companies	At-Home Tests	Clinical Tests
Nutrient status	**Vitamin panel Mineral panel**	**Coenzymes**; **Vitamin D** - **regulator** of whole metabolism, **Folate** - DNA **methylation**, Cell **division**	⇩ **Decreased**	**Wellness Multiple**		medichecks, JINFINITI	Alcala, LIFE, US BioTek
Endocrine system	**Thyroid Panel Sex hormones Growth hormone Cortisol**	**FSH, TSH*, T3, T4** - **Metabolism** regulators **Estradiol, Testosterone** - **Multisystem** control **Tissue and organs renewal** **Stress** hormone, **increases blood glucose**	⇩ **Decreased** **Decreased** **Decreased** ⬆ **Increased**	**Wellness Multiple**	LANDMARK, veMico	medichecks, ZRT, FLUIDS iQ	LIFE, Alcala, US BioTek
Glucose metabolism	**Insulin Glucose HbA1C***	Regulator of **glucose utilisation**, cell **growth** High glucose **accelerates Aging** Retrospective marker of high glucose	⇩ **Decreased** ⬆ **Increased** **Increased**	**Cardiovascular Metabolic Dementia**	Roche, LANDMARK, San Diego State University	122 L BS, JINFINITI	LIFE, PORTEA
Cardiovascular system	**Lipid panel**	**Cholesterol, LDL, TAGs*** increase, and **HDL*** decreases **Atherosclerosis risk**	⬆ **Increased** ⇩ **Decreased**	**Cardiovascular Metabolic Dementia**	Roche, LANDMARK, San Diego State University	ZRT, JINFINITI, medichecks	LIFE
Liver	**Liver panel**	**Albumin** - Marker of chronic **liver dysfunction**; **ALT*** - liver cells **destruction** markers; **ALP*** - **cholestasis** marker	⇩ **Decreased** **Decreased** ⬆ **Increased**	**Wellness Liver**		medichecks, JINFINITI	PORTEA
Renal system	**Kidneys panel**	**eGFR*** (based on **cystatin C, creatine**) - marker of **renal filtration ability**	⇩ **Decreased**	**Wellness Kidneys**		medichecks	ARUP, PORTEA
Nervous tissue	**Ng, NPTX2, Aβ. tau, P-tau, GFAP** **α-Syn***	**NPTX2, Ng** support **synapse integrity** **Aβ 42** - major component of **Amyloid plaques** **Tau, P-tau*** - major components of **NTF*** **GFAP** - risk of **AD** **α-Syn** is linked to the presence of **Lewy bodies**; key protein involved in **PD pathology**	⇩ **Decreased** ⬆ **Increased**	**AD Dementias** **PD**	ADx, IOWA, BIOARCTIC, Lilly, sysmex		C2N DIAGNOSTICS

* *FSH - follicule stimulating hormone; TSH - thyroid stimulating hormone; HbA1C - Glycated hemoglobin; LDL - low densuty lipoprotein; HDL - high density lipoprotein;*

TAGs - triacylglycerols; eGFR - estimated glomerular filtration rate; Aβ - beta-amyloid; P-tau - phosphorylated Tau protein; GFAP - Glial fibrillary acidic protein; Ng - Neurogranin; NPTX2 - Neuronal Pentraxin 2; NTF - neurofibrillary tangles; α-Syn - Alpha-synuclein; AD- Alzheimer's disease; PD - Parkinson's disease

Informativity of Inflammation Biomarkers

Process	Biomarker	Biological Role	Changes in Ageing	Indication	BioTech Companies	At-Home Tests	Clinical Tests
	C-Reactive Protein	Activates **innate immune system**	⬆ Increased	Aging, CD		JINFINITI	LIFE
	IL1-β	Key **proinflammatory cytokine**	⬆ Increased	Aging		JINFINITI	
	IL-6	**Promotes Inflammation**	⬆ Increased	Aging, CD	San Diego State University	JINFINITI	
General inflammaging	IL-8	Attracts neutrophils in **inflammatory** regions	⇩ Basal decreased ⬆ Induced increased	Aging,		JINFINITI	
	IL-18	Potent **proinflammatory cytokine**	⬆ Increased	Aging, CD	San Diego State University		
	TNFα	**Proinflammatory cytokine**	⬆ Increased	Aging		JINFINITI	
	IgG glycans	**Promote Inflammation** contributing to Ageing	⬆ Increased	Aging		GlycanAge	
Autoimmune inflammaging	Rheumatoid Factor	Autoantibody contributing to **RA pathology**	⬆ Increased	RA			
	ANA	Strongly age-dependant autoantibodies	⬆ Increased	ATI disorders			
	Anti-DGP antibodies	DGP activates immune system to **destroy gut epithelium**	Unclear	Celiac disease			US BioTek
Leaky gut-induced Inflammaging	Zonulin	Modulates **permeability of the gut**	⬆ Increased	Welness Food sensitivities		FLUIDS iQ	
	Histamine / DAO ratio	Histamine - mediator of **inflammation**, DAO inactivates histamin	⬆ Increased	Gut inflammation		FLUIDS iQ	
	GTAs	Digestion	Unclear	Aging Inflammation			

CD - Cardiovascular diseases; IL - interleukin; TNFα - Tumor necrosis factor alpha; RA - rheumatoid arthritis; ANA - antinuclear antibodies; ATI - autoimmune; anti-DGP antibodies - antibodies specific for deamidated gliadin peptide; DAO - Diamine oxidase; GTAs - Gastrointestinal tract acids

Informativity of Mitochondrial Biomarkers

Biological Pathways	Biomarker	Biological Role	Changes in Aging	Indication	Biotech Companies	At-Home Tests	Clinical Tests
	PDH* activity	PDH converts Pyruvate to acetyl-CoA for subsequent oxidation in mitochondria	⇩ Decreased	MD, Metabolic, Cancer, AD			MAYO CLINIC LABORATORIES
	Organic acids	Lactate, Pyruvate - glycolysis end-products, and Beta-Hydroxybutyric Acid - markers of MD	⬆ Increased	MD, Metabolic, Cancer, AD			MAYO CLINIC LABORATORIES, ARUP, US BioTek
Mitochondrial energy metabolism	NADH levels	NADH is a marker of ETC* dysfunction	⬆ Increased	MD, Metabolic, Cancer, AD			Children's Hospital of Philadelphia
	Complex I, II, III and IV of ETC	Transport of electrons and formation of proton gradient for OXPHOS*	⇩ Decreased	MD			
	Carnitine	Transport of FFA* to mitochondria for oxidation	⇩ Decreased	MD, frailty			ARUP
Mitochondrial synthetic function	Creatine	Partly produced in mitochondria; reflects mitochondrial dysfunction	⬆ Increased	MD			ARUP
Mitochondrial translation	FGF21*	Marker of mitochondrial transfer-RNA mutations and mtDNA deletions	⬆ Increased	Mitochondrial myopathies			NHS
	GDF15*	Marker of mitochondrial translation defects or mtDNA deletions	⬆ Increased	Mitochondrial myopathies			MAYO CLINIC LABORATORIES
	ROS*	Harmful, formed as side product by ETC	⬆ Increased	Aging, AD, PD		JINFINITI	
	F2-Isoprostanes	Oxidative stress marker	⬆ Increased	Aging, OS, Diabetes, CD, AD	LIFE EXTENSION		
Oxidative stress	Oxidative damage markers	8-OHdG - oxidative DNA damage marker oxidised LDL-C, MDA* - oxidative damage	⬆ Increased	Aging	San Diego State University	FLUIDS iQ, JINFINITI	
	Antioxidant capacity	Blood glutathione, total antioxidant capacity - markers of ability to neutralise ROS	⇩ Decreased	MD		JINFINITI	Children's Hospital of Philadelphia

** PDH - pyruvate dehydrogenase ; MD - mitochondrial diseases; ETC - electron transport chain; OXPHOS - oxidative phosphorylation ; FFA - free fatty acids; FGF21 - fibroblast growth factor 21; GDF15 - growth differentiation factor 15; ROS - reactive oxygen species; 8-OHdG - 8-hydroxydeoxyguanosine; LDL - low density lipoprotein; MDA - malondialdehyde; AD - Alzheimer's disease; PD - Parkinson's disease; CD - cardiovascular disease; OS - oxidative stress*

Relationship between single and multiomics data analysis

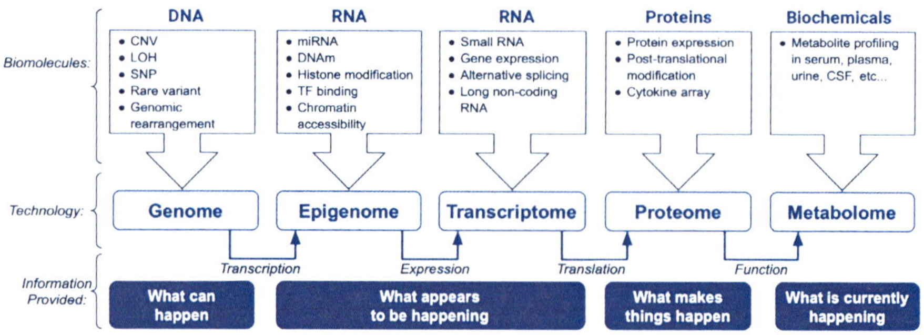

For the **evaluation of ageing and age related diseases**, in traditional diagnostics, patients are classified into groups according to their clinical characteristics (i.e., phenotypes). However, these classifications do not provide insight into the functional or pathobiological mechanisms of the disease within the individual (i.e., endotype). **Omics is the comprehensive assessment** of the molecules that **constitute a cell, tissue, or organism**. The integration of multiomics data, such as genomics, proteomics, and metabolomics, with clinical data **allows for a better understanding of age-related disease pathogenesis** and will be **important for prognosis of ageing** and for **predicting, diagnosing, and treating disorders**.

Gradation of Individual Omics Technologies for Studying Ageing

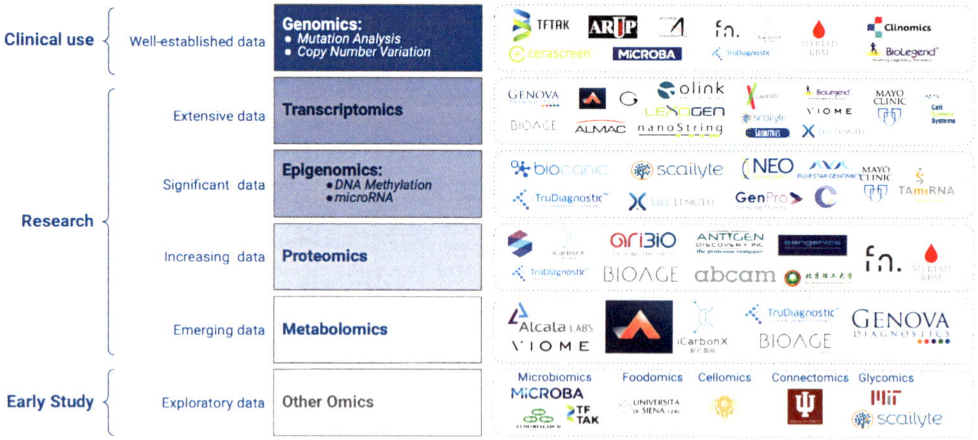

Specific UP- or DOWN-regulated Proteins Associated with Ageing

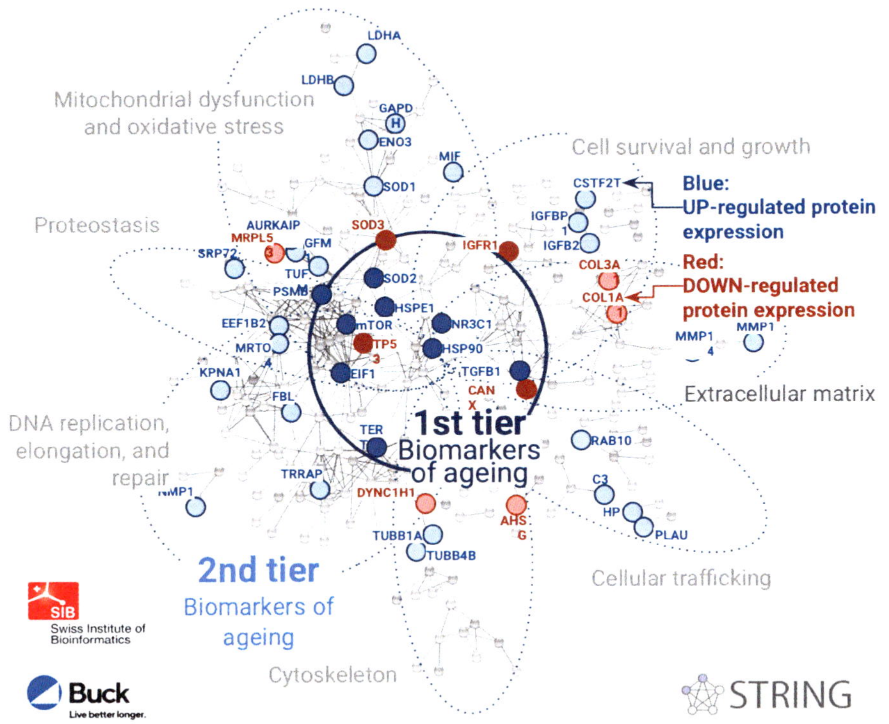

Note: *Analysis of protein-protein interaction is based on the biomarkers reported in literature and* **STRING** *database. The interactions include direct (physical) and indirect (functional) associations; they stem from computational prediction. Interaction score: > 0.07.*

Analysis of **biomarker expression in human cells** shows the several **pathways that lead to senescence**, including **mitochondrial dysfunction and oxidative stress**, **persistent inflammation**, and **deregulated proteostasis**.

In addition, senescence can be induced by cellular replication, which resulted from **telomere shortening.** Taken together, it is possible to draw a common pathway connecting ageing to age-related diseases, and the central point of this process, senescence, can be the target for new therapies, which may result in the Healthspan matching the lifespan.

The **senescence-associated secretory phenotype (SASP)** has recently emerged as a **driver of and promising therapeutic target for multiple age-related conditions**, ranging from neurodegeneration to cancer. The complexity of the SASP, typically assessed by a few dozen secreted proteins, **has been greatly underestimated**, and a small set of factors cannot explain the diverse phenotypes it produces in vivo.

Proteome analysis show potential biomarkers of ageing in human lung fibroblasts. Some of them are **UP-regulated** (SOD2, HSP90, mTOR, TGFB1 etc.) while others are **DOWN-regulated** (TP53, IGFR1, CANX, COL1A1 etc.). On the chart the proteins with significantly altered (q-value <0.05 and **>2-fold change**) secretion by senescent compared with quiescent cells following genotoxic, oncogenic, or ATV treatment stress are shown.

Prices vs. Gradation of Individual Omics Technologies

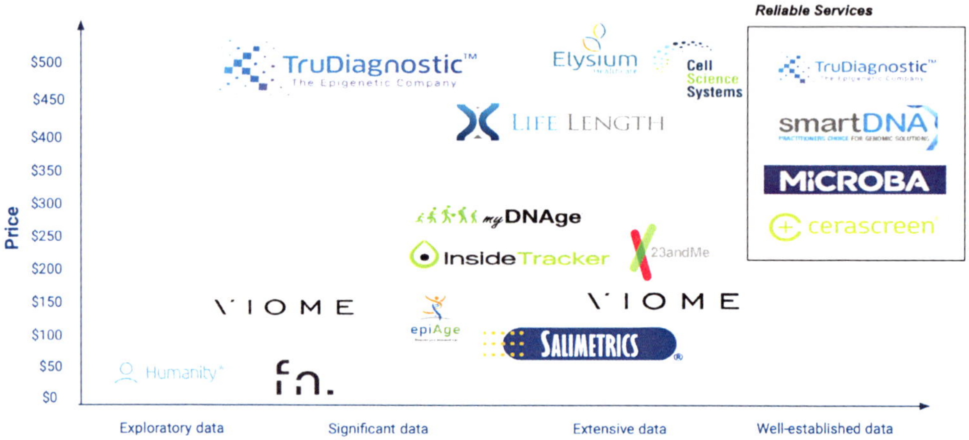

Physiological biomarkers are popular in the market as useful tests in **Sports physical examination**, **Occupational health management**, **Health-care services**. The great benefits of physiological markers are their **reliability**, **informativity**, and **non-invasive procedure of measurement**. Some companies provide solutions for assessment and tracking physiological biomarkers to evaluate **Aging rate**.

Scope of Companies with Physiological Biomarker Approaches

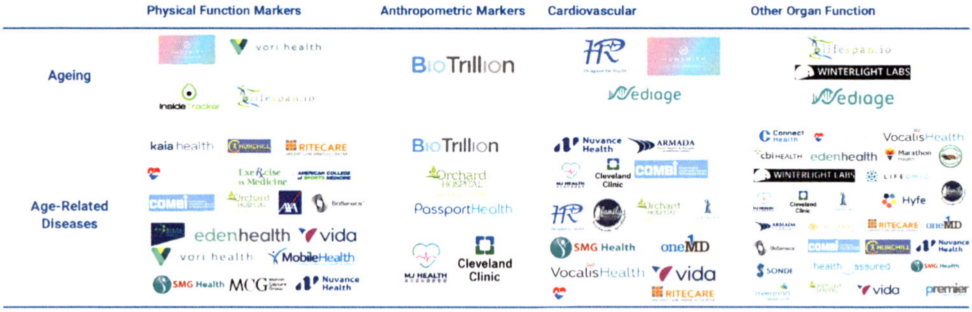

Markers of **Physical function** of body (balance, lift ability, walking speed) as well as **Cardiovascular** system markers (blood pressure, heart rate, EKG) are routinely used and prevail in the market. At the same time, innovative markers (facial features, vocal markers) are perspective to be used both as clinical assays and at-home tests for purposes of **Longevity medicine**.

Physiological Biomarkers in Practical Use for Aging Rate Evaluation

Marker type	Biomarker	What is assessed	Digital platform	Application	At-home tests
Physical Function	**Walking speed** **Fitness activity**	Marker of **Frailty**	*Humanity App* *InsideTracker App* *Mediage Mobile Health* *Vori Health*	Ageing rate evaluation Wellness	InsideTracker Mediage vori health
	Muscle movement time	Marker of **Age-associated impairments** in motor control systems	*AgeMeter® App*	Ageing rate evaluation	lifespan.io
	Alternate button tapping	Marker of **Muscle coordination**	*AgeMeter® App*	Ageing rate evaluation	lifespan.io
	Digital eyes & face biomarkers	Real-time motor, cognitive, and emotional abilities of **Brain**	*BioEngine4D*	Ageing rate evaluation Neurology	BioTrillion
Anthropometric Markers	**BMI**	Indicator of the **Energy metabolism balance**	*InsideTracker App* *Mediage Mobile Health*	Ageing rate evaluation	InsideTracker Mediage
	Height & weight	**Abdominal obesity**	*Mediage Mobile Health*	Ageing rate evaluation General Check up	Mediage
	Waist circumference	**Abdominal obesity**	*Mediage Mobile Health*	Ageing rate evaluation	Mediage

Organ	Biomarker	What is assessed	Digital platform	Application	At-home tests
Cardiovascular system	**Heart rate**	Indicator of **Heart functioning**	*Humanity App* *InsideTracker App*	Ageing rate evaluation	InsideTracker
	Blood pressure	Index of **Cardiovascular activity**	*Mediage Mobile Health*	Ageing rate evaluation	Mediage
	Pulse rate	Marker of **increased arterial stiffness / stress / autonomous nervous system** function	*Health Reviser*	Ageing rate evaluation	HR
Ear	**Highest audible pitch**	**Ability to hear sounds**	*AgeMeter® App*	Ageing rate evaluation	
Eyes	**Visual reaction time**	**Ability to respond quickly to visual stimuli**	*AgeMeter® App*	Ageing rate evaluation	lifespan.io
	Visual perception	**Ability to perceive light stimuli**	*AgeMeter® App*	Ageing rate evaluation	
Lungs	**Forced vital capacity**	**Lung function** test	*AgeMeter® App*	Ageing rate evaluation	lifespan.io
	forced expiratory volume	**Lung function** test	*AgeMeter® App*	Ageing rate evaluation	Mediage
Brain	**Decision reaction time**	**Brain activity**	*AgeMeter® App*	Ageing rate evaluation	lifespan.io
	Memory	**Ability** to to **reconstruct the experience** and **make relevant judgments**	*AgeMeter® App*	Ageing rate evaluation	

The physical capability and physiological function biomarkers are mostly implemented in clinics for the **overall health assessment of middle-aged and older adults**. However, such biomarkers are detectable only in the late stage of aging, so they are not helpful for early diagnostics.

Additionally, physical capability tests are used in **population-based studies**. These objective measures complement self-reports, improve validity and reproducibility.

Key Companies Digital Biomarkers

Increasing necessity of digital biomarkers in research arises from many advantages brought by it. **The main advantage** of digital biomarkers is that they are **cheap, publicly accessible, and do not require the assistance of medical professionals.** Digital biomarkers utilize objective, quantifiable physiological and behavioural data. These products **have created new opportunities,** enabling remote monitoring for biomedical research, decentralised clinical trial designs, and routine patient care. Additionally, **digital biomarkers can predict some diseases**, and they are commonly used in aging clocks.

Key Companies Imaging Biomarkers

The **trends driving** the development of **Medical Imaging sector** are arising of **an aging population** and the increasing **need for early-stage diagnostic** among elderly and youths.The high-yield niche in preventive and personalised diagnostics will be occupied by AI medical imaging due to the nature of its algorithms - neuronal networks, deep-learning able to identify patterns that clinicians may not be thought to look for or even to detect an initial disease trays more accurately than clinicians.

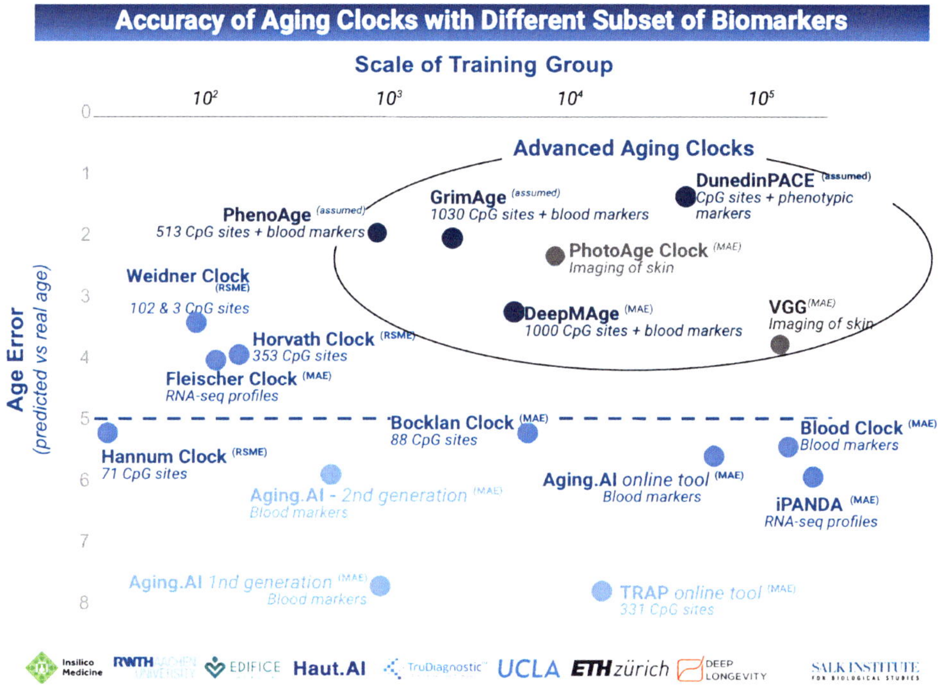

To determine biological age and predict life expectancy, a set of similar or dissimilar biomarkers is most often used. **The DNA methylation (DNAm) clocks** combined with blood markers are currently the most accurate biological clocks. Their developers claimed 2-3 times higer accuracy than conventional DNAm Clocks like Horvath and Hannum.

Cross-Correlations of Different Age-Clocks *(shown above the diagonal)*

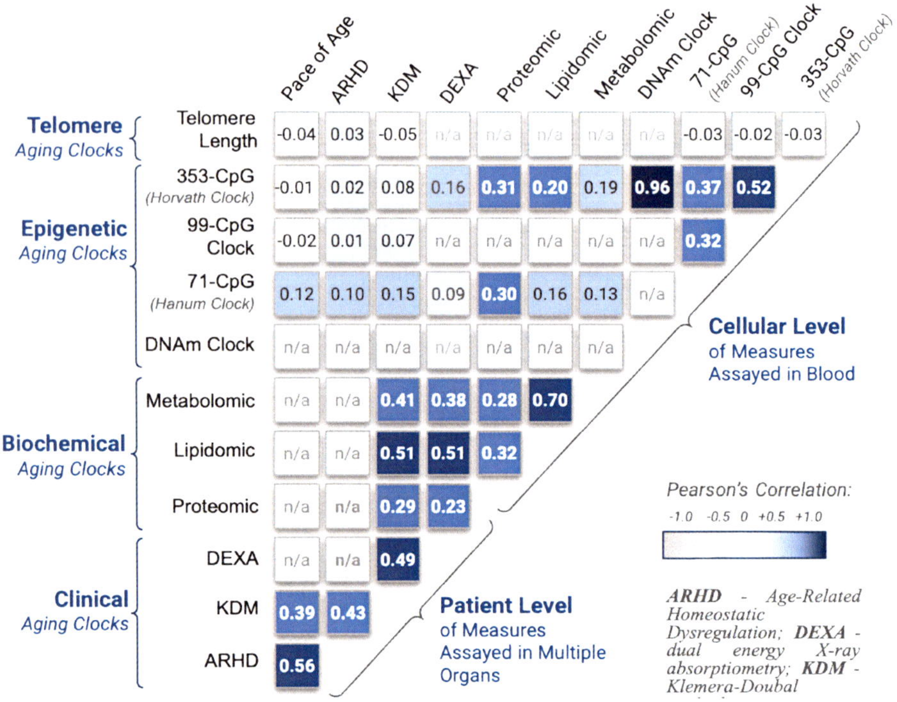

Epigenetic Clocks were correlated with each other in the **r = 0.3–0.5** range *(P < 0.001 for all)*. **Clinical Aging Clocks** algorithm measures were correlated with one another in the **r = 0.4–0.6** range (P < 0.001 for all). However, **Telomere Clocks** was **not significantly correlated** with estimates from epigenetic clocks or clinical-biomarker algorithms (**r = -0.05–0.03**; P > 0.05 for all), and correlations of epigenetic clock measures with clinical-biomarker-algorithm measures were generally low.

Digital Transformation of the Health and Longevity Industry

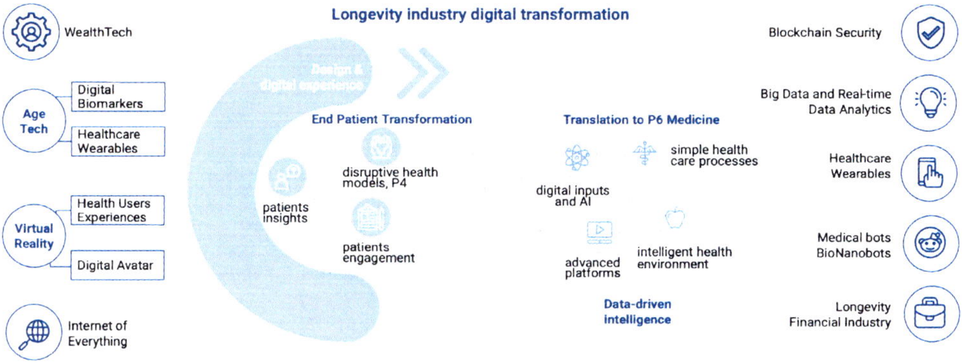

The future of health will be driven by digital transformation enabled by radically interoperable data and open, secure platforms. Radically interoperable data and AI can empower consumers in ways that are difficult to visualize today. Data about individuals, populations, institutions, and the environment will be at the heart of the future of health.

Source: Image: Modified from ***everis***[1]

AI-Driven Biomarkers and P4 Medicine

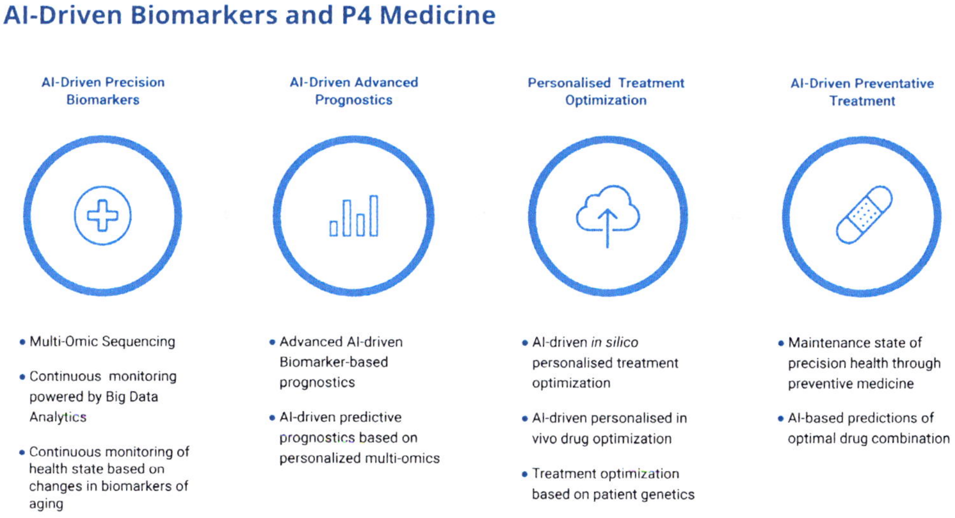

[1] www.everis.com/about-us

AI and Digital Platforms:

Toward a "Digital Avatar"

AI-driven Mobile Apps with Aging Biomarkers

YOUNG.AI Young.AI is a product of Deep Longevity company. It allows to upload your selfie or recent blood test and find out your biological age. Based on uploaded data, their AI algorithms will provide a scientifically based personalized longevity plan. The tactic is divided into easy-to-do tasks that help isolate harmful habits.	Haut.AI Haut.AI provides skincare brands with a B2B SaaS tool, available as an API and SDK, for AI analysis of their customers' skin. This product facilitates R&D, provides essential tools for building skin analysis apps, and allows brands to develop a better understanding of their clients' needs. The platform is trained on millions of face images.	ME The Me Biosciences solution combines the iterative power of artificial intelligence and machine learning to match the body, digestion, and activities with the correct set of nutritional supplementation and body care products to maximize performance, optimization, and longevity	LIFE Extend helps track eating, physical activity, intermittent fasting, slipping, stress reduction activities to improve health and increase longevity.
Facelytics Facelytics is a face recognition solution that is able to detect peoples' morphological criteria such as age and gender, by analyzing the video feed in real time.	RYNKL RYNKL is a wrinkle analysis app. It is an artificial intelligence that cares about looks and helps adjust lifestyle to look younger.	Eterly Eterly calculates personal longevity score daily using a bespoke algorithm that looks at thousands of different data points. With the help Eterly's interactive chatbot, which prompts for information and offers suggestions concerning diet, fitness, supplements.	G Gero is a free research app that helps build a healthier lifestyle for a longer life. Gero analyses the data from Apple Health to calculate biological age and predict healthspan on a daily basis.

Non AI Biomarkers Condition Stage

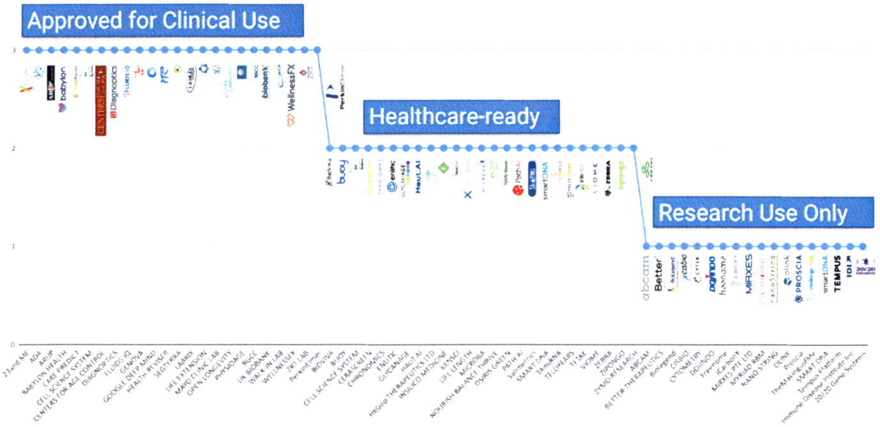

AI Biomarkers Conditioning Stage

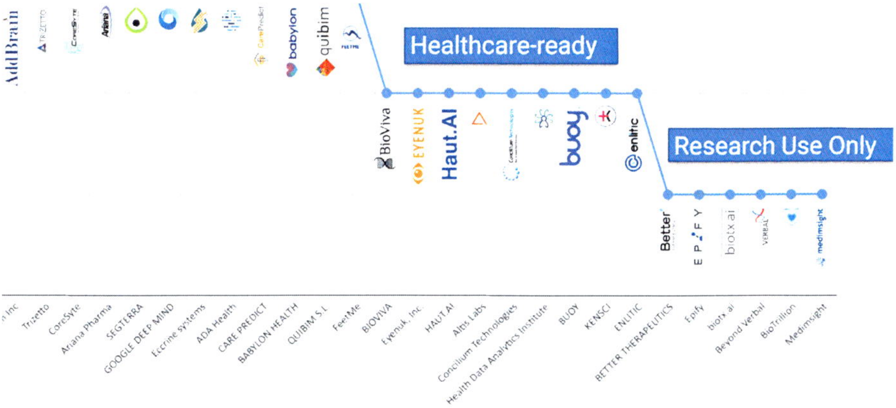

Prices for Biomarker Panels, $ US

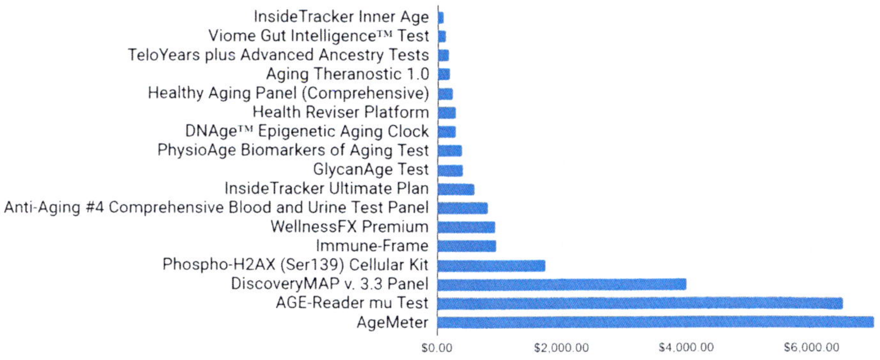

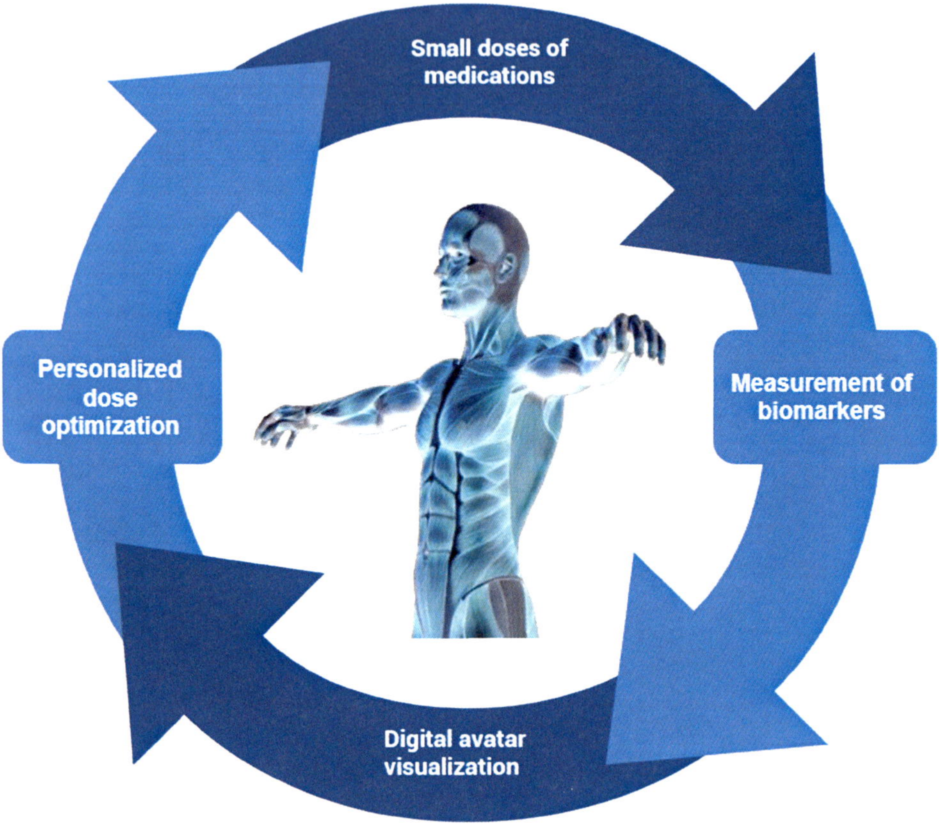

Not only do new methods of standard industry benchmarking and forecasting need to be developed to combat the issues of overcomplexity and multidimensionality in the Longevity Industry, but new methods of testing the basic safety and efficacy of Longevity and Precision Health prevention, diagnostics, prognostics and therapeutics need to be adapted as well, moving away from the use of model organisms, towards a more human-centric approach.

Digital biomarkers satisfy all these new industry requirements: they can be continuously tested on all users, notifying adverse micro-effects and ultra-stratifying patients.

Precision Diagnostics

Digital avatar visualizes a combination of Biomarkers and other diagnostic results

A large part of health information is digitized, which allows us to compile enormous amount of data, access global servers, and compare patient information: a dynamic repository of information that is constantly being updated. The massive advance in these databases facilitates doctors in their diagnostic process, their ability to measure, analyze, compare patients, and produce medical reports that are more accurate and personalized, that will, in turn, lead to the best available therapy or treatment of the time.

Collect your data today:

- Biological specimen
- Biomarker analysis
- Database of personal biomedical data stored on blockchain

Future benefits:

- Data driven analysis of Biomarkers dynamics over time
- Analyse the changes in your digital avatar
- Personalized interventions

The intensive application of AI to all stages of Longevity and Preventive Medicine R&D, and healthcare, has the potential to rapidly accelerate the clinical translation of experimental, validated and non-validated biomarkers toward diagnostics, prognostics and therapeutics, to empower patients to ultimately become the CEOs of their own health through continuous AI-driven monitoring of minor fluctuations in biomarkers, and the rapid development of the global Longevity Industry to scale.

Index Values for Single Aging Biomarkers and Aging Panels

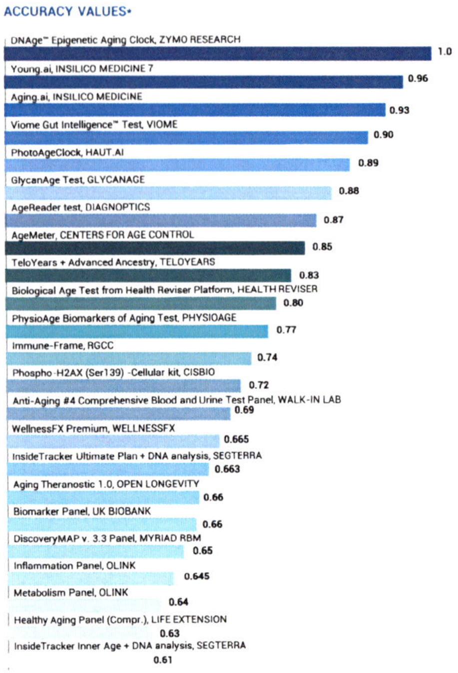

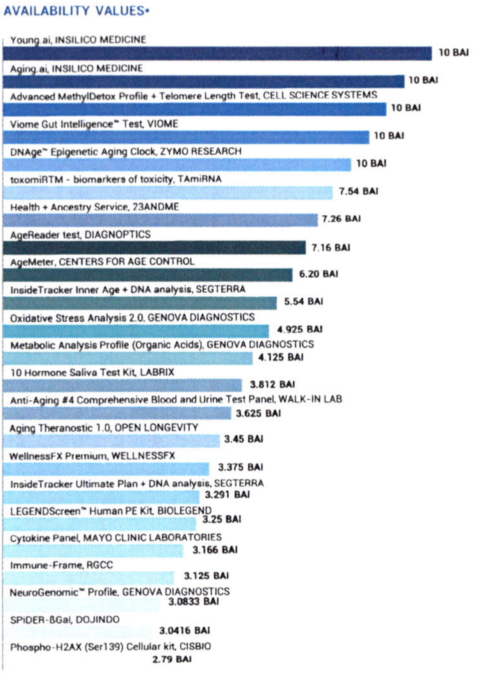

*Illustrative only; non-real values.

Aging Analytics Agency uses detailed mathematical procedures to assign a value (or numerical factor) to each of the three variables taken into account in the evaluation of single biomarkers and whole biomarker panels. Many qualitative considerations are made when assigning scores to these three variables: firstly the CONDITIONING STAGE CATEGORY of the single biomarker or the panel; that is, if they are approved or not for clinical use. Second, its OPERATIONAL CATEGORY; if they are offered in the market as laboratory research kits, as medical tests; if the company gives access to the prices; if it is a theoretical panel or one employed in epidemiological surveys, in which case it is not materially offered to the market by the entity; if it is a real-time evaluation technology for a set of biomarkers, which could decrease availability due to the increase in cost but actually increases it because it allow large-scale assessment of vast amounts of patients, and so on. Thus, the numerical evaluation is to some extent subordinated to a multiplicity of qualitative factors considered by our highly specialized professionals in the field of biotechnology and pharmaceutical intelligence.

Open Access and Proprietary Metrics

Open Access Metrics

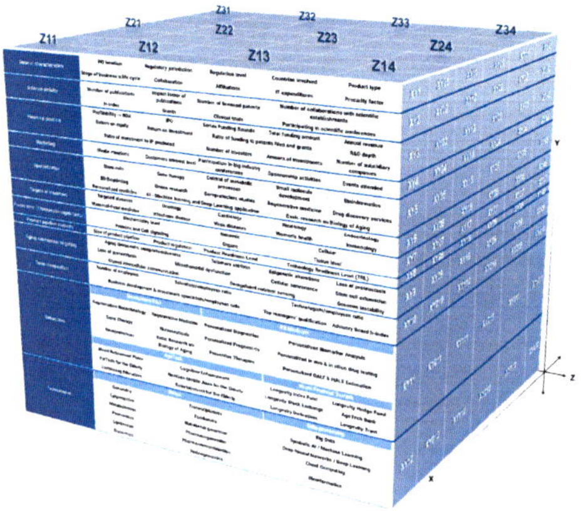

Proprietary Metrics

The methodology and metrics featured in this appendix for this contracted analytical case study are **public** and **could have been used in a number of other open-access Aging Analytics Agency reports**, whereas a large portion of the analytical frameworks used for the report's benchmarking are proprietary, available to potential partners via NDA. These include both **absolute values** (quantitative or qualitative) and **dynamic parameters** to analyze metrics as they change over time. The following are examples of parameters used in assessment of P4 Medicine clinics, so many of them may concern this report while others may not.

Appendix B

Longevity Industry 1.0: Defining the Biggest and Most Complex Industry in Human History

Book Teaser

Evolution of the Longevity Industry from Zero to 1.0;
The Industrialization of Longevity;
The Current State of Longevity Science, Business, Finance, and Practical Applications

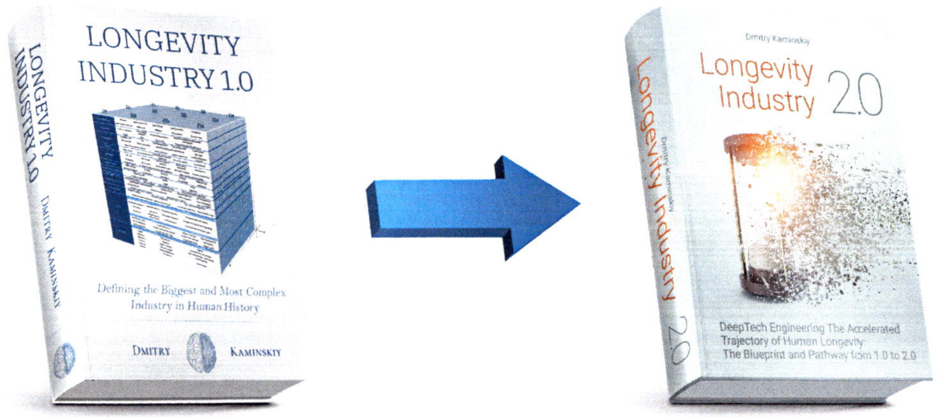

2010-2020: Evolution of the Longevity Industry from Zero to 1.0

- The Industrialization of Longevity
- The Current State of Longevity Science, Business, Finance, and Practical Applications
- Longevity Becomes National Priority Item for the Strategic Agenda of Progressive Governments
- Transforming the Challenge and Deficit of Aging into the Opportunity and Asset of Longevity
- Defining and De-Risking: Hype vs. Reality

2023-2025: DeepTech Engineering The Accelerated Trajectory of Human Longevity – The Blueprint and Pathway from 1.0 to 2.0

- Global Industrialization of Longevity to Scale
- The Evolution from Longevity Start-ups to Multitrillion Dollar Longevity Corporations
- How AI-Driven Preventive Medicine will Disrupt the BioTech and Healthcare Industries
- Novel Financial Instruments and InvestTech Solutions
- The Rise of Progressive Longevity MegaHubs

About the Book:
Introducing Longevity Industry 1.0

PART I.
Longevity Industry Mega-Complexity:
Challenges, Issues, Opportunities

- The Unprecedented Complexity, Intersectionality and Multidimensionality of the Longevity Industry
- Why Traditional Methods of BioTech Analytics, Forecasting, Benchmarking and Investment Strategy Will Fail For Longevity
- The Need for New Analytical Methodologies and Frameworks Equal to the Complexity of the Longevity Industry
- Longevity.Capital: Specialized Hybrid Investment Fund for Longevity-Industry De-Risking and Diversification

PART II.
Longevity Policy and Governance:
Longevity Technocracies & MegaHubs

- The Rise of Progressive Longevity MegaHubs and Technocracies
- Longevity as a Major National Priority Item of Progressive Governments' Strategic Agendas
- The Emergence of Longevity-Focused Parliamentary Groups and Government Bodies
- National Longevity Industrial Strategies and Healthy Longevity Development Plans
- How Longevity Will Determine the Outcome of National Elections by the Year 2025

About the Book: Introducing Longevity Industry 1.0

PART III.
The Longevity Financial Industry:
Novel Financial Instruments and Health as New Wealth

- Longevity Embraced by the World's Biggest Financial Corporations: Investment Banks, Insurance Companies, Pension Funds, Asset Management Firms
- Integrated AgeTech, FinTech, WealthTech, InvestTech Solutions
- Longevity Stock Exchanges, Financial Marketplaces and AgeTech-Longevity Banks
- New Financial Instruments and Derivatives: New Business Models and Novel Financial Instruments Tied to the Rising Longevity Industry

PART IV.
Longevity Industry:
Science, Technology and Biomedicine

- Global Industrialization of Longevity to Scale
- Precision Health and the Patient as CEO
- How AI-Driven Preventive Medicine and Novel InvestTech Solutions Will Disrupt BioTech and Healthcare Industries by 2025
- Longevity FemTech
- Biomarkers of Longevity 2.0: The Need for Maximally Actionable Biomarkers of Aging, Health and Longevity

About the Book: Longevity Financial Industry

In *"Longevity Industry 1.0: Defining the Biggest and Most Complex Industry in Human History,"* seasoned Longevity Industry professionals Dmitry Kaminskiy and Margaretta Colangelo distill the complex assembly of deep market intelligence and industry knowledge that they have developed over the past 5 years into a full-scope understanding of the Global Longevity Industry, showing the public exactly how they managed to define the overwhelmingly complex and multidimensional Longevity Industry for the first time, and how they created tangible frameworks for its systematization and forecasting.

The book features first of its kind coverage of entirely new segments and sectors of the rising Longevity Industry, including Longevity Politics and Governance, the Longevity Financial Industry (including coverage of AgeTech, WealthTech, FinTech, and the coming rise of new financial instruments and derivatives), the current state and forecasts on the Global Industrialization of Longevity to Scale, and an overview of the near-future trajectory of the Longevity Industry's evolution between 2020 and 2025.

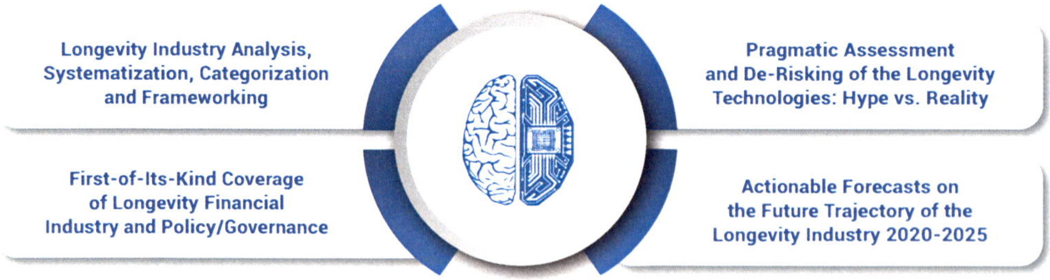

Longevity Industry – A Multitrillion Opportunity

While many analysts and investors consider the Longevity Industry to be of a much more limited scope, consisting of core anti-aging therapies at the forefront of advanced biomedicine, **Deep Knowledge Group**[1] has been working for the past 5+ years to landscape and understand the global multifaceted Longevity Industry through the work of its analytical subsidiary **Aging Analytics Agency**[2], which has gained a deep comprehension through tens of thousands of pages worth of public and proprietary Longevity Industry analytical reports.

World Longevity Economy Size Projections, current US $*

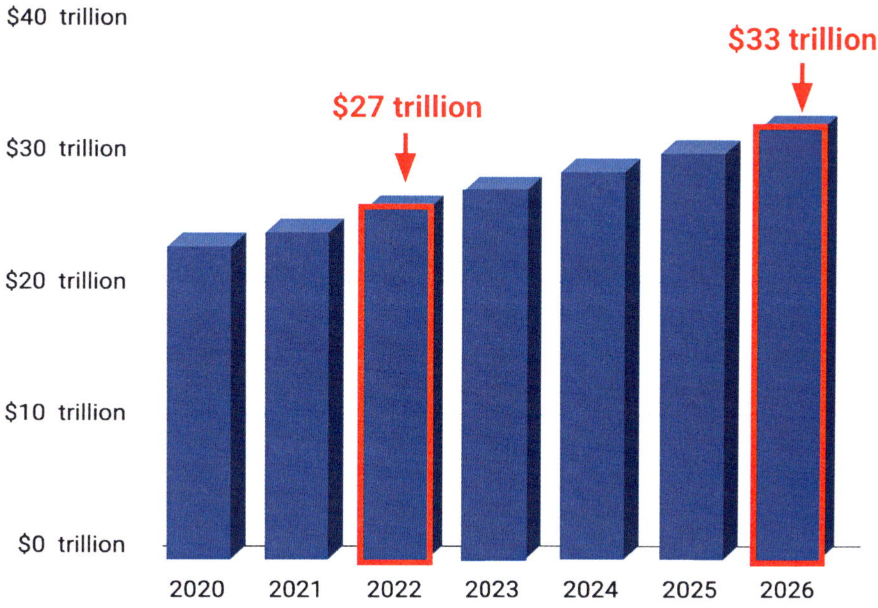

** projections are made under the assuption that the share of the Longevity Economy is 20 % of the world'd GDP*

[1] www.dkv.global

[2] www.aginganalytics.com

Longevity Industry – A Multitrillion Opportunity

The perception of the Longevity Industry, in both its scope and its main trends, in the USA in general and the Silicon Valley in particular, most often is conflated with biomedical moonshots – very advanced biomedicine not yet on the stage of human clinical trials – and this is an angle that severely underestimates technologies which we consider to be within the scope of the Longevity Industry, which are actionable and closer to market readiness.

Our work in this area has led us to define the industry much more broadly, and to include sectors that lie completely outside of biomedicine – sectors like AgeTech, WealthTech, FinTech and the emerging Longevity Financial Industry, which are also much closer to market readiness and to market capitalization than the anti-aging biomedicine sector.

As a result of this unprecedented complexity and multidimensionality, the typical approaches used for analytics, investment target identification and due diligence that worked for the BioTech Industry are destined to fail when applied to the Longevity Industry.

This has created an unmet need for analytical methodologies and de-risking techniques of equal complexity and multidimensionality as the industry itself. And this has been a large part of Deep Knowledge Group's mission to create a robust and sophisticated market intelligence and analytics base that can survive under the enormous weight of the Longevity Industry's complexity.

The 7th Continent - 1 Billion People in Retirement

Older people are the fastest growing demographic group (with around 2.5% annual population growth vs. 0.7% for the overall population). With growing numbers, older adults represent a dynamic emerging market and human capital resource. The multitrillion market of 1 billion people currently in retirement can be thought of as the world's 7th continent. The Longevity Industry includes the products and services satisfying the needs of the cohort aged 50+. The substantial increase in the population of this group is the main driver of the industry.

The Global Longevity Economy reached $17 trillion in 2019 and is showing stable growth to achieve $27 trillion in 2026. According to the most conservative estimates, it makes up 20% of the global GDP. While the global Longevity Economy is projected to reach $27 Trillion in 2026, the AgeTech segment alone will reach $2.7 trillion by 2025. This would imply a 21% annual growth in the global Age-

The 7th Continent - 1 Billion People in Retirement Globally

The 7th Continent – 1 Billion People in Retirement

Tech market. This growth is driven by the general development of the elderly care sector enhanced by advancing IT, FinTech and other digital technologies and solutions.

This creates both opportunities and threats for the economy as a whole and for the Financial Industry in particular. As for the last one, we believe aging to have rather a positive impact on the Financial Services Industry as there will be a greater need for financial products over the longer term, but of newer types and kinds — newer asset classes, newer investment strategies, and longer-dated bonds/securities will have to be developed. This will require participation and collaboration across both the public and private sectors to have a positive effect for all long-livers.

Formulating the Longevity Industry Framework

"A mere three years ago the current rise of the Longevity Industry would have been unthinkable. Now, a few years later, it has become unthinkable that biomedicine should not have healthy Longevity among its objectives. The time has come to establish a framework for the rising Longevity Industry." - Dmitry Kaminskiy, in Longevity Industry Landscape Overview Volume II: The Business of Longevity (Aging Analytics Agency, 2018)

Key points:

- The Longevity Industry has seen rapid evolution since 2014
- Unprecedented diverse scope of sectors and domains
- Ongoing trend of growth and diversification
- Extreme levels of complexity, multidimensionality and intersectionality

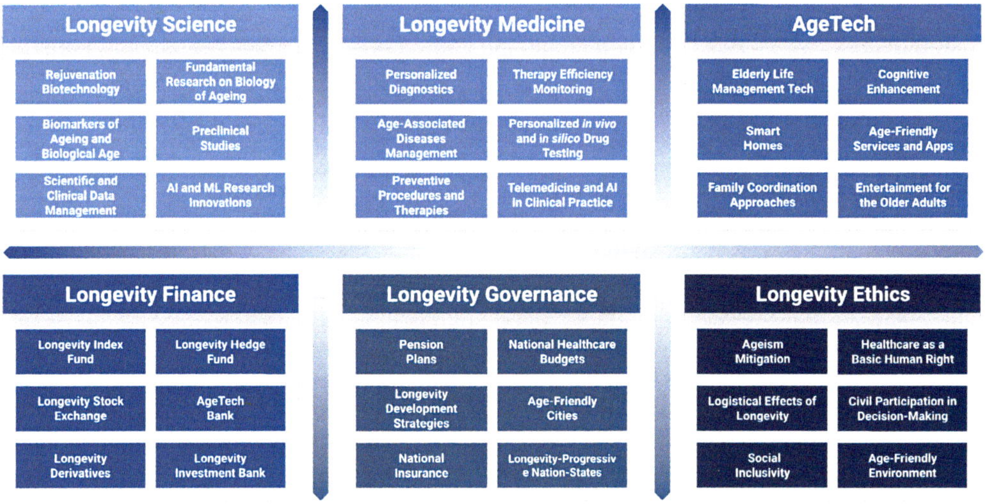

Formulating the Longevity Industry Framework

- Requires novel sophisticated techniques for analysis and due diligence and modern de-risking methods
- Unprecedented opportunities for ethical profit generation
- The Longevity Industry will be the biggest and most complex of all time

Aging Analytics Agency was the first entity to formulate a comprehensive and inclusive Longevity Industry Framework that encompassed the full range of its sectors, domains and technologies, which have only recently been recognized by the wider scientific, business, finance and investment communities as legitimate aspects of the industry. Utilizing this framework, Aging Analytics Agency was also the first entity to predict in 2014 the substantial rise in Longevity Industry investments that occurred during 2017-2018, three years in advance.

Preface by Dmitry Kaminskiy

Approximately seven years ago I became acquainted with a number of scientists working in the field of aging research. The more I learned from them, the prospect of successfully intervening in aging in the next few years seemed less a mystery like time travel and more a challenge like space exploration. At the beginning of the 20th century, the sciences necessary for space travel — rocketry, Newtonian physics, mathematics and basic astronomy — were already centuries old. It was the additional powers of foresight and precision, brought about by digitization, alongside government initiatives and funding, which opened the door to another world. So it was with space travel last century, and so it is with Longevity this century.

This realisation brought with it my personal eureka moment: I understood that the extension of Healthy Longevity is scientifically feasible, and that it just requires an engineering mindset to structure, formulate and format its practical realization — that all necessary components are in place, and that these components just need to be assembled, integrated and synthesized in an optimal way.

Having understood this, I also realized that the tangible and practical implementation of Longevity technologies could create value and generate wealth far beyond any profit and wealth itself, outperforming the opportunities and impact existing in any other sector and industry, and at the same time delivering the most valuable thing that you could provide for humanity, for your family and for yourself.

In 2013, I founded ***Aging Analytics Agency***[1] in order to keep track of relevant technological, social and political developments as they unfolded, which I now can proudly claim has since evolved into the undoubtable leader in the Longevity Industry analytics sphere, producing a great quantity of open access and proprietary analytical reports, and being capable to monitor, analyze, forecast and ultimately understand, on the deepest level, the entire scope of the current state of Longevity science, technology, business and policy, and to predict the future trajectory of the industry, taking into account its unprecedentedly diverse scope and extreme levels of complexity, multidimensionality

[1] www.aginganalytics.com

and intersectionality. The Longevity sphere attracts some of the brightest, most innovative and most committed talents in the world, being on the very forefront of human progress. As a result of this, it was natural and logical that during my journey along this pathway I met Margaretta Colangelo, who has since become my closest friend, business partner and confidant.

Deep Knowledge Group's activities in the Longevity field became a magnet for the talented and forward-thinking experts and specialists, and we found ourselves blessed with a stellar team of like-minded people with us, including Eric Kihlstrom, Stefan Hascoet, Kate Batz, Richard Siow, Franco Cortese, Sergey Balasanyan and Robin Farmanfarmaian, each of whom is a bright star in our constellation, and each of whom contributed a wealth of valuable content and specific chapters to this book.

This brings us to a quick note I must make on the structure of this book. The science behind the industry is highly advanced and involves a great many technical concepts, many of which are already quite well documented in other relevant books. Our concern here is presenting a broad, industry-level conceptual frame work in order to provide a

AGING ANALYTICS AGENCY

Aging Analytics Agency was established in 2013 as a Longevity-focused analytical subsidiary of Deep Knowledge Group, and began producing analytics before the Longevity Industry itself emerged. The company is exclusively focused on Longevity, Geroscience, AgeTech and Preventive Medicine, has been developing its methodology since 2015, and is the main source of market intelligence in the field.

DEEP KNOWLEDGE ANALIYTICS

Deep Knowledge Analytics focuses on DeepTech and frontier technology industries using sophisticated multidimensional analytical frameworks and algorithmic methods that combine hundreds of specially designed metrics to deliver sophisticated market intelligence, pragmatic forecasting and tangible industry benchmarking.

general overview of the industry with the degree of understanding of specific domains necessary to discern threats versus opportunities, fiction versus facts, hype versus reality, and realistically predict future scenarios. As such, in this first edition we have restricted the science to 50 pages. But as true participation in the Longevity sphere will inevitably require some mastery of advanced scientific concepts, ***interested readers of the first edition can request a free of charge second edition after January 2021, which will contain hundreds of additional pages on science and technology.*** We are confident that most readers would be interested to read about the science, technology and advanced biomedicine no less than they would about the investment, financial and governance sides of the Longevity Industry.

The science and technology of Longevity is much more multifaceted than in the current biotechnology industry in general, requiring much more ***sophisticated analytics***[1]. This is due to the highly complex and cross-sector nature of the technologies, which range from advanced biomedicine, Data Science and Artificial Intelligence to finance, investment, governance, policy, socioeconomic domains and others, and this complexity requires much more sophisticated multi-tier analytical frameworks, the integrated synthesis and assembly of many frontier domains of knowledge, and novel de-risking techniques because of the wide scope for risk and failure, much higher compared with the current BioTech Industry.

We at ***Deep Knowledge Group***[2] value knowledge as the most precious of all assets, and knowledge in the Longevity space as double precious deep knowledge. Deep Knowledge Group has an analytical GigaFactory, comprised of two major analytical arms (***Aging Analytics Agency***[3] and ***Deep Knowledge Analytics***[2]), and seven analytical divisions within them, each focused on specific domains (***AI-Pharma***[4], ***e-GovTech***[5], ***NeuroTech***[6], ***Longevity-GovTech***[7] and ***Longevity-FinTech***[8], Systemic Risk Forecaster, and InvestTech Laboratory).

---

[1] www.aginganalytics.com/our-approac
[2] www.dkv.global
[3] www.aginganalytics.com
[4] www.ai-pharma.dka.global
[5] www.govtech.global
[6] www.neurotech.com
[7] www.aginganalytics.com/policy-governance
[8] www.aginganalytics.com/financial-longevity-industry

We also are in possession of a specific strategy, involving first acquiring an extreme breadth of knowledge in particular domains, producing tens of thousands of pages of open-access and proprietary analytics on specific sectors, created with the primary purpose of gaining a broad understanding of the entire scope of such industries, in order to lay the groundwork for identifying all relevant players in the field, their activities and progress. We then utilize this extreme breadth of knowledge to cultivate an equally extreme depth of knowledge, sufficient to hone our strategic business agenda in each specific sector. Acquiring sufficient knowledge of the Longevity Industry was the greatest challenge, but we can feel safe to say that we have now accomplished the first level of it. And as such, we will work proactively in the following domains within the scope of Longevity Industry: Biomarkers of Longevity, Geroscience R&D, P4 Medicine, Regenerative Medicine, Advanced Cosmetics, Longevity FinTech, Longevity NeuroTech, Longevity FemTech, AgeTech and first of all AI for Preventive Medicine, Precision Health and Longevity.

Two looming global changes, one demographic, one technological, are about to shake up the

DEEP KNOWLEDGE VENTURES

Deep Knowledge Ventures is a data-driven investment fund focused on the synergetic convergence of DeepTech, frontier technologies and technological megatrends, renowned for its use of sophisticated analytical systems for investment target identification and due-diligence. Major investment sectors include AI, Precision Medicine, Longevity, Blockchain and InvestTech. While it's major short-term interests are AI and DeepTech, it's long term vision is focused on Longevity and Precision Health.

Longevity.Capital

Longevity.Capital is a specialized hybrid investment fund specifically focused on the Longevity Industry; it has unparalleled investment and exit strategies, structuring its portfolio based on sophisticated industry intelligence and advanced analytics provided by the specialized analytical firm Aging Analytics Agency.

global financial system. One is the Silver Tsunami, the economic devastation resulting from the global aging demographic. The other is the advancing Longevity Industry, the complex interplay of intersecting technologies which together will give healthy productive life back to this population.

The nations who succeed to thrive rather than decline in the face of the tsunami will have reformed themselves beyond recognition, having been forced to adapt and refashion around novel business models, new types of financial institutions and entire national financial systems, supported by progressive initiatives of technocratic governments.

Government will be responsibl[1]e for driving forward the development of many facets of the Longevity Industry, ranging from social care to financial reforms, and also for developing and supporting the missing technological synergies, such as the integration of Big Data and healthcare with the focus on AI for preventive medicine, the current lack of which further slows down industry growth. As such, Aging Analytics Agency is systematically producing ***Longevity Policy Proposals***[2] for a number of international and government entities, including the World Health Organization and the United Nations.

The Longevity Industry will be the biggest and most complex industry of all time, generating enormous wealth, and representing the most ethical way to conduct business, and its end product will be the most valuable and real asset ever conceived — Healthy Longevity. But it also inevitably carries an unprecedented number of risks. As such, Margaretta, myself and our team feel that it is our duty to demystify this sphere and to inform all future participants in this industry of potential benefits and pitfalls in equal measure.

Indeed, we would like to see a greater number of participants in the industry, an industry so big and diverse, calling upon so many forms of talent from so many sectors, from so many world regions, that there is enough space for unlimited cooperation, and zero competition, unlike any other industry before.

We want to welcome as many people as possible into this industry, which is why we decided to publish this book, bringing the actual deep knowledge of the Longevity Industry to the public.

---

[1] www.aginganalytics.com/longevity-development-plans

[2] www.aginganalytics.com/policy-governance

PART I - Longevity Industry Mega-Complexity

Key Points:

- *The Longevity Industry is poised to become the biggest and most complex industry in all of human history*
- *As it is situated at the intersection of the most advanced domains of science, medicine, technology, finance and policy, it possesses unprecedented levels of complexity, which turn the strategic decision-making within its scope into a challenge*
- *As a result of this complexity, entirely novel methods of analytics are required for intelligent and relevant forecasting, benchmarking and strategic investing*

Besides having immense opportunities for ethical profit, the Longevity Industry will also deliver an entirely new kind of ROI more valuable than money: health as wealth, and additional years of healthy, functional, productive life for both citizens and national economies.

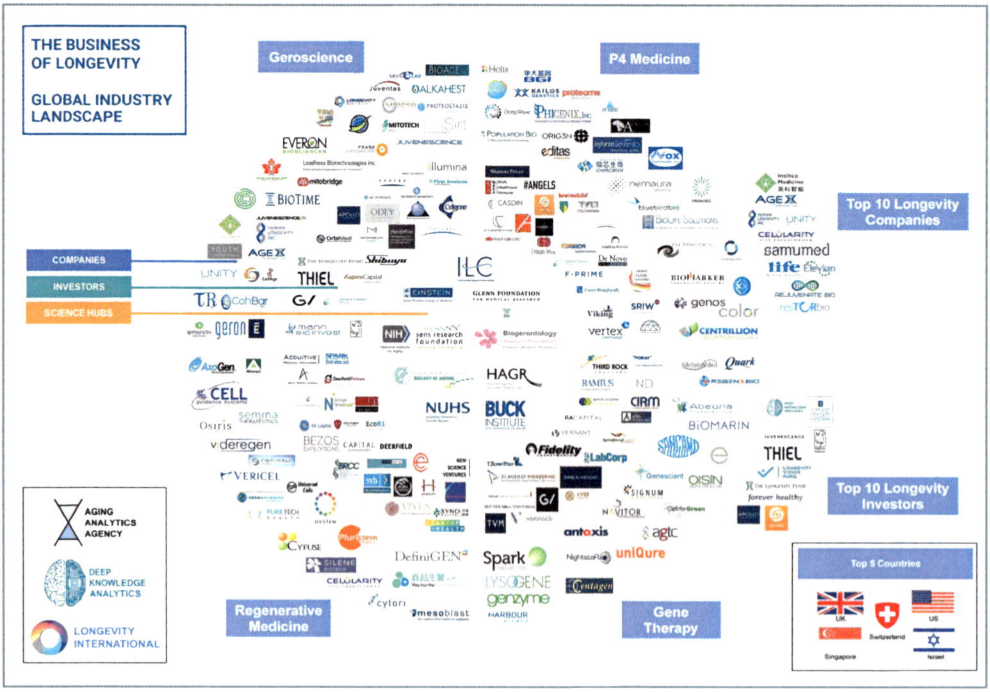

PART I - Longevity Industry Mega-Complexity

Developed nations now face the oncoming collision of two opposed Longevity MegaTrends: aging population and advancing biomedicine. Whether they stagnate or prosper when this collision hits depends entirely on how proactively and progressively they act today.

The Longevity Industry has the potential to transform the challenge and deficit of aging into the opportunity and asset of Healthy Longevity, turning a source of stagnation into a thriving wellspring of growth, prosperity and stability. However, being located at the intersection of the most advanced domains of science, technology, medicine, finance and governance in human history, it possesses unprecedented levels of complexity and multidimensionality, necessitating entirely new methods of industry analytics, benchmarking, forecasting and investing in order to leverage and manifest its opportunities and neutralize its risks.

PART II – Longevity Policy and Governance

Key Points:

- *The Rise of Progressive Longevity Technocracies and MegaHubs*
- *National and international government policy can help to shape cooperation between sectors and facilitate the growth of the Longevity Industry*

Part I stated that the Longevity Industry means more than just the science of aging, it is very multi-faceted with regard to technological sectors (Geroscience Research and Development, Preventive Medicine, AgeTech, and Novel Financial System), and only a dedicated, synergistic focus on all four of these domains at once can escort the global population to a longer and healthier life, and turn Healthy Longevity into an asset.

Part II explains why the future structure of global finance, if it is to withstand the Silver Tsunami, must necessarily consist of a Novel Financial System: Longevity VC funds, AgeTech Banks, Longevity Hedge Funds and Longevity Trusts, culminating in a Longevity Stock Exchange and Longevity Index. Each of these is designed either to utilise the increased working productive life of their user population, or to extend it. The increased liquidity provided by these combined structures would set in motion a self-perpetuating cycle of growth in Longevity Finance: as soon as Longevity becomes a dividend, then the greater the progress in achieving Healthy Longevity, the more the owners of wealth will want to reinvest in the technologically reinvigorated labour force endowed with a greater healthspan, leading to further growth and a greater Healthy Longevity.

The nations that initiate these reforms would be capable of attracting several trillions of potential wealth that is currently inaccessible and locked away, and succeed in building a bridge between the currently separate spheres of the Longevity Industry on one hand, and the conservative investment community on the other hand.

PART III – Longevity Financial Industry

Aging and the Silver Tsunami should be considered one of the most acute problems of our time, whereas progress in biomedicine and increasing Health Longevity, one of the most promising opportunities in human history.

The upcoming collision of these opposed MegaTrends has created a massive multitrillion market for AgeTech, Longevity WealthTech and novel financial products targeted toward 1 billion people in retirement globally. The market size of the global Longevity economy reached $17 trillion in 2019, and is expected to grow to $27 trillion by 2026.

Financial institutions such as investment banks, pension funds, insurance and reinsurance companies, asset management firms and others can either utilize the opportunity of Healthy Longevity and thrive, or stagnate and decline when hit by the oncoming Silver Tsunami.

Whether they succeed at riding the wave or drown under it will depend not only on their willingness to deploy new business models adapted to the aging population but also on the quality of analytics and forecasting methods underlying them.

The traditional financial industry needs to adapt its products and services to respond to the challenges of an aging society. This implies creating new innovative products in the fields of AgeTech, Longevity WealthTech, Health Insurance and new financial products for clients aged 60+ and aiming to live up to 100 years and beyond.

Financial institutions that don't adapt will not survive in their present form! *After the tsunami has passed, such economies will have reformed themselves beyond recognition, absorbed by novel financial institutions and systems, or transformed by government initiatives of progressive technocratic countries.*

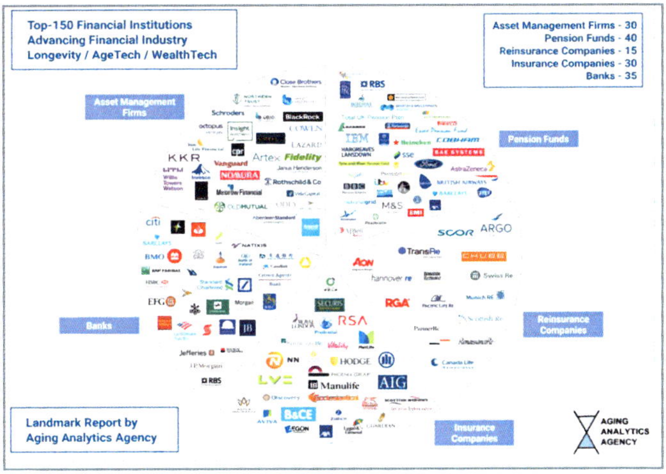

PART IV – Longevity Industry (Science, Technology and Biomedicine)

Introducing the scientific sector from which the multi-sector Longevity Industry emerged

Key points:

- *The 20th century has greatly increased our understanding of the mechanisms of aging, but has also revealed the limits of our ability to intervene in bodily processes.*
- *The advent of the "repair approach" to aging, not presuming to intervene in bodily processes, and the discovery of the hallmarks of aging at the turn of the 21st century have opened up new doors and created new sectors to which the old limits no longer apply.*

The Rise of Geroscience

The war on human illness has evolved from an invasion into a counter-insurgency operation.

With infectious diseases now under biomedical control, the stubborn illnesses inflicted from within — cancer, Alzheimer's, arthritis, osteoporosis, not to mention more subtle defects such as progressive hearing loss and cognitive decline — are the next (and perhaps the final) challenge for the future of biomedicine.

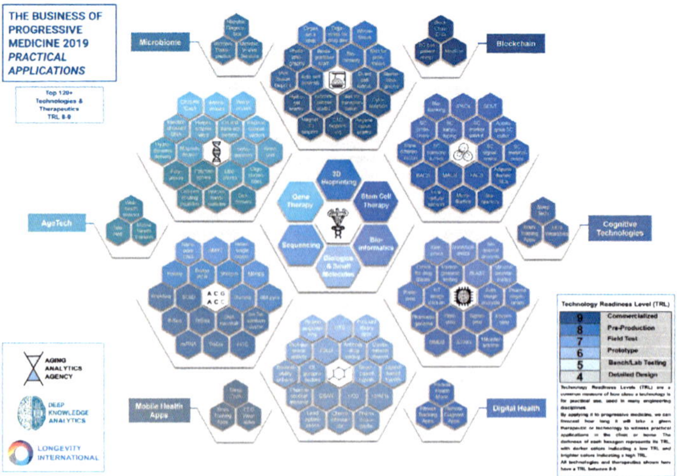

Introduction to the Next Book – Longevity Industry 2.0

The completion of the book Longevity Industry 1.0 does not bring you to the end of the Longevity story, but rather to its beginning. In Longevity Industry 1.0 we look at the factors involved in bringing us to this point.

The Longevity Industry 1.0 book is a summary of the current state of affairs and the forces at work, the challenges and opportunities that have made the industrialization of Longevity a national priority for the world's most forward thinking and progressive governments.

The Longevity Industry 2.0 book will be focused on the state of the industry in the year 2027, and the transformation of the global Longevity Industry from 1.0 to 2.0. It will take a deeper analytical, more specified look at some of the mechanisms and practical techniques that would encompass a fully realized Longevity Industry. This sequel will explore in detail how today's Longevity start-ups will become tomorrow's multi-trillion dollar corporations; how AI will guide preventive medicine and how human health and vitality will be transformed in a still unimagined way.

www.longevity-book.com

contact@longevity-book.com

It will also discuss the important role that Longevity focused MegaHubs will play as well as painting a detailed and knowledge rich picture of the full scope and scale of the global potential of the Longevity Industry, as

Introduction to the Next Book – Longevity Industry 2.0

BioTech and current Healthcare models are disrupted by AI driven preventive medicine and novel InvestTech solutions. If Longevity Industry 1.0 explains how we got to this point, 2.0 will give a detailed map of where we will be by the year 2025, and the hows and whys of the processes involved in every sector, from the practical applications and processes of precision medicine to a more detailed analysis of de-risking strategies for Longevity investment. In short, Longevity Industry 1.0 has brought us to the birth of not just a new asset class, but also, in its novel approaches, a new knowledge class. In Longevity Industry 2.0, we will explore this deeper realm of knowledge to its fullest extent.

Longevity Industry 2.0

2023-2027: DeepTech Engineering The Accelerated Trajectory of Human Longevity: The Blueprint and Pathway from 1.0 to 2.0

- Global Industrialization of Longevity to Scale
- The Evolution from Longevity Start-ups to Multitrillion Dollar Longevity Corporations
- How AI-Driven Preventive Medicine will Disrupt BioTech and Healthcare Industries
- Novel Financial Instruments and InvestTech Solutions
- The Rise of Progressive Longevity MegaHubs

Appendix C

Longevity Financial Industry

Book Teaser

Health as New Wealth

Engineered Solutions to Bridge the Longevity Liquidity Gap

Rise of Longevity Investment Banks, Stock Exchanges and Financial Instruments

2010-2023: Evolution of the Longevity Industry from Zero to 1.0

- The Industrialization of Longevity
- The Current State of Longevity Science, Business, Finance, and Practical Applications
- Longevity Becomes National Priority Item for the Strategic Agenda of Progressive Governments
- Transforming the Challenge and Deficit of Aging into the Opportunity and Asset of Longevity
- Defining and De-Risking: Hype vs. Reality

Health as New Wealth, Engineered Solutions to Bridge the Longevity Liquidity Gap, and the Rise of Longevity Investment Banks, Stock Exchanges and Financial Instruments

- Engineered InvestTech and FinTech Solutions to Bridge the Big Liquidity Gap
- Longevity Futures, Shorting, Exchange Traded Funds and Technology Marketplaces
- Longevity Stock Exchange, Investment Bank, Index Fund and Financial Derivatives
- Financial Futurism: Derivatives Tied to National Healthcare Systems, Pension Systems, National Healthy Longevity & Longevity Economies

About the Book:
Introducing Longevity Financial Industry

PART I.
Defining the Longevity Financial Industry

- Introduction
- What is Longevity Finance?
- Major Financial Corporations Re-Tuning their Business Models to Neutralize the Risk of Ageing Population and Capture the Opportunity of Healthy Longevity
- The Need for Longevity Finance: Health as the New Wealth (Where Healthspan Meets Weathspan)
- Longevity Finance Big Data Analytical Dashboard
- Top Longevity Challenges and Opportunities for Financial Corporations

PART II.
The Present State of the Longevity Financial Industry (2021)

- The Rise of Longevity Finance at the Intersection of Population Ageing vs. Longevity
- The State of Global Longevity Finance in 2021
- Longevity Finance Benchmarking: Top Longevity Financial Corporations in 2021
- Longevity FinTech Benchmarking: Top Longevity FinTech Companies in 2021
- Longevity Banking Card (Integrated AgeTech/HealthTech/WealthTech Solution)
- Biomarkers of Human Longevity as Catalyst for Investment De-Risking, Reliable Clinical Forecasting, and Tangible Metrics for Company Valuation and Due Diligence
- The Big Longevity Liquidity Gap
- Longevity Investment Digest

About the Book:
Introducing Longevity Financial Industry

PART III.
The Near Future of Longevity Finance (2023-2030)

- Forecast on the Near Future of Longevity Finance: 2023-2025
- Longevity Investment Bank
- The Next 5 Years Will Determine the Fate of Financial Corporations at the Intersection of Ageing Population vs. Health Human Longevity
- Longevity Banking: the 1+ Trillion Opportunity of 1 Billion People in Retirement
- Longevity Innovation Marketplace
- Longevity Stock Exchange
- Longevity Exchange Traded Funds
- Longevity Financial Instruments and Derivatives 1.0 (2021-2025)

PART IV.
The Future of Longevity Finance (2025 - 2030)

- Longevity Financial Instruments and Derivatives 2.0 (2025-2030)
- The Paradigm Shift From National Economies to National Longevity Economies
- Longevity Financial Industry as the Key Mechanism for Securing the Future of Longevity Industrialization (and the Stability of National Economies)
- The Concept of the Longevity Finance Valley: Global Benchmarking of Top Contenders to Create the First Municipal Longevity Finance Hub by 2025
- The Evolution of the Longevity Finance State: Global Benchmarking of Top Contenders to Create the First National Longevity Finance Hub by 2030
- Forecast on the Deep Future of Longevity Finance: 2025-2030
- Only Longevity Finance Can Secure the Health & Wealth of Citizens & Economies

About the Book: Longevity Financial Industry

In *"Longevity Industry 1.0 - Defining the Biggest and Most Complex Industry in Human History,"* we distilled the complex assembly of deep market intelligence and industry knowledge that Deep Knowledge Group has developed over the past 5 years into a full-scope understanding of the global Longevity Industry, showing the public exactly how the international consortium of commercial and non-profit entities managed to define the overwhelmingly complex and multidimensional Longevity Industry for the first time, and how they created tangible framework for its systematization and forecasting. One of that book's four major parts was devoted to the recent rise, present state and near future of the Longevity Financial Industry.

Now, as the pace of innovation and diversification in Longevity Finance continues to grow, with increasing numbers of Financial Corporations onboarding the topic of Longevity, the time has come for a full book dedicated to this topic. In "Longevity Financial Industry: Health as New Wealth, Engineered Solutions to Bridge the Longevity Liquidity Gap, and the Rise of Longevity Investment Banks, Stock Exchanges and Financial Instruments," Deep Knowledge Group General Partner Dmitry Kaminskiy charts the near and far future of Longevity Finance, charting the emerging paradigm of Health as New Wealth and the increasing need for weathspan extension to match the pace of increasing healthspans, forecasting the rise of Longevity-focused Retail Banks, Investment Banks, Stock Exchanges, ETFs, Technology Marketplaces and Modern Financial Instruments and Derivatives, and offering a glimpse at

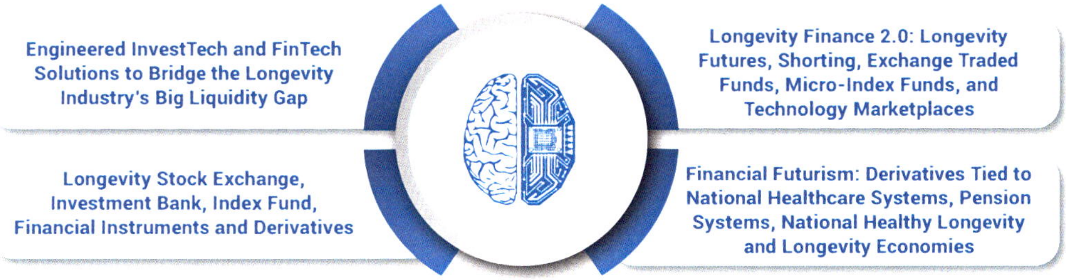

Longevity Financial Industry – Foreword by Dmitry Kaminskiy

Deep Knowledge Group's blueprint for modern, engineered Invest-Tech solutions to bridge the Big Longevity Liquidity Gap.

In our 2020 book, '*Longevity Industry 1.0: Defining the Biggest and Most Complex Industry in Human History*[1]', we distilled the complex assembly of deep market intelligence and industry knowledge that Deep Knowledge Group[2] and its Longevity-focused subsidiaries (including Longevity.Capital[3] and Aging Analytics Agency[4]) have developed over the past 7 years into a full-scope understanding of the global Longevity Industry.

One of the four sections of that book was dedicated to the topic of Longevity Finance, charting the ways in which increasing numbers of Financial Corporations are re-tuning their business models to neutralize the challenge of ageing population and leverage the growth potential and opportunity of Longevity. Now, in 2021, given the unrecognized importance of this sector to the health and stability of the Global Longevity Industry as a whole, the time has come to devote an entire topic of the present state and near-future of this rapidly evolving sector.

While the majority of practical outcomes in healthy human Longevity will be driven by precision health technologies, they will also require an assembly of other, non-biomedical components, including financial products and services such as InsurTech, AgeTech, HealthTech, and WealthTech. Progressive financial corporations are already onboarding the topic of Longevity in significant ways. Meanwhile, increasing numbers of HWNIs and members of the global investment community are embracing the emerging paradigm of Health as New Wealth, and recognizing Health as the Most Precious Asset; in 2018 UBS bank conducted a survey among its HNW clients and found that 50% planned to live past the age of 100 years. It stands to reason that such clients

[1] www.longevity-book.com

[2] www.dkv.global

[3] www.longevity.capital

[4] www.aginganalytics.com

will require a far longer horizon of financial planning and specifically designed financial products.

Against this backdrop of activities, the continued positive trajectory of growth and stability of Global Longevity Industrialization faces a major bottleneck and risk factor: Longevity start-ups require increasingly large levels of funding to move forward, and appetite for exposure to Longevity among both individual and institutional members of the conservative investment community continues to mount, but an extreme lack of innovation in the technologies, institutions, and instruments used to direct and manage Longevity investment and financial activities has created a critical Big Gap preventing the two sides of the Longevity Liquidity equation from being bridged.

Having been actively conducting market intelligence in this space for several years, and having originally coined the term Longevity Financial Industry in 2017, Deep Knowledge Group has been systematically applying its DeepTech Engineering mindset to this issue, developing and designing the blueprint of frameworks, institutional structures, technologies and financial innovations required to bridge this big gap and achieve a massive acceleration in the rate of Longevity Industrialization. Now, for the first time, we are unveiling parts of this blueprint to the larger investment and finance communities, outlining the major challenges and opportunities in this space for Investment Banks, Insurance Companies, Pension Funds, Asset Management Firms and Retail Banks, and forecasting the rise of Longevity-focused Investment Banks, Stock Exchanges, ETFs, Technology Marketplaces and Modern Financial Instruments and Derivatives, and offering a glimpse at Deep Knowledge Group's ongoing work to create, validate and deploy the exact modern DeepTech tools and engineered solutions needed to unleash the Future of Longevity Finance years ahead of schedule.

PART I:
Defining the Longevity Financial Industry

- *Introduction*
- *What is Longevity Finance?*
- *Major Financial Corporations (Insurance Companies, Pension Funds, Asset Management Firms, etc) Beginning to On-Board Topic of Longevity*
- *The Need for Longevity Finance: Health as the New Wealth (Where Healthspan Meets Weathspan)*
- *Longevity Finance Big Data Analytical Dashboard*

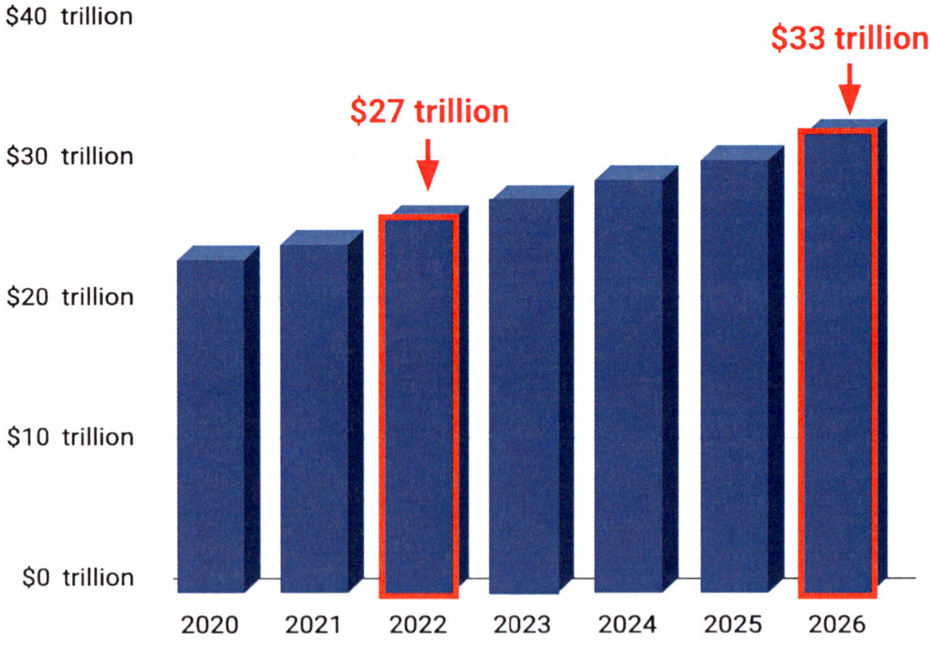

www.aginganalytics.com/investment-digest

Longevity Industry: Multi-trillion Dollar Opportunity

Longevity, AgeTech & WealthTech Market

Globally
1 Billion in Retirement

In the UK
10 Million in Retirement

"The one billion retired people globally are a multi-trillion dollar opportunity for business"

~ Dmitry Kaminskiy, interview in the Financial Times

FT FINANCIAL TIMES

"The global spending power of those aged 60 and over will reach $15 trillion annually by 2020"

~ Bank of America Merrill Lynch

Longevity Financial Industry Framework

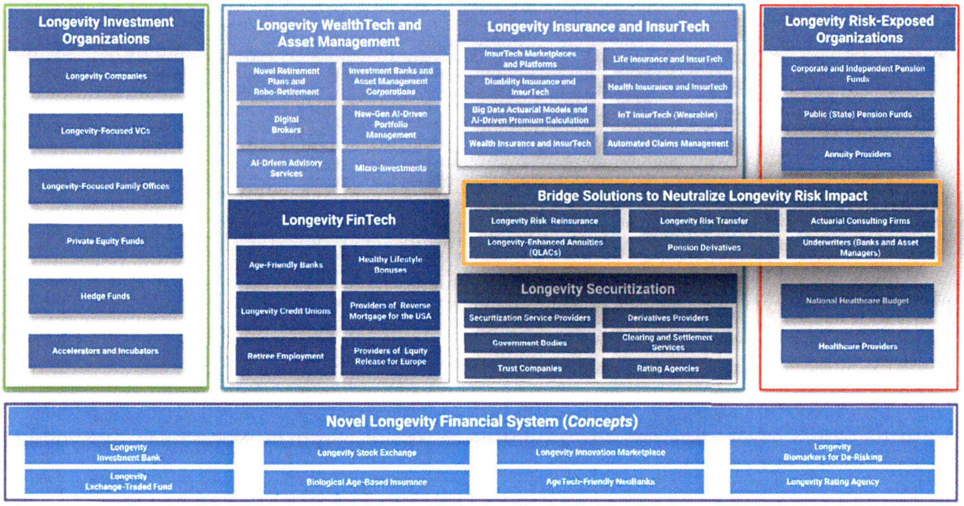

Longevity Finance: Defining the Sector

Longevity has become a major focus of some of the largest financial institutions in the world, with many major institutional investors seeking opportunities to contribute to the development of the Longevity industry by investing in AgeTech, Longevity Fintech, Longevity Biomedical companies, and start-ups.

Aging has become more than a challenge at the intersection of many of the most acute problems of our time - it also presents one of the most promising opportunities.

Financial institutions that will be able to adapt their business models to the new industries of AgeTech, WealthTech, and Longevity Finance will have a chance to outperform their competitors and grow rapidly. **High-quality analytics on how Longevity can be on-boarded into financial institutions' business models will help make this opportunity a reality.**

We define 5 categories of companies that contribute to the development of the Longevity Financial Institutions industry: Asset Management Firms, Banks, Insurance, Reinsurance Companies, and Pension Funds. We expect the Longevity Finance market to grow at an average rate of 6% CAGR, reaching $28,532 billion by 2025.

Global Financial Services Market Volume (Forecast, $ Billions)

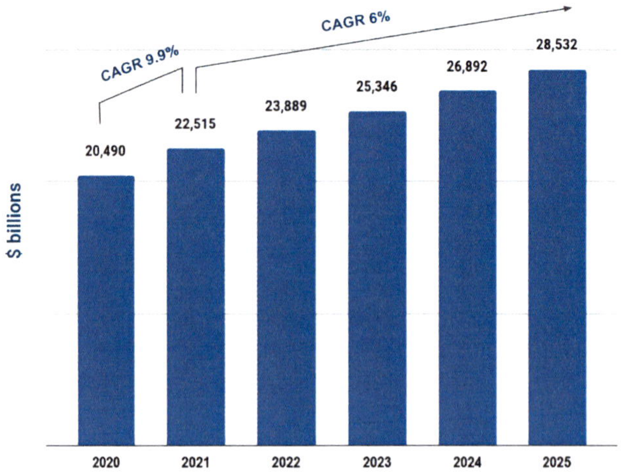

www.aginganalytics.com/investment-digest

Longevity Finance: Defining the Sector

Deep Knowledge Group's Longevity focused analytical subsidiary Aging Analytics Agency was the first to coin the term "Longevity Financial Industry" in their May 2018 analytical report "Longevity Industry Landscape Overview Volume II: The Business of Longevity," forecasting the rise of Longevity Finance institutions (such as Longevity Stock Exchanged, Index Funds and AgeTech Banks) and instruments (specialized financial derivatives tied to the Longevity Industry) as far back as Q2 2018, and was the first to produce a dedicated analysis of the sector, and a specific analytical framework capable of defining the sector and forecasting developments within it, through their July 2019 special analytical case study "Advancing Financial Industry: Longevity / AgeTech / WealthTech."

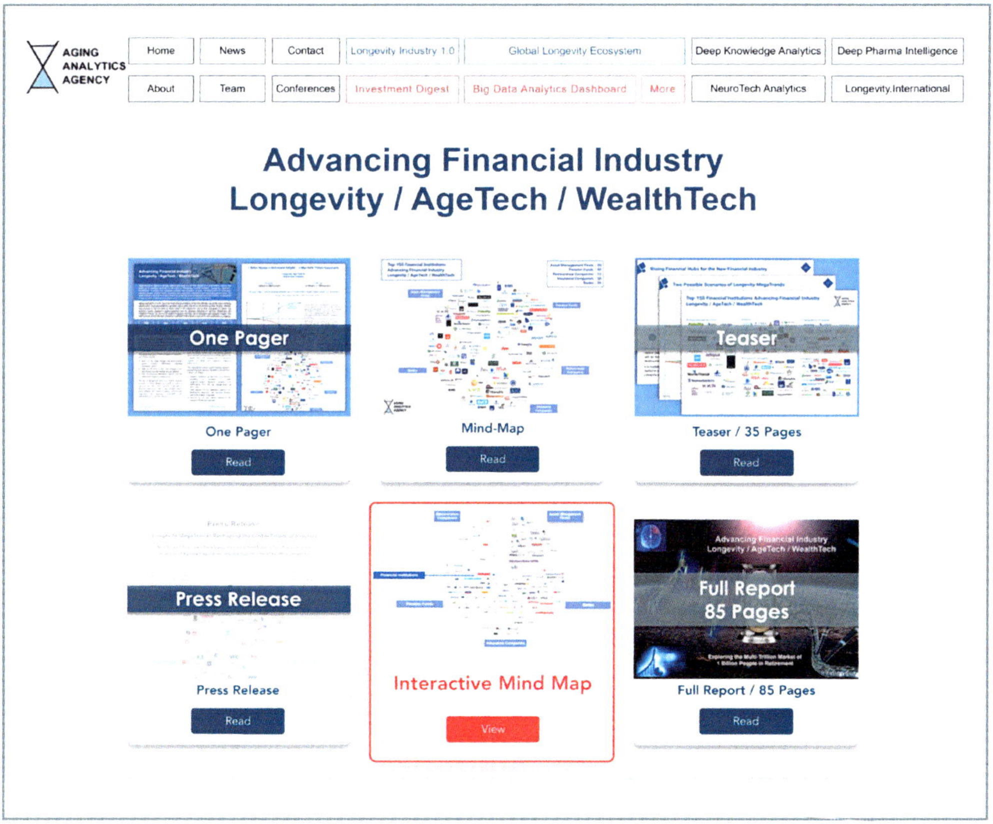

Top-20 Financial Institutions Advancing the Longevity Industry

NYSE

US-Based

Berkshire Hathaway (BRK-B)
Market Cap: $605B

Citigroup (C)
Market Cap: $151B

JPMorgan Chase & Co.(JPM)
Market Cap: $470B

Goldman Sachs Group Inc. (GS)
Market Cap: $118B

Bank of America Corp.(BAC)
Market Cap: $342B

Wells Fargo & Co. (WFC)
Market Cap: $165B

American Express (AXP)
Market Cap: $118B

Morgan Stanley (MS)
Market Cap: $150B

UK-Based

HSBC Holdings PLC (HSBC)
Market Cap: $120B

China-Based

China life (LFC)
Market Cap: $157B

Canada-Based

Royal Bank of Canada (RY)
Market Cap: $132B

Toronto-Dominion Bank (TD)
Market Cap: $118B

Hong Kong (HKEX)

China-Based

China Construction Bank (939)
Market Cap: $211B

China Merchants Bank (3968)
Market Cap: $199B

Bank of China (3988)
Market Cap: $139B

AIA Group(1299)
Market Cap: $149B

China (SSE)

China-Based

Ping An Insurance (601318)
Market Cap: $279B

Agricultural Bank of China (601288)
Market Cap: $176B

Australia (ASX)

Australia-Based

Commonwealth Bank of Australia(CBA)
Market Cap: $118B

NASDAQ

Switzerland-Based

AstraZeneca pension fund(AZN)
Market Cap: $131B

300 Longevity Finance Institutions: Regional Proportion

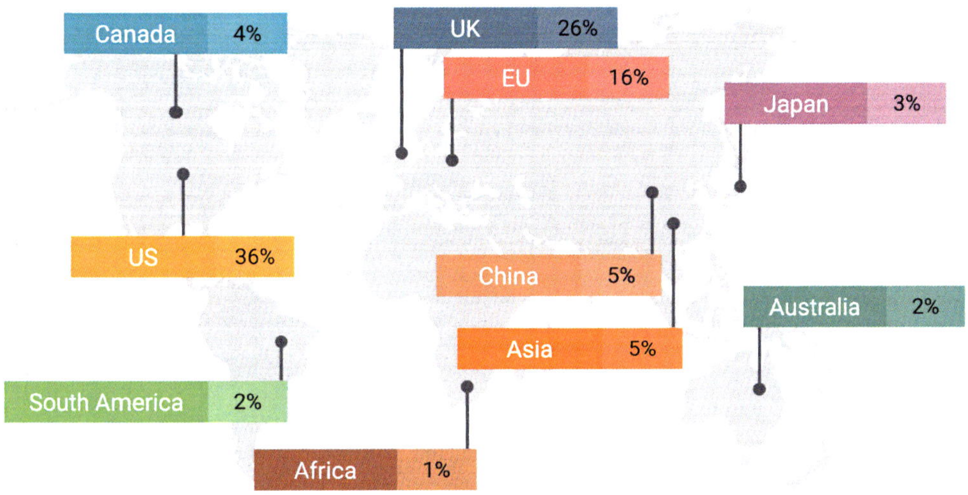

www.invest-solutions.tech/access-to-Longevity-finance

The above chart represents the distribution of Longevity-focused financial institutions by country. On the one hand, the largest number of the selected companies have their headquarters in the US because of the high availability of developed financial markets there. On the other hand, quite a few companies are headquartered in the UK and in the EU for similar reasons.

Financial Institutions Advancing the Longevity Industry

Since the Q1 2020 global financial market downturn, 194 financial institutions have fully recovered by year-end and continued to grow in 2021.

The largest institutions involved in the Longevity Finance sector by market capitalization include **Berkshire Hathaway, JPMorgan Chase & Co., Industrial and Commercial Bank of China, and Shell Contributory Pension Fund.**

Given that many Longevity-advancing financial institutions are included in the S&P 500, this chart represents the approximate dynamics of the entire stock market.

Cumulative Capitalization Dynamics in 2016-2020

www.invest-solutions.tech/access-to-Longevity-finance

PART II: The Present State of the Longevity Financial Industry (2021)

- The Rise of Longevity Finance at the Intersection the Opposed MegaTrends of Population Ageing vs. Healthy Human Longevity
- The Rise of the Silver Economy: The Multi-Trillion Opportunity of 1 Billion People in Retirement
- The State of Global Longevity Finance in 2021
- Global Longevity Finance Benchmarking: Top Longevity Financial Corporations in 2021
- Global Longevity FinTech Benchmarking: Top Longevity FinTech Companies and Start-ups in 2021
- Longevity Banking Card (Integrated AgeTech/HealthTech/WealthTech Solution)
- Biomarkers of Human Longevity as the Major Catalyst for Investment De-Risking, Reliable Clinical Forecasting, and Tangible Metrics for Company Valuation and Due Diligence
- Biomarkers of Human Longevity for InsurTech: Spotlight / Case Study
- The Big Longevity Liquidity Gap
- Longevity Investment Digest 2021

Longevity Financial Industry Threats and Opportunities

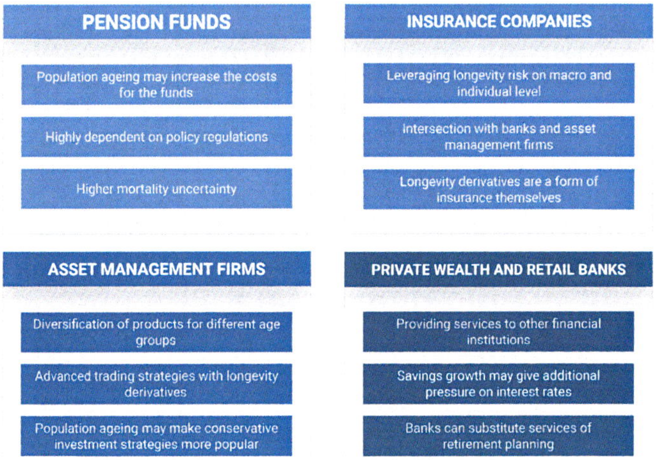

Top-300 Financial Institutions Advancing Financial Industry

www.invest-solutions.tech/access-to-Longevity-finance

Number of Financial Institutions Engaged in the Longevity Sector by Region

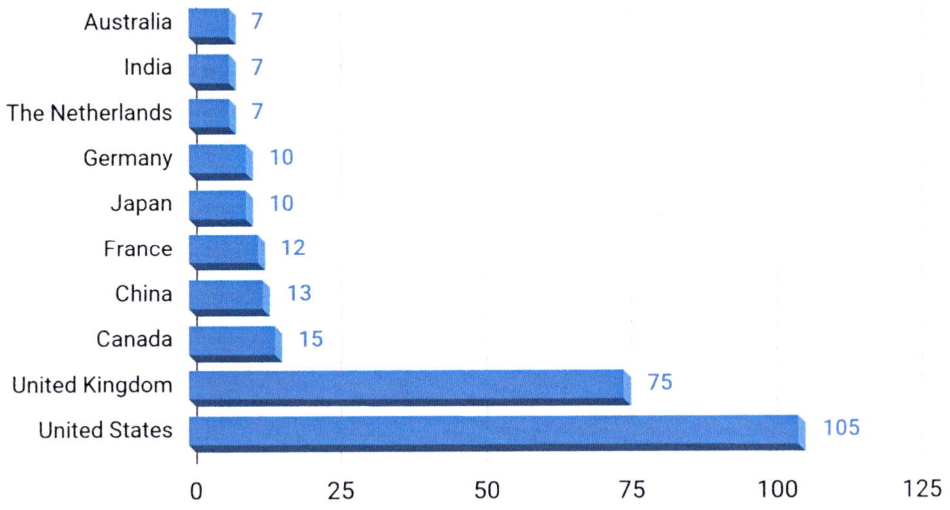

www.invest-solutions.tech/access-to-Longevity-finance

Total Market Capitalization of Financial Institutions Engaged in the Longevity Sector in 2020: Regional Distribution

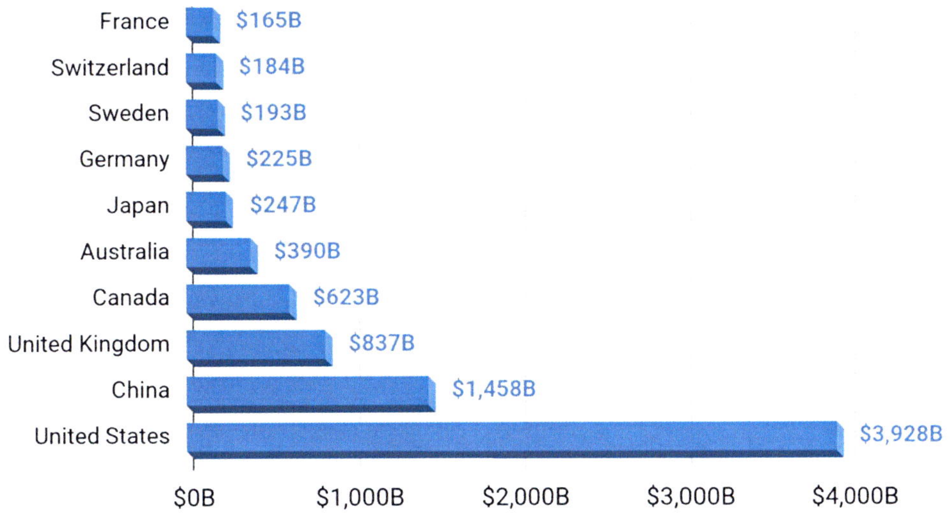

www.invest-solutions.tech/access-to-Longevity-finance

Top 10 Financial Institutions Advancing the Longevity Industry in 2020 (by Market Capitalization)

www.invest-solutions.tech/access-to-Longevity-finance

The Biggest Financial Institutions Advancing the Longevity Industry

Four of the five biggest financial institutions (by market capitalization) advancing the Longevity Industry are represented by banks. Nevertheless, the first place is taken by Berkshire Hathaway asset management firm. Despite the COVID-19 financial crisis that occurred in the beginning of 2020, nearly all financial institutions have recovered their market capitalization to pre-pandemic figures and some of them such as Berkshire Hathaway and JPMorgan Chase & Co. have reached even higher positions by the beginning of 2021.

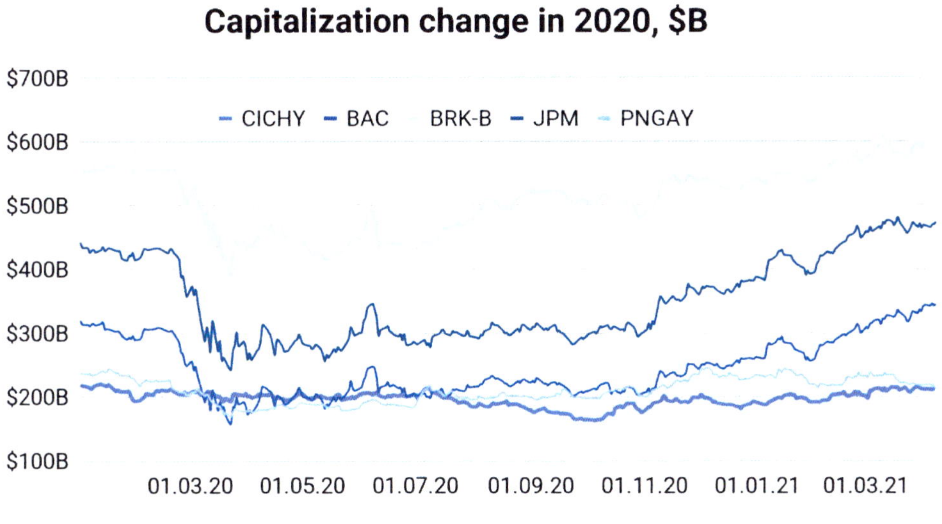

Name	Country	Market Cap	Annual Revenue	Net Income	Volume	Total Return Level	ROE	ROA	PE Ratio	Beta (5Y)
Berkshire Hathaway	USA	611.04B	245.51B	42.52B	6.2M	260.91	9.83%	5.15%	24.37	0.87
JPMorgan Chase & Co.	USA	473.85B	102.06B	27.41	16.79M	582.16	10.78%	0.96%	17.62	1.225
Bank of America Corp.	USA	345.00B	86.86B	17.89B	58.09M	102.03	7.34%	0.68%	21.38	1.572
Ping An Insurance Group	China	214.09B	161.25B	20.75B	30.9M	31.06	17.32%	1.51%	10.11	0.89
China Construction Bank	USA	212.06B	109.35B	39.27B	367.52M	N/A	11.83%	1.02%	5.41	0.53

Biomarkers of Human Longevity for Investment De-Risking, Reliable Clinical Forecasting, and Tangible Metrics for Company Valuation and Due Diligence

- The Longevity Industry's Systemic and Most Fundamental Risk: Reliance on Animal Model (Mice) Data for Company Valuation
- Lack of tangible frameworks for human validation is leading to unstable market capitalization dynamics for public Longevity companies, and risks de-stabilizing long-term investor confidence Biomarkers of Human Longevity as the only market-ready tool to de-risk Longevity investing
- Consensus frameworks of Biomarkers of Human Longevity to bridge the translational gap of successful results in model organisms (mice) vs. replication of results in humans (clinical trials)
- Tangible biomarker-based frameworks for company valuation and due diligence, as the tool for reliable investor expectations and forecasts
- Digital Human Avatars as the platform for safe self-experimentation and real-time therapeutic validation, personalization and fine-tuning
- Longevity Economy 2.0 — Biomarkers of Human Longevity as Major Catalyst for Longevity Investment Bank, InsurTech, Longevity Stock Exchange and Financial Instruments and Derivatives

www.longevity-book.com/biomarkers

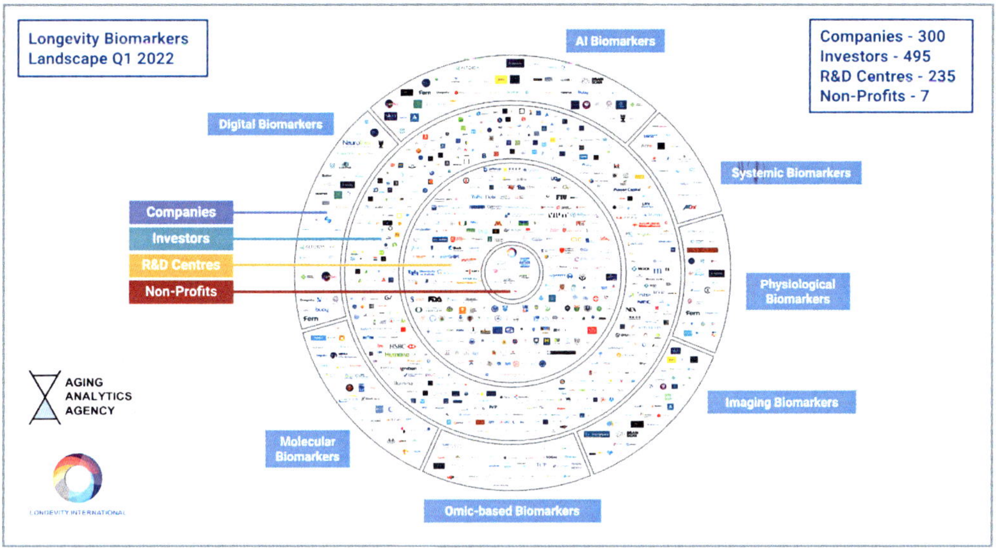

www.longevity.international/biomarkers

Digital Human Avatar Framework: 1.0, 2.0, 3.0 and 4.0

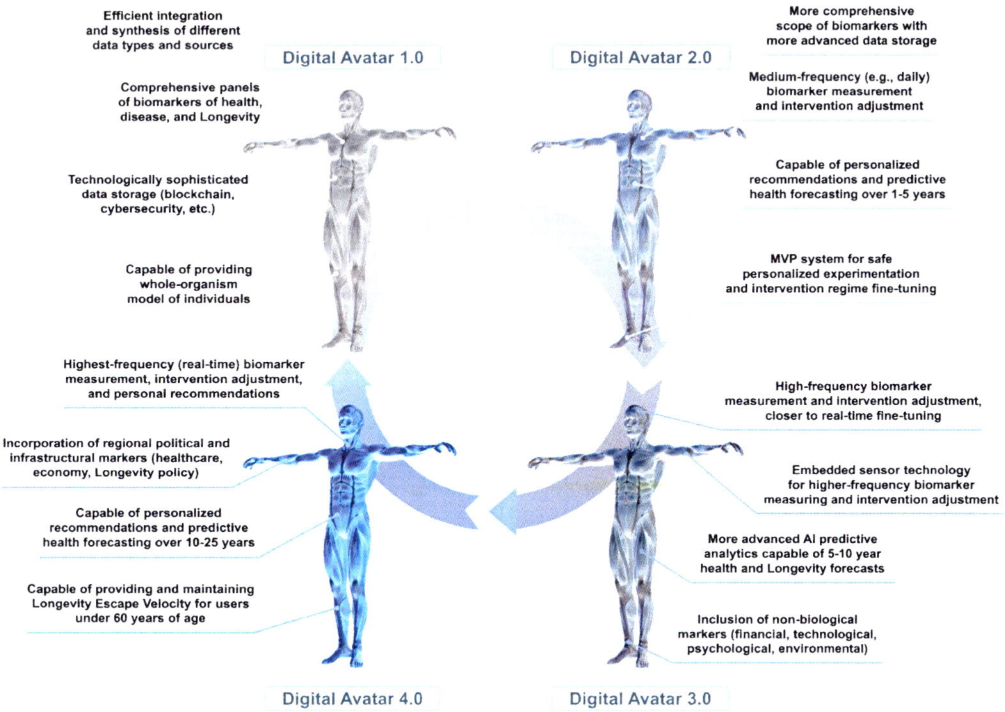

www.longevity.international/biomarkers

PART III: The Near Future of Longevity Finance (2021-2025)

- Forecast on the Near Future of Longevity Finance: 2021-2025
- What the Opposed MegaTrends of Aging Population and Advancing Biomedicine Means for the Finance Sector
- The Rise of the Silver Dollar: 1+ Billion Global Retired Hold the Majority of Global Purchasing Power, and Recognize Health as the New Wealth, and Longevity as Most Precious Asset Class
- Forecasting Major Challenges and Opportunities for Pension Funds, Insurance Companies, Retail Banks, Investment Banks, Asset Management Firms, and Financial Advisory Services
- Longevity Financial Corporations vs. Longevity FinTech Companies: Competitive Landscape Analysis
- Longevity Innovation Marketplace
- Longevity Stock Exchange
- Longevity Exchange Traded Funds
- Longevity Financial Instruments and Derivatives 1.0 (2021-2025)

Longevity Finance Big Data Analytical Dashboard

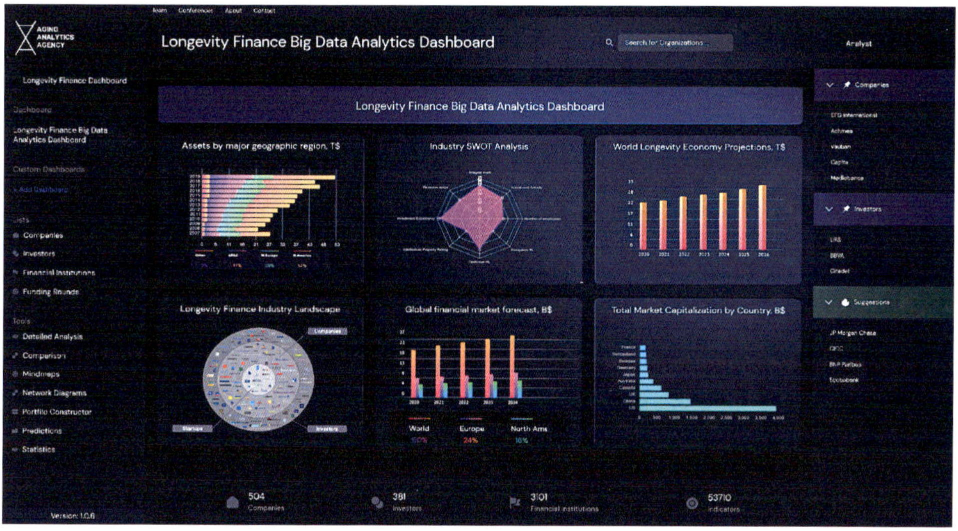

www.deep-innovation.tech/longevity-finance-dashboard

AI-driven Big Data Analytics System IT Architecture

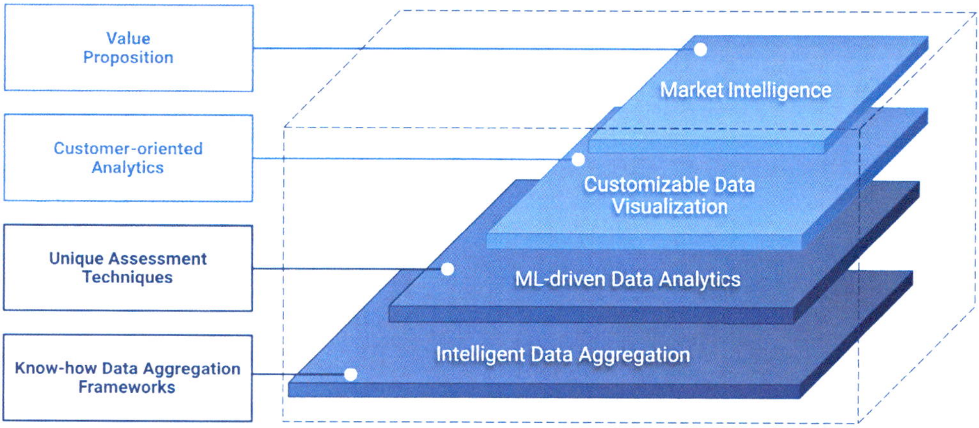

Longevity Banking Card: Overview of Feature

UK & EU IBANS

MULTI CURRENCY ACCOUNTS

DEBIT CARD

MONEY TRANSFERS

CASH LOAD

LONGEVITY MARKETPLACE

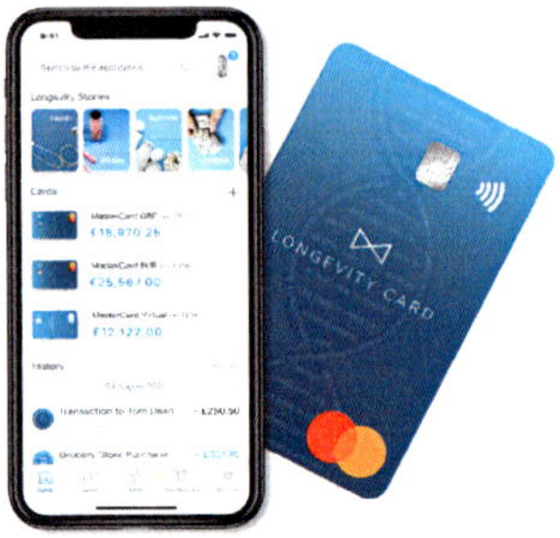

PERKS

FAST SIGN UP

REAL-TIME NOTIFICATIONS

TRAVEL INSURANCE

BUSINESS ACCOUNTS

24/7 SUPPORT

Longevity Banking Card: Marketplace Components

FITNESS & WELLNESS

HEALTHCARE

BEAUTY & COSMETICS

HEALTH & TRAVEL INSURANCE

WELLNESS TOURISM & SPA

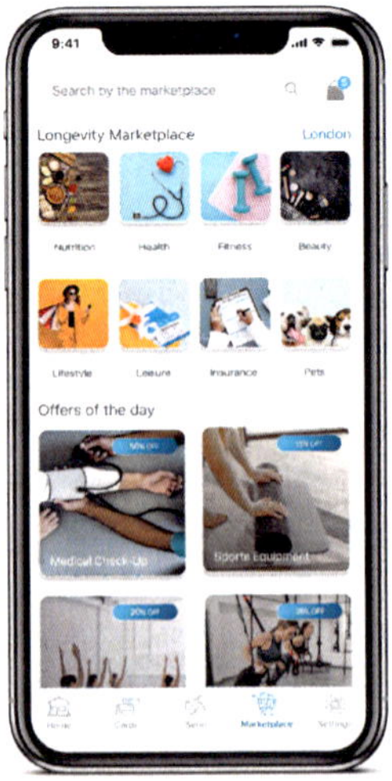

HEALTHY LIFESTYLE

PRIVATE CLINICS

MEDICAL CHECK-UPS

HEALTHY NUTRITION & FOOD

PET CARE

www.longevity.cards

HealthTech Integration and Healthspan Gamification

HEALTHSPAN AND FITNESS GAMIFICATION

LONGEVITY POINTS

DAILY ACTIVITY ANALYSIS

SLEEP ANALYSIS

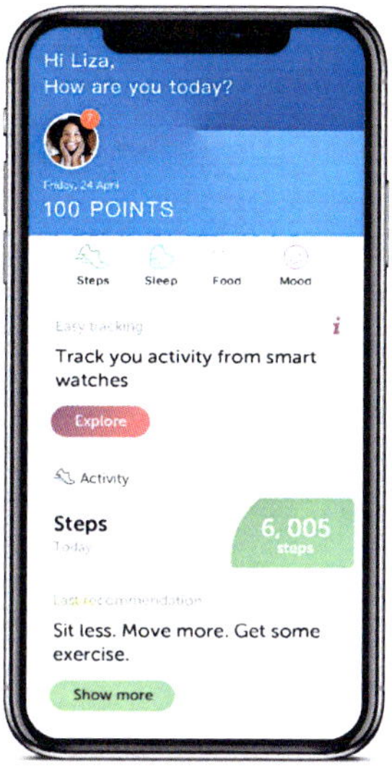

AI POWERED PERSONALISED RECOMMENDATIONS

WEARABLES COMPATIBLE

LONGEVITY STORIES

NUTRIENT TRACKING

www.longevity.cards

Longevity Membership Cards

Longevity Club is a one stop solution for members providing a wide range of unique advantages for:

- Club Members
- Corporate Club Members
- Ambassadors

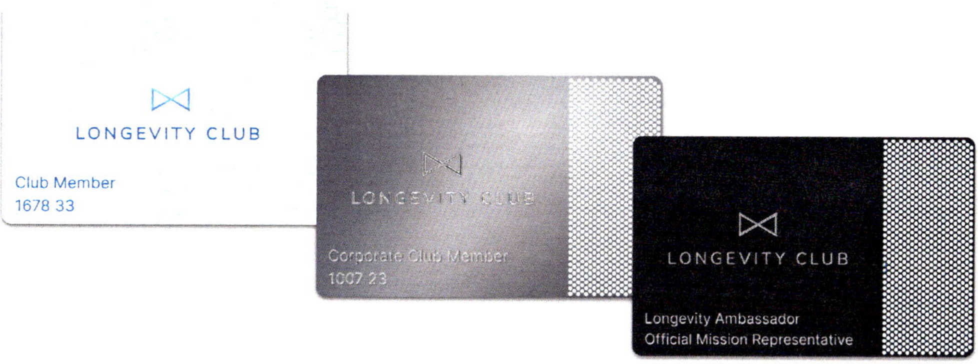

www.longevity.cards

PART IV: The Future of Longevity Finance (2025-2030)

- Longevity Financial Instruments and Derivatives 2.0 (2025-2030)
- The Paradigm Shift From National Economies to National Longevity Economies
- Longevity Financial Industry as the Key Mechanism for Securing the Future of Longevity Industrialization (and the Stability of National Economies)
- The Concept of the Longevity Finance Valley: Global Benchmarking of Top Contenders to Create the First Municipal Longevity Finance Hub by 2025
- The Evolution of the Longevity Finance State: Global Benchmarking of Top Contenders to Create the First National Longevity Finance Hub by 2030
- Forecast on the Deep Future of Longevity Finance: 2025-2030

Deep Knowledge Group Longevity Finance Activities Roadmap

Appendix **D**

Longevity Politcs

Book Teaser

Longevity Technocracy;
Modern Approaches to Policy, Governance, and National Industrial Strategies;
and Longevity as the New Political Priority of the 21st Century

2010-2023: Evolution of the Longevity Industry from Zero to 1.0

- The Industrialization of Longevity
- The Current State of Longevity Science, Business, Finance, and Practical Applications
- Longevity Becomes National Priority Item for the Strategic Agenda of Progressive Governments
- Transforming the Challenge and Deficit of Aging into the Opportunity and Asset of Longevity
- Defining and De-Risking: Hype vs. Reality

2023-2030: The Solidification of Longevity as the New Political Priority of the 21st Century

- The Rise of Longevity Technocracies as Global Competitors in the Creation of Full-Scope Longevity Industry Ecosystems and the Optimization of National Healthy Longevity
- Modern Approaches for Formulating and Executing National Longevity Industrial Strategies
- The Set of Factors That Will Make Citizens Demand the Delivery of Additional Health-Adjusted Life Expectancy (HALE) and Quality-Adjusted Life Years (QALY) From Their Governments and Politicians as a Basic Resource and Human Right

About the Book: Introducing Longevity Politics

In *Longevity Industry 1.0 – Defining the Biggest and Most Complex Industry in Human History*, we distilled the complex assembly of deep market intelligence and industry knowledge that Deep Knowledge Group has developed over the past five years into a full-scope understanding of the Global Longevity Industry. We showed the public exactly how Deep Knowledge Group managed to define the overwhelmingly complex and multidimensional Longevity Industry for the first time, and how they created tangible a framework for its systematization and forecasting. That book devoted one of its four sections to the topic of Longevity Policy and Governance and the future rise of Longevity Politics.

Now, in 2023, that time has come, as governments begin to prioritize the concept of Longevity from both a population health and an industrialization standpoint, and developments are accelerating to such an extent that an entire book is now necessitated to gain an actionable, deep and comprehensive understanding of the challenges and opportunities at play on the new frontier of Longevity Governance. In ***Longevity Politics: Longevity Technocracy, Modern Approaches to Policy, Governance & National Industrial Strategies, and Longevity as the New Political Priority of the 21st Century***, Dmitry Kaminskiy charts the rise of progressive Longevity MegaHubs and Technocracies, the concept of the Longevity Valley and its evolution toward the Longevity State, global benchmarking of national and local Longevity Industrial Strategies, international benchmarking of National Healthy Longevity, Longevity Ethics (how potential negative societal outcomes resulting from Longevity can be proactively optimized to ensure maximized socioeconomic benefit), and how Longevity will become central, essential and fundamental to developed nations' political agendas by 2030.

Longevity Politics
Foreword by Dmitry Kaminskiy

In our 2020 book, Longevity Industry 1.0: Defining the Biggest and Most Complex Industry in Human History[1], we distilled the complex assembly of deep market intelligence and industry knowledge that Deep Knowledge Group[2] and its Longevity-focused subsidiaries (including Longevity.Capital[3] and Aging Analytics Agency[4]) have developed over the past 7 years into a full-scope understanding of the global Longevity Industry.

One of the four sections of that book was dedicated to the topic of Longevity Policy and Governance, charting the rise of Longevity as a topic of increasingly prominent recognition and prioritization among progressive governments, and the emergence in the past several years of Longevity Politics – a state in which the concept of optimized National Healthy Longevity becomes an actual political talking point and tangible goal among politicians and between national and municipal governments. In 2023, we have reached the point at which the realm of Longevity Politics is emerging, and developments have accelerated to such an extent that we now feel compelled to dedicate a full book to this topic. We have done this in order to provide a comprehensive and actionable overview of the present state and near future of Longevity Politics. It is with this agenda in mind that I have written *Longevity Politics: Longevity Technocracy, Modern Approaches to Policy, Governance & National Industrial Strategies, and Longevity as the New Political Priority of the 21st Century*.

In the next few years, several technologically advanced, small, smart states will emerge as global competitors in the development of unified Longevity Industry ecosystems. Some will focus on specific sectors, while others will seek to create so-called Longevity States:

[1] www.longevity-book.com

[2] www.dkv.global

[3] www.longevity.capital

[4] www.aginganalytics.com

fully integrated hubs encompassing the entire scope of the industry (including biomedicine, tech, and finance). Five years from now we can expect to see a "new normal" of small, technocratic nations selectively promoting both personal and institutional citizenship to individuals, companies, and investors committed to advancing Longevity-related technologies in exchange for access to some of the world's most sophisticated and progressive healthcare, life insurance, MedTech, HealthTech, Longevity FinTech, AgeTech, and WealthTech ecosystems. These products and services will enable the simultaneous maintenance of Healthspan and Wealthspan and ensure the highest possible quality of life, social activity, mental wellness, and overall functionality.

As increasing numbers of governments are beginning to formulate and execute national industrial strategies to facilitate the synergetic development of four main sectors and driving forces of Longevity Industrialization (Advanced Biomedicine, AgeTech, Artificial Intelligence (AI), and the Longevity Financial Industry), with the goal of transforming the challenge of population aging into the opportunity of Healthy Longevity, it is clear that the age of Longevity Politics has already begun. The political, economic, and industrial capital that governments control is larger than any other industry stakeholder, and the stakes are proportionately higher, given that they are tasked with maintaining and optimizing the health and wealth of their nations. It is now indisputable that the Longevity Industry has reached a critical inflection point where politics becomes the most important driver of progress in the continued positive trajectory of Longevity Industrialization, as well as the delivery of prolonged and optimized Health, Wealth, and Practical Human Longevity to the citizens around the world.

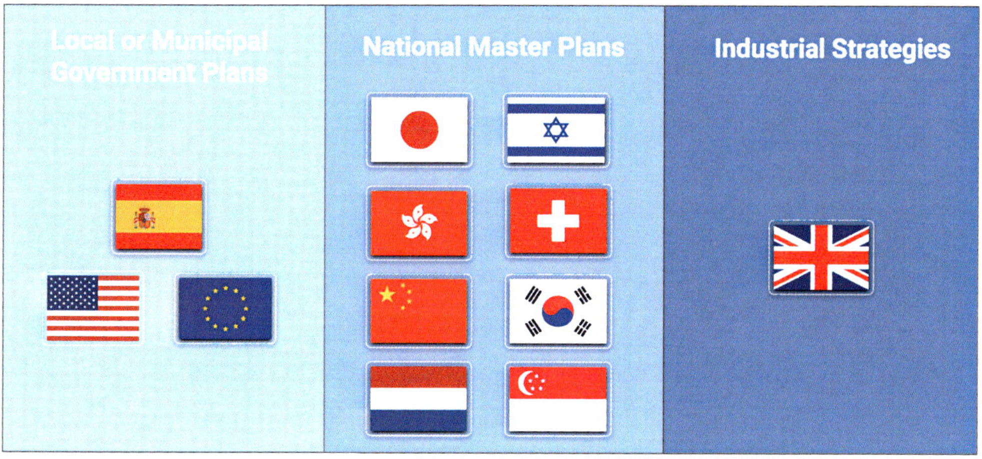

Chapter I. Longevity Policy Will Define Global Humanity by 2030

- The Challenges and Opportunities of Aging Population and Human Longevity Lie at the Direct Intersection of Society's Most Pressing Problems and Important Values.
- Longevity is a Direct Responsibility of Governments, which are Obligated to Maintain and Optimize the Health and Wealth of their Nation (their Citizens and Economy)
- National Healthy Longevity and Longevity Industrialization is the Most Powerful Tool at Governments' Disposal to Achieve Economic Stability and Societal Prosperity
- Just as Advances in Longevity Expanded Past the Point of Scientists, Necessitating Involvement from Industry and Investors, It Has Now Reached its Next Inflection Point, Requiring Dedicated and Proactive Governmental Involvement to Progress

Chapter II. The Big Longevity Mindset Gap: Why Governments Are Focusing on the Problem of Aging Instead of the Opportunity of Longevity

- The Mindset that Aging Should Be Mitigated Versus the Mindset that Longevity Should Be Perceived As An Opportunity for Healthy Life Extension
- Understanding the Opportunities of Longevity
- Some governments are pursuing policies and programs to implement Healthy Longevity. However, while some of these actions serve to mitigate the consequences of population aging, others are less effective due to underfunding.

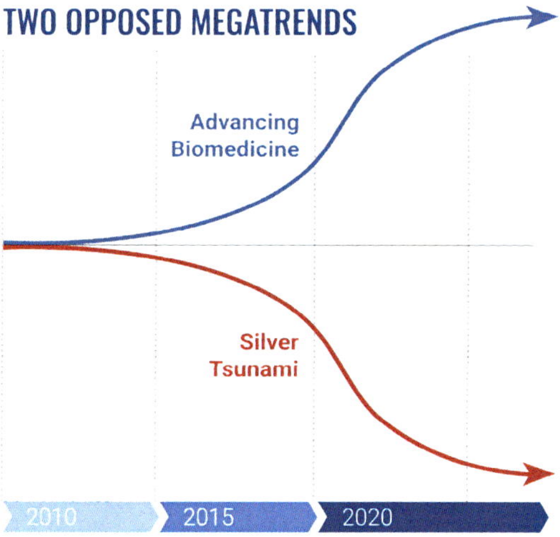

The acceleration of biomadicine has been mainly spurred by advances in the collection, gathering, and analysis of data. The industry is poised to witness a quantum leap in the near future, particulary because of the impact of Artificial Intelligence in biomedicine R&D and the upcoming paradigm shift from tratment to prevention.

At the same time, the inevitable Silver Tsunami (demographic aging) poses major economic burdens not just for the healthcare systems of developing nations but also for major financial institutions, including pension funds, insurance companies, asset management firms, and retail/private wealth banks. An increase in the costs associated with old age is expected.

Chapter III. The Rise of Longevity Technocracies

- The Concept of the Longevity Valley: The World's First Longevity-Friend Cities and Integrated Longevity Hubs
- Benchmarking Top Contenders (Cities and Municipalities) for Establishing the First Longevity Valley by 2025
- The Evolution of the Longevity Valley into Longevity State: The World's First Truly Progressive Longevity Nation, Prioritizing Human Longevity at the Center of their Industrial and Innovation Economy
- Benchmarking Top Contenders for Establishing the First Longevity State (AKA Longevity Nation) by 2030
- The Non-Linear Rate of DeepTech Progress Makes it Inevitable that Longevity Will Emerge as a New and Fundamental Political Asset Class of the 21st Century
- How Increased Health-Adjusted Life Expectancy (HALE) and Quality-Adjusted Life Expectancy (QALY) Will Become a Fundamental Citizen Right by 2030

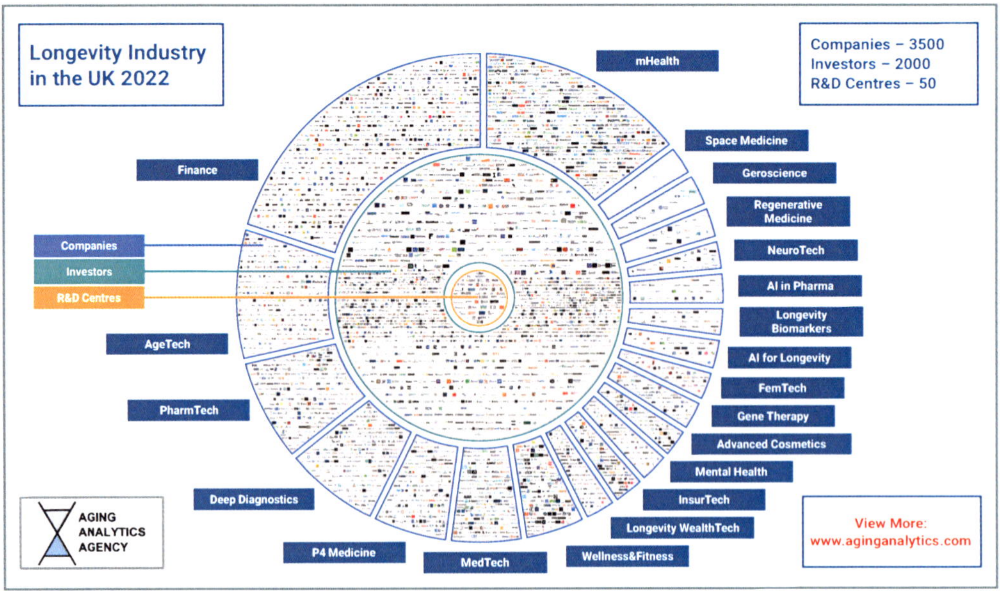

Chapter IV. Longevity Policy in Practice 2021: Progressive vs. Regressive Approaches to Longevity Governance

- Comparison Between Progressive and Regressive Approaches to Longevity Governance
- Comparatively Independent Municipalities or Sub-national Regions May Be Ideally Suited to Be Good at Strategy Formulation and Execution and We Can Expect to See Much Progress on This Front Between Now and 2030
- A Need for an Optimal Strategy in Regards of the Longevity Governance

Life Expectancy of Males and Females in 2019

Males		Females
82.2	Hong Kong	88.1
81.5	Japan	86.9
81.8	Switzerland	85.1
80.3	South Korea	86.1
80.7	Spain	85.7
81	Singapore	85.5
80.8	Israel	84.4
80.4	Netherlands	83.1
79.8	United Kingdom	83
75	European Union	82
76.3	USA	80.7
74.7	China	80.5

Chapter V. The Big Longevity Analytical Gap: Governments Lack the Systems and Technologies Required for Effective Planning, Forecasting and Decision Making

- Current Systems of Government Are Outdated and Ineffective
- The Need for True Big Data Analytics Systems with AI Components and Predictive Analytics
- The Need to Leverage the Power of AI and Its Predictive Analytics in Response to the Lack of Governmental Systems and Technologies

Chapter VI. Longevity Policy and Governance Analytics: The Need for Data-Driven Systems to Understand and Forecast the Complex Dynamics of Longevity Societies and Economies

- The Importance and Unprecedented Complexity of Longevity Policy and Governance Analytics
- Longevity Governance Big Data Analytics Dashboard as a Data-Driven System for the Complex Dynamics of Understanding and Forecasting Longevity

Chapter VII. Longevity Policy and Governance Big Data Analytical Dashboard (Deep Comparative Multiparametric Analysis of National Healthcare Systems)

- Overview of Longevity Governance Big Data Analytics Dashboard
- Big Data Analytics Application of 50 Countries and Recommendations for Their National Healthcare Systems

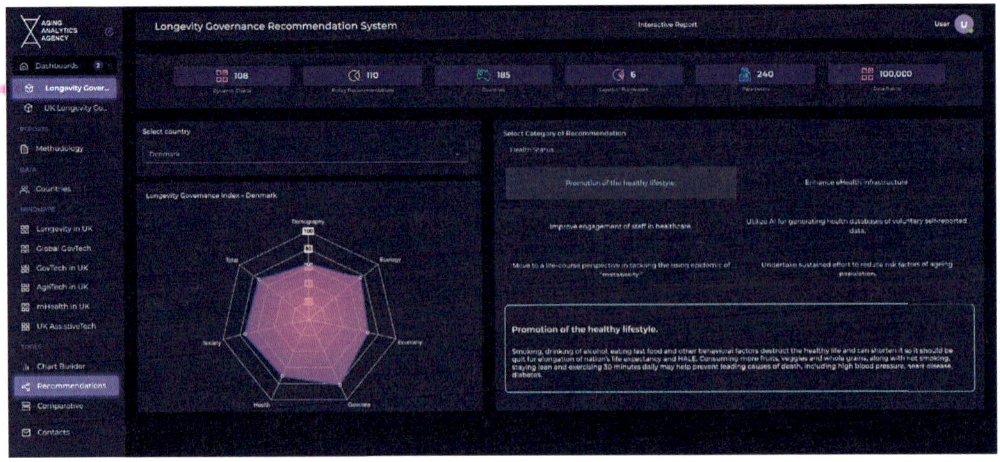

Chapter VIII. AI-Driven Analytical Systems for Semi-Automated SWOT Analysis, Actionable Long-term Forecasting, and Practical Recommendations for Good Governance

- SWOT Analysis on Ai Technology
- Good Governance Today Requires Ai-Driven Analytical Systems as Evidenced in Our COVID-19 Response
- Methodology for Long-Term Forecasts and Practical Recommendations for Good Governance
- AI-Driven Analytical Systems Can Have Profound Impact on Policy and Governance

Chapter IX. Semi-Automated Longevity Governance SWOT Analysis and Practical Recommendation System: UK Special Case Study and Proof of Concept

- The Methodology of the Semi-Automated SWOT Analysis Algorithm
- Visualizations of the Semi-Automated SWOT Analysis Algorithm

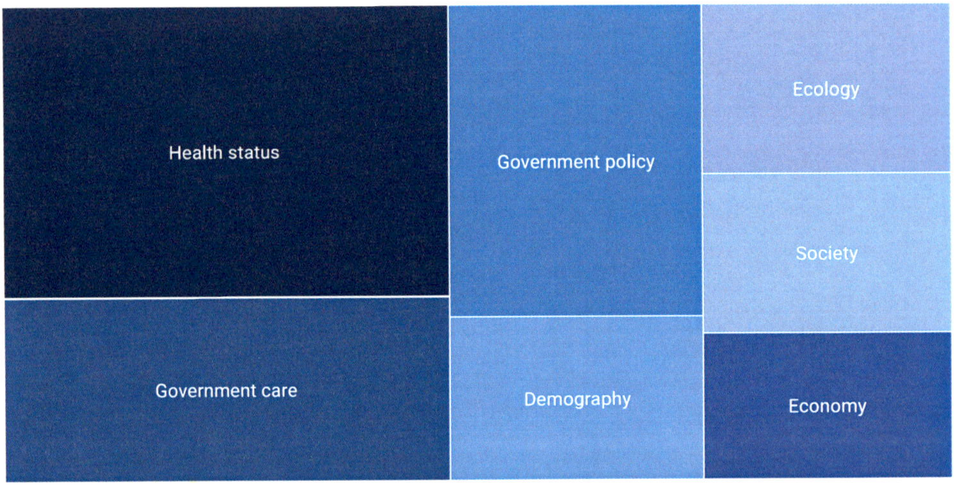

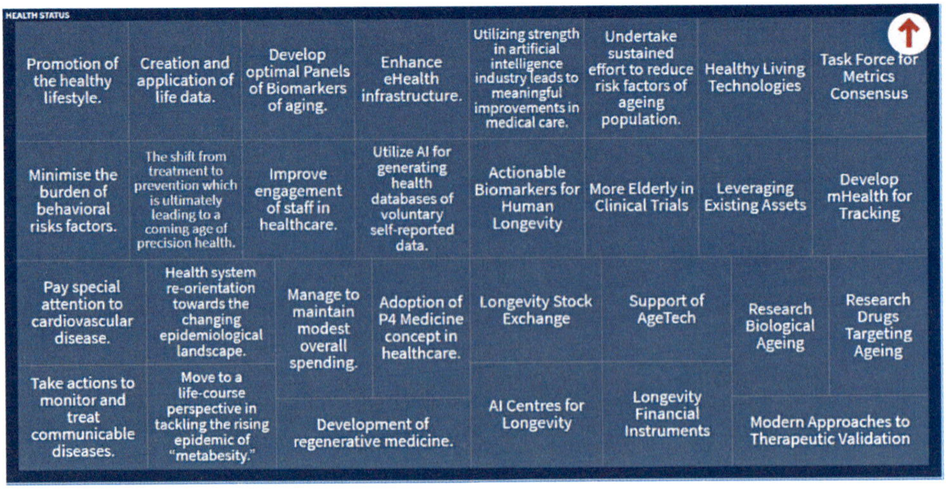

Chapter X. Semi-Automated Longevity Policy Analytics and Benchmarking: UK House of Lords and Healthy Ageing Industrial Strategy Special Case Study

- The Aging Population of the United Kingdom
- Vision of the House of Lords on the Problem of Aging in the UK
- Benchmarking of House of Lords Recommendations
- Benchmarking of Aging Analytics Agency Recommendations and a Brief Summary of Their Response to the House of Lords
- Comparison of Aging Analytics Agency Recommendations vs. Those Made by House of Lords Science and Technology Select Committee
- The Need for a National Strategy and Our Focus on Government Strategy
- Recommended List of Metrics for a National Longevity Strategy
- Establishment of Leading AI UK Centres for Longevity
- Launching the Task Force Focused on Initiatives to Reduce Psychological and Social Aging

House of Lords Recommendations

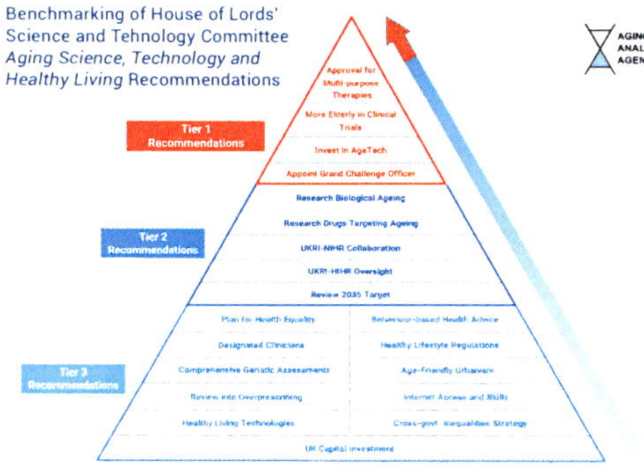

Aging Analytics Agency Recommendations

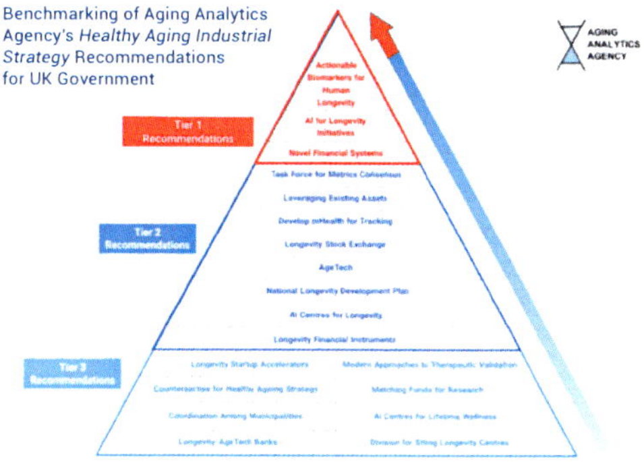

Chapter XI. Municipal Longevity Policy and Governance Analytics: Manchester (UK) Special Case Study

- Population Aging Is One of the UK's Four Grand Challenges in the Government's Industrial Strategy
- Manchester as an Age-Friendly City
- The Leading Longevity Companies and Investors in Greater Manchester
- Main R&D Longevity Hubs of Greater Manchester
- Trends and Prospects of the Longevity Industry in the Greater Manchester Area

2003	2009	2010	2016	2017
Manchester's ageing programme launched under the name of Valuing Older People	The city's ageing strategy, Manchester: A Great Place To Grow Older, is launched	Manchester becomes first UK city to join World Health Organisation Global Network of Age Friendly Cities and Communities	Department for Business, Energy and Industrial Strategy (BEIS) Founded	• Healthy Ageing Industrial Strategy Grand Challenge defined • Manchester age-friendly city strategy updated

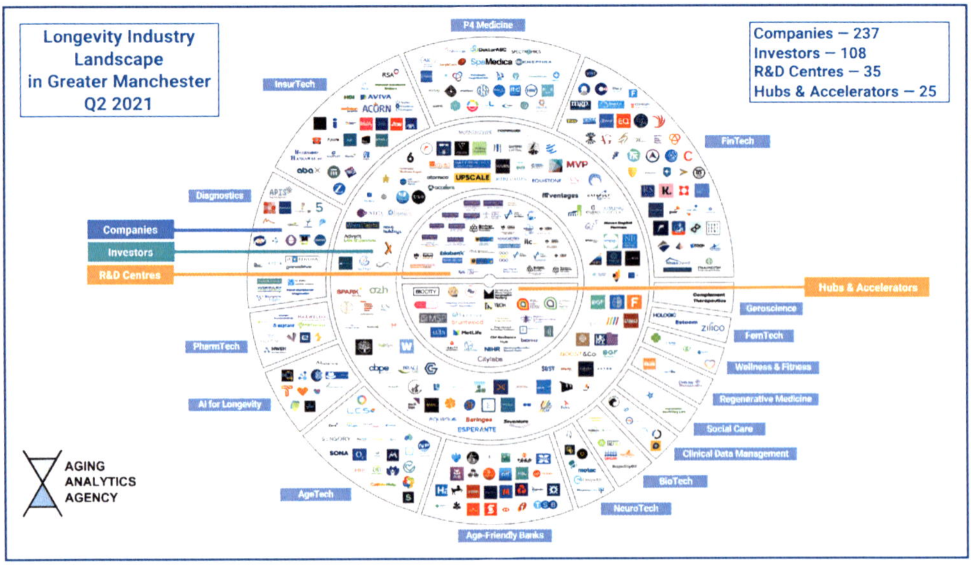

Chapter XII. The Big Longevity Hub Gap: Many Age-Friendly Cities, Zero Longevity-Friendly Hubs

- Age-Friendly Cities as a Way to Address Demographic Aging
- Age-Friendly Cities in South Korea, Japan, and Europe
- The Emergence of Longevity Hubs
- The Evolution from Age-Friendly Cities to Longevity-Friendly Cities

Age-Friendly Cities and Communities

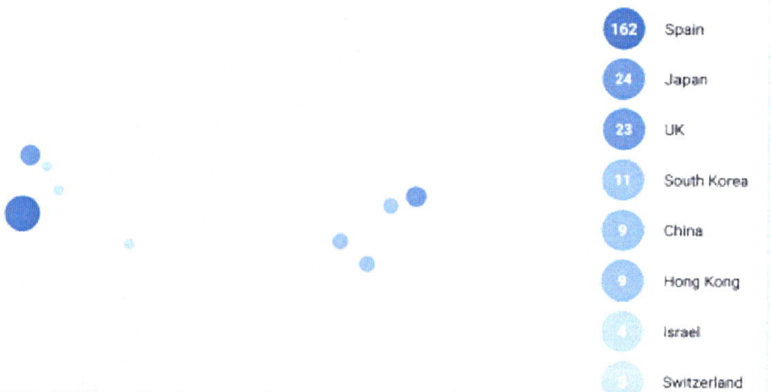

319 out of 833 of WHO age-friendly cities and communities are located in Europe. Among them, 162 are in Spain. Other countries, including ones located in Asia, contain significantly less number of WHO age-friendly cities/communities.

A number of these cities, such as Seoul in South Korea or Akita in Japan, are the products of detailed government master plans. Such master plans are recorded as instances of government initiatives in this document.

Source: WHO Global Network

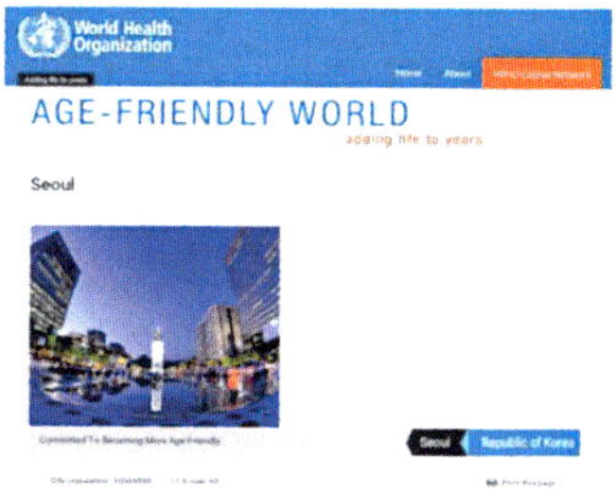

Chapter XIII. The Concept of the Longevity Valley: Full-Scope Longevity Industrialization to Combine the Health and Wealth of a Nation

- Comfortable Retirement vs. Empowered Lifelong Activity
- The Paradigm Shift of Health as New Wealth, and the Transition from Age-Friendly Cities to Longevity-Friendly Cities: Comfortable Retirement vs. Empowered Lifelong Activity
- Special Benchmarking Case Study: Top Contenders for Establishing First Longevity Valley by 2025 (Manchester, Geneva, Tel Aviv, Abu Dhabi, Hong Kong, California)
- Small Technocratic Municipalities Will Emerge as Global Competitors in the Race to Create the First Unified Longevity Ecosystem and True Longevity Valley by 2025
- How Longevity Will Become a Significant Factor Driving Municipal Company, Investor, and Hnwi Relocation; Inflow of External Investments; and Socioeconomic Prosperity

Chapter XIV. The Evolution From the Concept of Longevity Valley to the Concept of Longevity State

- The Rise of the First True Longevity State (Longevity Nation) by the Year 2030
- Full-Scope Longevity Industrialization (Biomedicine + Tech + Finance) is National Governments' Most Powerful and Efficient Tool for Boosting Economic Growth, Maintaining Socioeconomic Stability, Preventing Economic Decline, and Maximizing Societal Benefits and Dividends
- The Evolution of the Longevity Valley into Longevity State: The World's First Truly Progressive Longevity Nation, Prioritizing Human Longevity at the Center of Its Economic Strategy and Vision
- Special Benchmarking Case Study: Top Contenders for Establishing the First Longevity State by 2030 (Switzerland, Singapore, Israel, United Kingdom, and Japan)

Specialized Government Led Longevity Initiatives

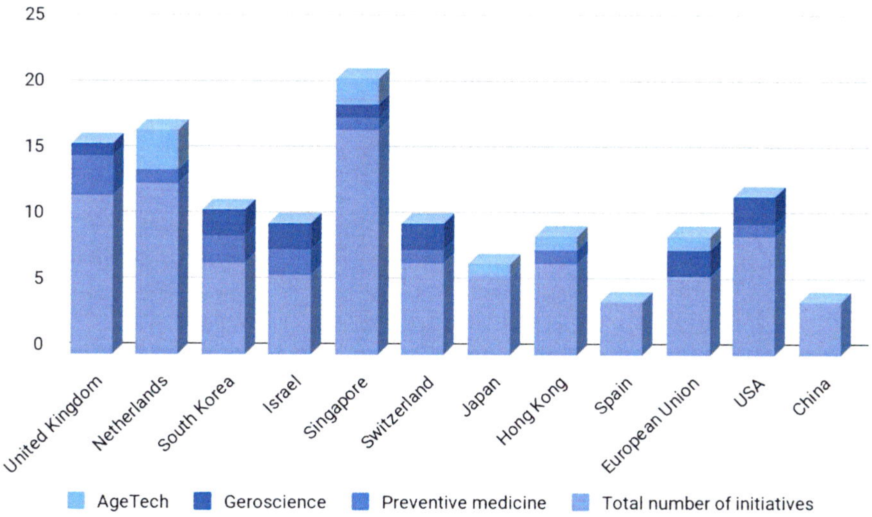

Chapter XV. The Current State of Longevity Policy (Theory) in 2021: Global Benchmarking

- Updated Global Overview and International Benchmarking (Ranking) of National Longevity Industrial Strategies, Policies, and Initiatives 2021
- Bridging the Gap Between Theory (National Longevity Plans and Industrial Strategies) and Practice (Tangible Optimization of National Healthy Longevity)
- The Longevity-Focused Vision and Ambition of National Governments Has Increased Dramatically
- Governmental Will and Commitment Are Secured in an Increasing Number of Progressive Nations
- Governments Remain Severely Limited by a Lack of Modern Approaches for Strategy Execution
- Data-Driven Approaches to Optimize the Practical Execution of National Longevity Policy Are Still Required

Number of Initiatives Foucused on Longevity

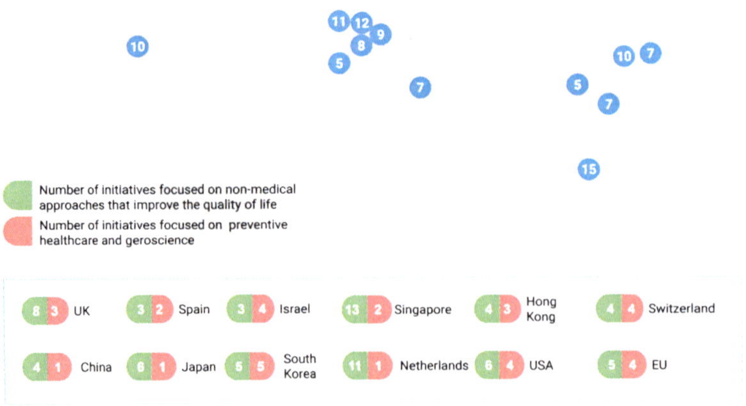

Chapter XVI. The Current State of Longevity Politics (Practice) in 2021: Global Benchmarking

- Updated Global Overview and International Benchmarking (Ranking) of Progress in Optimizing National Longevity, Health-Adjusted Life Expectancy, HALE Gap, and Quality-Adjusted Life Years
- Benchmarking (Ranking) of Region-Specific Success Rates in Translating Longevity Development Plans and Industrial Strategies (Theory) into Practical Results (Reality)
- Tangible Quantitative Frameworks for Improving the Execution of National Longevity Plans and Industrial Strategies: Data-Driven Approaches to Realistic Strategy Formulation, Tangible Progress Tracking, Execution, and Maximizing Societal Dividends
- Forecasting the Future State of Longevity Politics in 2025 and 2030

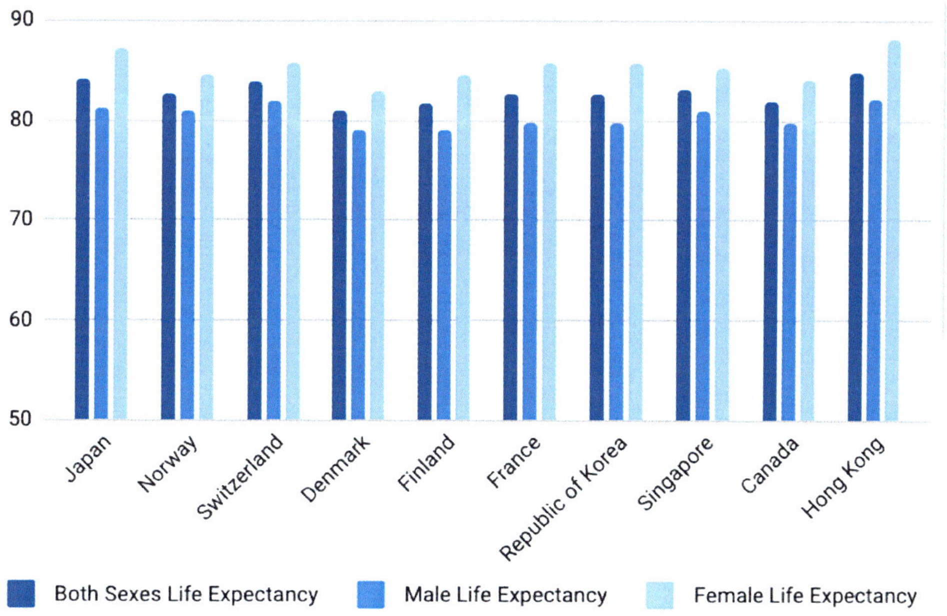

Chapter XVII. Regional Spotlight: Top Challenges and Opportunities for Longevity Policy in the UK

- The UK as the First Government to Prioritize National Healthy Longevity and Longevity Industrialization as Central to Its Economic Vision
- The UK Has the First Necessary Component of a Longevity Progressive Nation: Political Will and Government Commitment
- Ageing Society Grand Challenge Industrial Strategy Fund
- All-Party Parliamentary Group for Longevity
- 2035 Goal (Adding 5 Extra Years of Healthy Longevity to All UK Citizens by the Year 2035)
- The UK Also Has the Raw Resources Necessary to Fulfill its Vision: Competitive Longevity Industry Landscape, Prevention-Focused Healthcare System, Strong Position in International AI Race, Strong Financial Sector, and High GDP)
- Data-Driven Approaches to Translate the UK's Plan (Vision) into Practice (Reality)

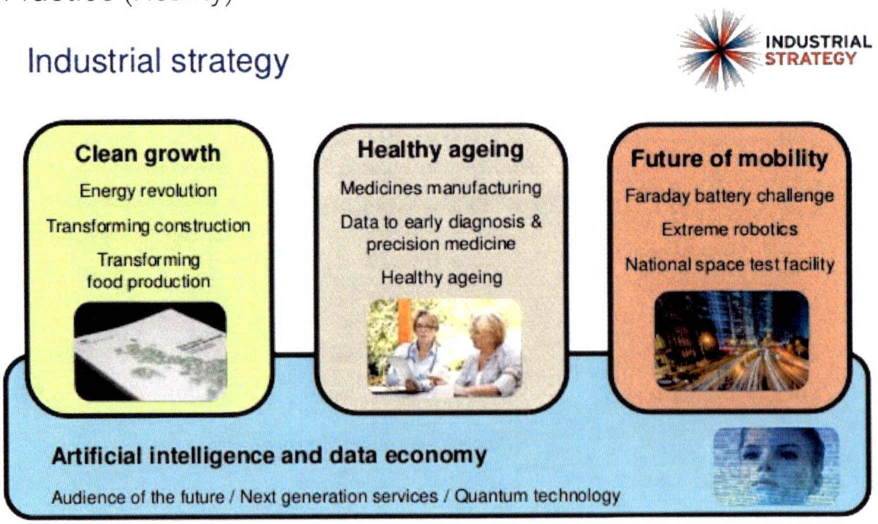

Chapter XVIII. Regional Spotlight: Top Challenges and Opportunities for Longevity Policy in the USA

- The United States Is the Leading Example of a Nation with the Untapped, Uncaptured, and Unrealized Potential to Become the First Longevity State by 2030.
- The USA's Raw Resources to Become More Longevity-Progressive Than Any Other Are Its:
  - High GDP;
  - Competitive Economy;
  - Robust Financial Sector;
  - Large Pool of Investment Capital;
  - Leading Position in the Global AI Race; and
  - Possession of the Largest Proportion of Longevity Start-ups, Corporations, Investors, R&D Hubs, Non-profits, Labs, and Scientists.
- But The United States Lacks the First Fundamental Component of a Longevity State: Top-Level Government Recognition of Longevity as Central to Its Political, Healthcare, and Federal Economic Agenda.
- The USA Can Only Become Longevity-Progressive via a Complete Strategic and Political Overhaul, Combined with a Strong Recognition of Longevity Prioritization by the Federal US Government.

Chapter XIX. Regional Spotlight: Top Challenges and Opportunities for Longevity Policy in Europe

- Europe at a Crossroads with Growing Economy and Accelerated Aging Population
- Can Europe Create the World's First Multi-Government International Longevity Development Plan, Leveraging Cross-nation Coordination for Rapid Strategy Implementation and Execution by Uniting Strengths and Neutralizing Disadvantages?
- Key Longevity Hotspots and Hubs in Europe: Switzerland, Liechtenstein, Germany
- Key Opportunities and Challenges for Europe at the Intersection of Aging and Longevity
- Can Europe Utilize Cross-Border Synergy and Cooperation to Become the Most Longevity-Progressive Continent by the Year 2030?

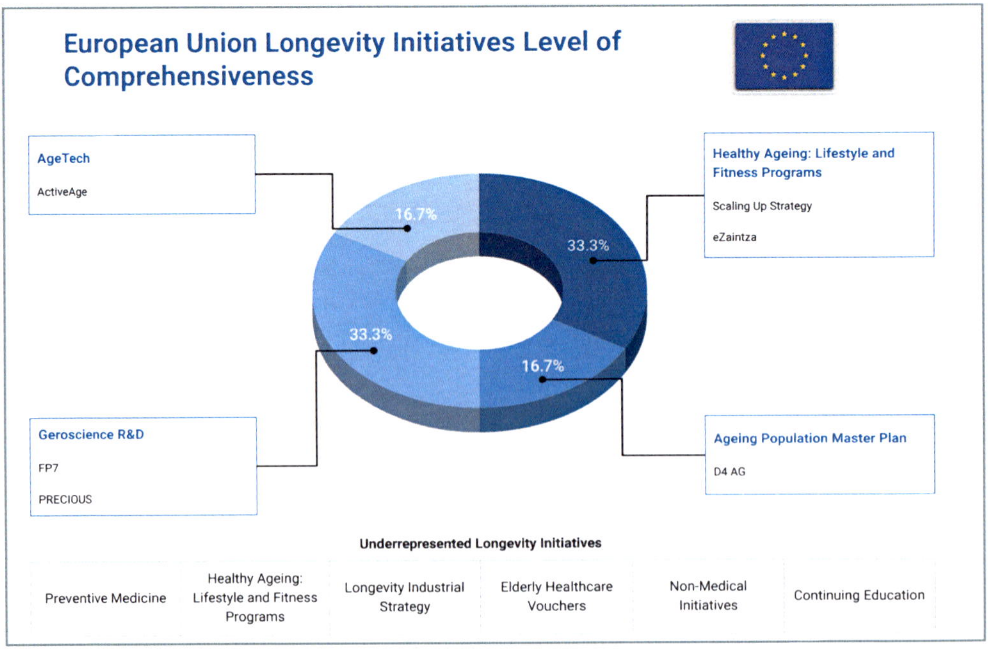

Chapter XX. Regional Spotlight: Top Challenges and Opportunities for Longevity Policy in Asia

- Asia's Unique Challenges: Aging Population and Wide Disparities Between Most Developed and Least Developed Regions and Territories in the Asian Continent
- Asia's Unique Opportunities: High Degree of DeepTech Commoditization and Societal Comfort with Emerging Technologies; Competitive Levels of Digitization and Innovation Economy; Strong Focus on Preventive Health and Digital Health
- Asia as the "7th Continent" and Epicenter of "Silver Wealth" (Greatest Proportion of Elderly Wealth in the World), and Its Basis as the Longevity Finance Hub of 2030
- Why the Future of Longevity Might Be Asian

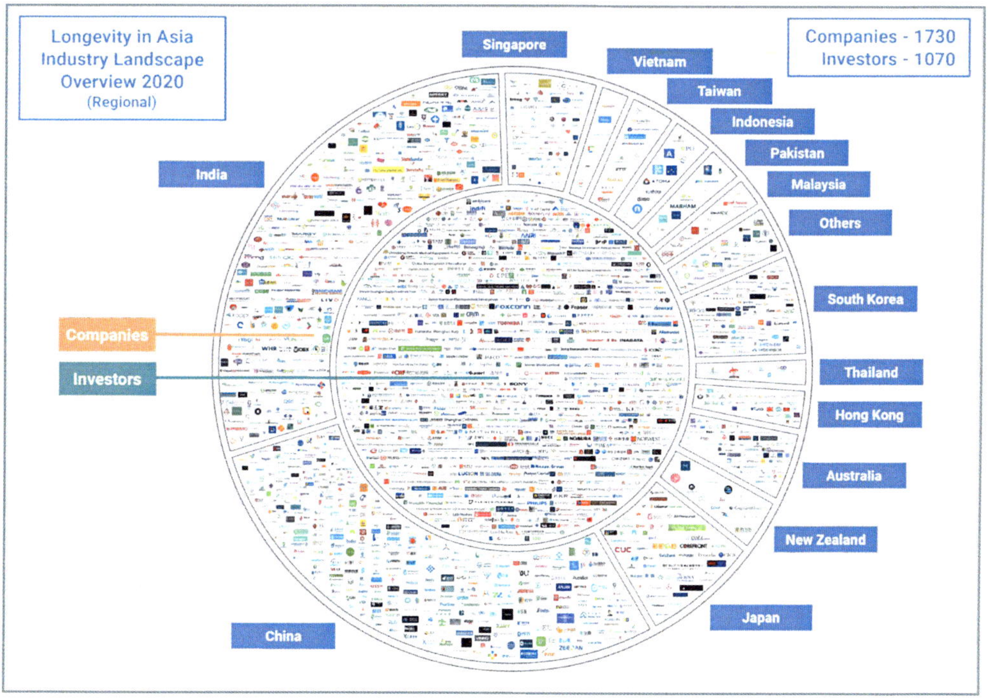

Chapter XXI. Regional Spotlight: Top Challenges and Opportunities for Longevity Policy in North America

- North America Is Home to the Most Significant Proportion of Longevity Start-ups, Corporations, Investors, Financial Institutions and R&D Hubs, Yet Has One of the Lowest Levels of Longevity Progressiveness
- An Overabudance of Raw Resources Gives North America the Chance to Establish the First Longevity State, But There Is a Complete Absence of Will and Recognition of This Among North American Governments
- How a Data-Driven and Tangible Analytical Overhaul of North America's Political Agenda, Economic Vision and Governmental Strategic Priorities Could Allow It to Become the First True Longevity Continent by the Year 2030

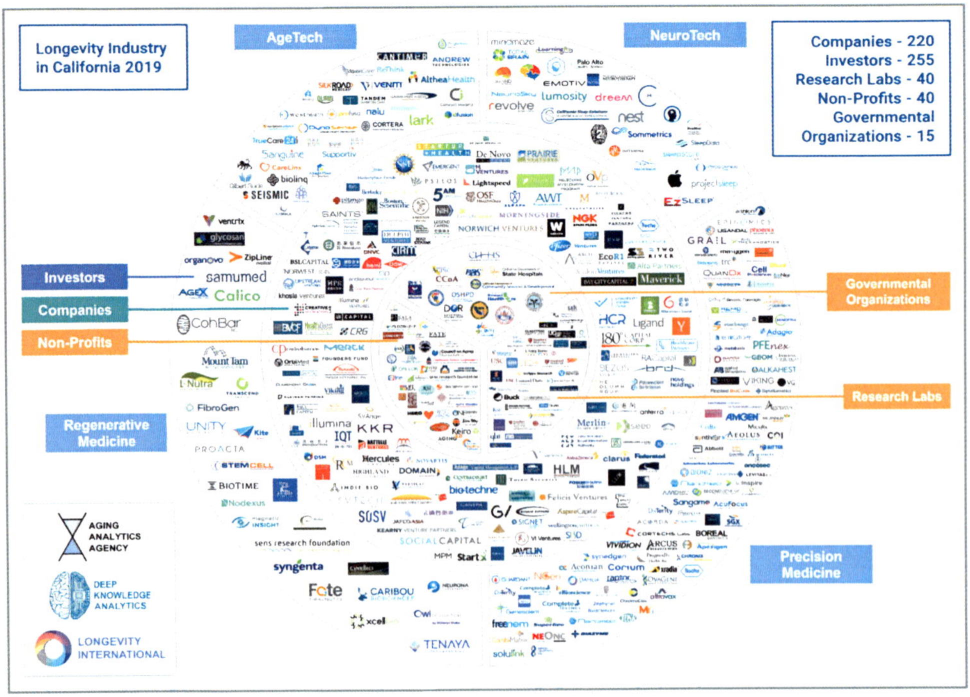

Chapter XXII. Regional Spotlight: Top Challenges and Opportunities for Longevity Policy in the Gulf and MENA Region

- Comparatively Lower Levels of Population Aging Mean the Gulf and MENA's Window of Longevity Opportunity Is Large
- High Incidence of Non-Communicable Diseases (NCDs) Among Population (and Diabetes in Particular) Creates National Population Health Deficit
- Large Economic Reserves, Shift Toward Digital and Innovation Economy Sectors, and Strong Prioritization of AI Place Gulf and MENA Territories at Strategic Advantage
- Major Focus and Support of Concept of "HealthTech Smart Cities" Could Allow for Easy and Rapid Transition to AI-Driven Longevity Valleys as Soon as 2025

Chapter XXIII. Regional Spotlight: Top Challenges and Opportunities for Longevity Policy in Latin America

- Population Aging in the Latin American Region: Is There a Window of Opportunity or Do Those Regions Need to Hurry?
- The Biggest Issues of the Latin American Region That Keep It From Becoming the Next Longevity Valley.
- What About Economic Readiness and Longevity Awareness?
- Is There Any Solution to the Region's Major Issues We at Aging Analytics Agency Can recommend?

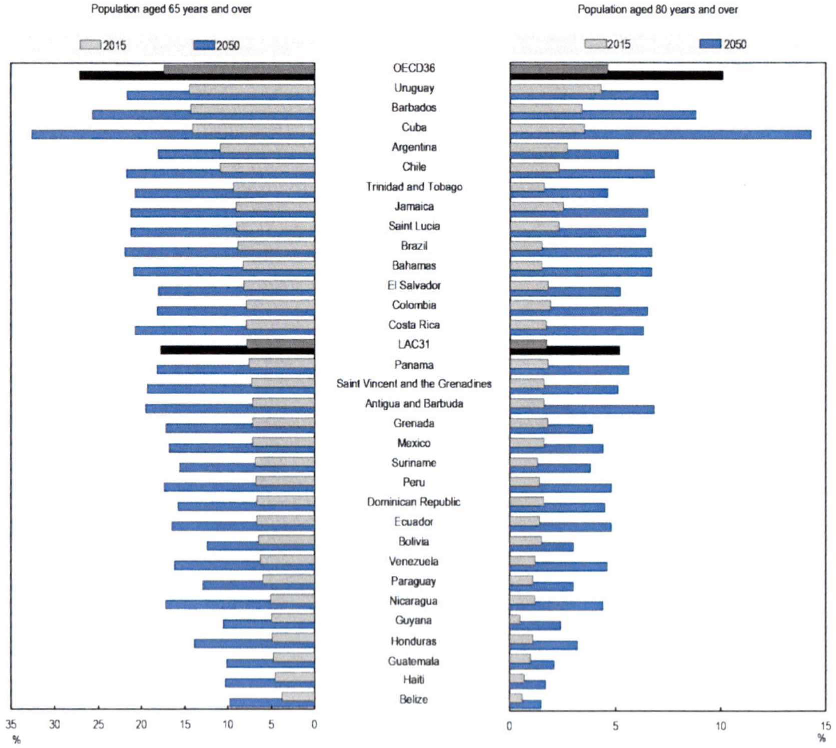

Source: UN World Population Prospects 2019[1]

[1] www.oecd-ilibrary.org/sites/6decc105-en/index.html?itemId=/content/component/6decc105-en#figure-d1e18923

Chapter XXIV. National Case Studies: Singapore (Governance), Switzerland (Finance), Israel (Technology)

- The World's First Truly Progressive Longevity Nation, Prioritizing Human Longevity at the Center of its Economic Strategy and Vision
- Singapore as a Tiger Economy with Diverse Government Programs and a Strong Focus on Continuing Education for the Elderly and AgeTech
- Switzerland's Strong Bioscience Initiative, and Diverse Policies for the Elderly in a Small Geographic Area
- Israel's National Master Plan on Aging, Strong Political Support for the Elderly, and Joint Geroscience Research Initiative with the UK

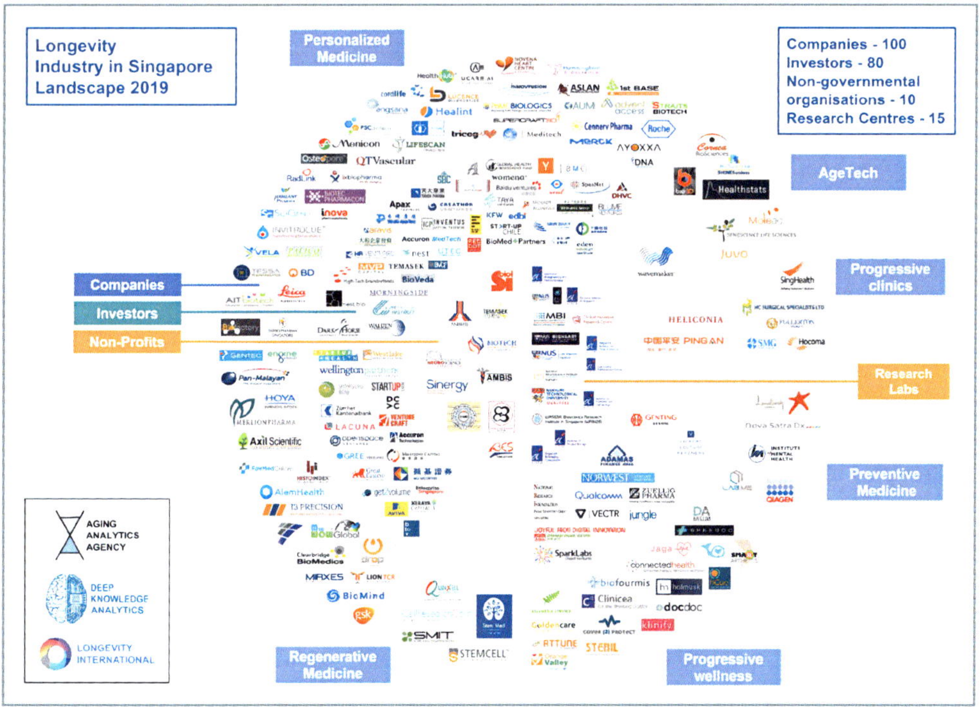

Chapter XXV. Longevity Ethics: How Potential Negative Societal Outcomes Resulting from Longevity Can Be Proactively Optimized to Ensure Global Longevity Harmonization

- Agism and AgeTech: the Elderly as the Most Politically Powerful but also Most Underrepresented, Unsupported and Marginalized Demographic of the Electorate
- Longevity Industrialization: An Unprecedented Challenge and Revolutionary Opportunity
- Participation of Senior Citizens in Policy-Making
- Longevity and Inequality: How Longevity Will Maximize QALY for all Demographics
- Lifelong Support and Maintenance of Health, Wealth and QALY Across Full Lifecycle Will Become a Basic Human Right Demanded by Citizens
- Governments Should Say No to War and Yes to Longevity
- Mitigating Detrimental Logistical Effects: Overpopulation, Famine, Pollution and Climate Change, Etc.
- Frameworks for Simultaneously Neutralizing Longevity-Related Challenges and Maximizing Longevity-Derived Dividends for Society
- Human Will and Ingenuity Can Refashion the Challenges Posed by Longevity into Opportunities

AgeTech Industry Market Capitalization

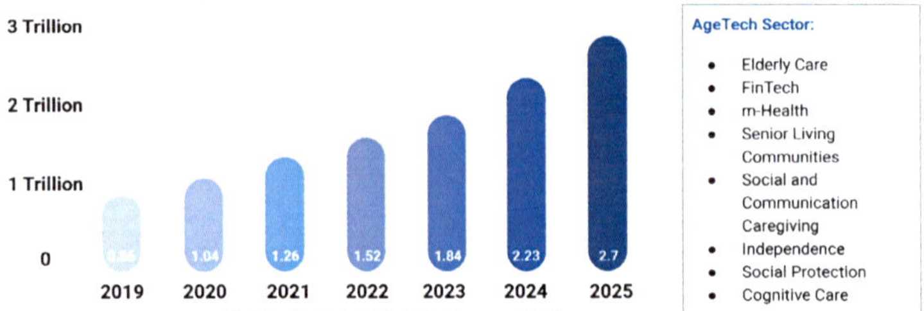

World AgeTech Industry Size Projections, current US $

Chapter XXVI. Health of Nations at the Intersection of Longevity and Pandemics

- Convergences Between Longevity-Progressiveness and COVID-19 Safety
- Is There a Correlation Between Longevity-Progressiveness and Pandemic Readiness?
- Longevity, Biodefense, Pandemics, and Health as New Wealth
- Intersections of Longevity and Biodefense Threats: Enhanced Immunity as a Critical Factor in Both Healthy Longevity and Preventive Defense Against Pandemics
- Aging Population and the Critical National Bottleneck of Demographic Groups at Extreme Risk of Pandemic-Related Illness, Prolonged Disease and Mortality
- COVID-19 Is Fueling BioTech and HealthCare Industry Growth and Investments, and Driving Recognition by Governments of Population Health as a Major National Asset
- COVID-19 Created a Manhattan Project Mindset for Governments, and Accelerated Practical Action; Why Aging Population and Longevity Need the Same Mindset Now

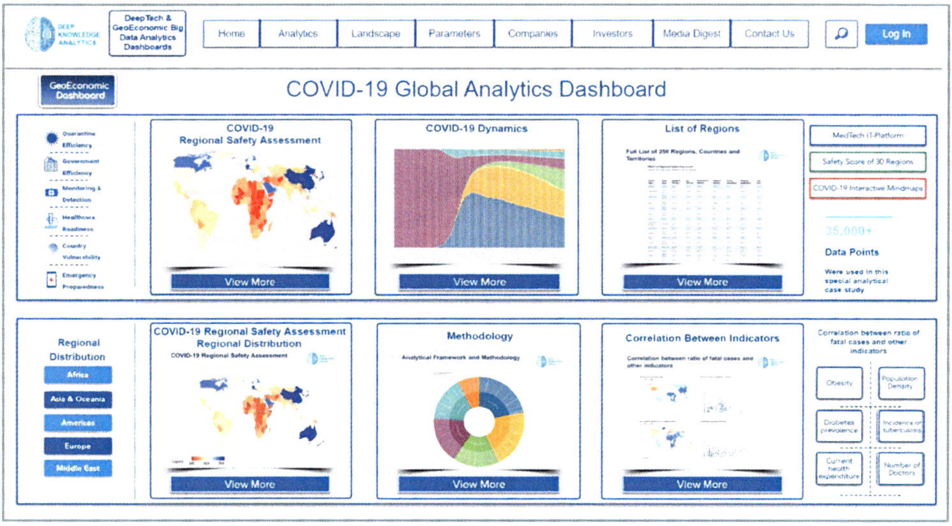

Chapter XXVII. FemTech, Female Health and Simplistic Competition (20th Century) vs. Sophisticated Competition (21st Century)

- The Rise of FemTech
- The Current State and Future of the FemTech Market
- Female Entrepreneurs Are Concerned About Equality
- The Gender Gap Issue
- FemTech Longevity

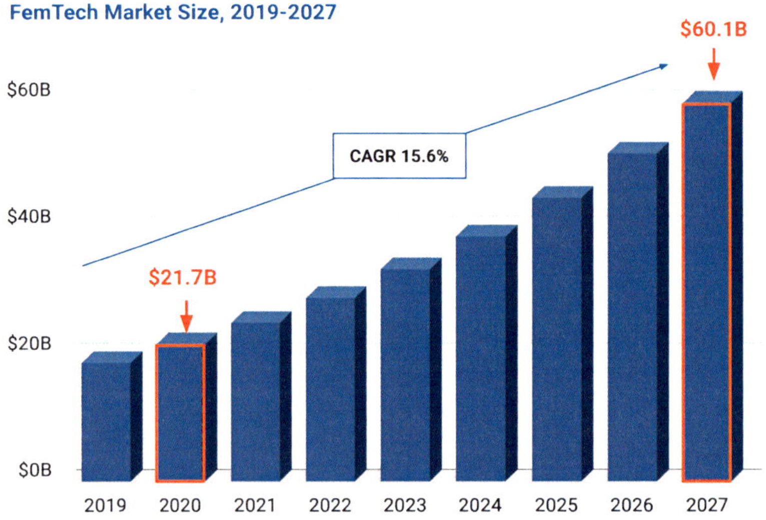

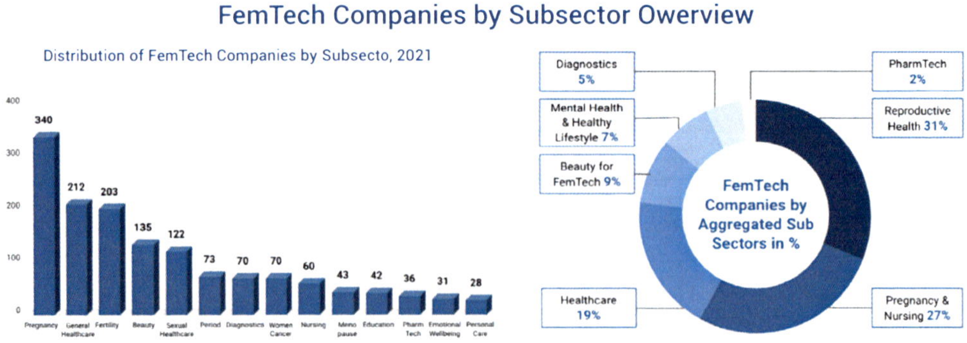

Chapter XXVIII. National Longevity Policy Platform Case Study: The Blueprint for and Actionable, Tangible and Responsible National Longevity Development Plan

- The Purpose of Longevity Envisioned and the World Engulfed in Longevity Revolution
- A Nation Molded and Led by the Best Longevity Paradigms and Industrial Strategies
- A Declaration of a World Where Longevity is at the Center of Politics

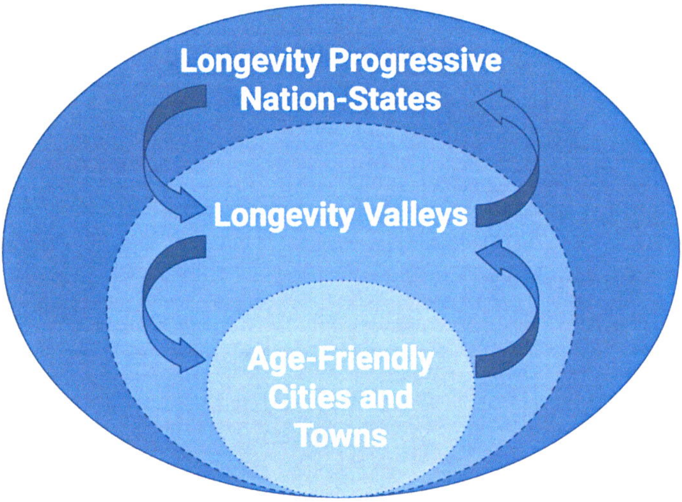

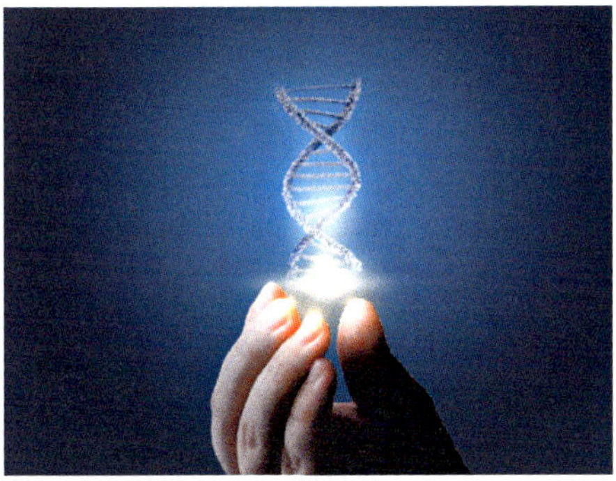

Chapter XXIX. Modern GovTech Driven Solutions for the Paradigm Shift From "Healthcare" (Reaction Sick-Care) to Precision Health (Prevention Health-Care)

- Overview of Healthcare Systems Around the World
- The Rise of GovTech and its Impact on Healthcare

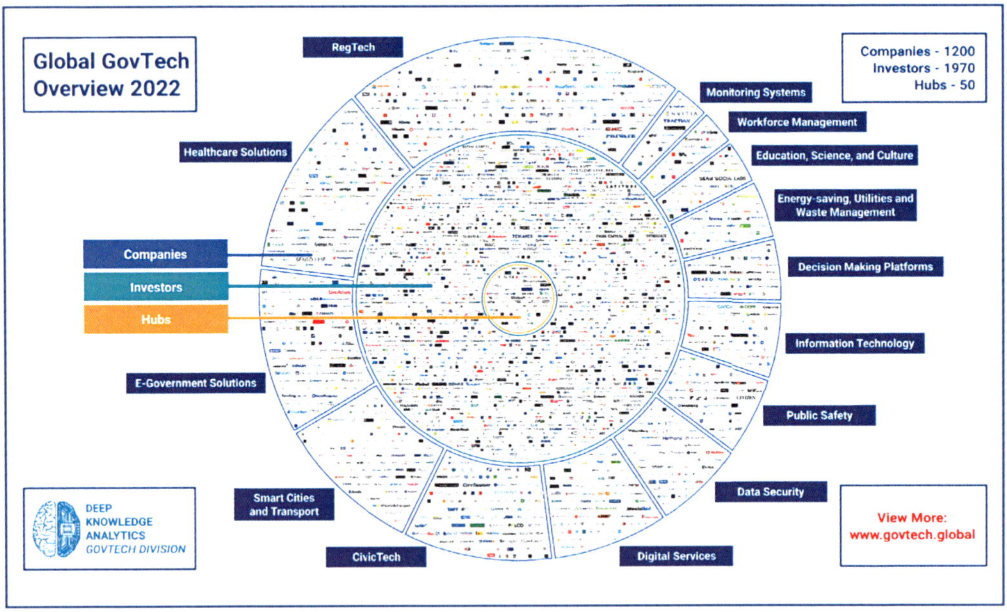

Source: www.analytics.dkv.global/govtech/govTech-global-mindmap.pdf

Chapter XXX. How Longevity Will Define and Decide the Fate of Nations in the Next 10 Years

- Longevity as a Fundamentally New Type of Political Asset Class at the Intersection of National Health and Wealth, the 5th Industrial Revolution, and National Economy 2.0
- Why Citizens Will Demand Proof of Added HALE and QALY From Their Politicians by the Year 2030
- How Longevity Will Become a Fundamental Pillar of National Economies and the Futurization Agenda of Developed Nations Starting From 2025
- Longevity Opportunities and Challenges Facing Developed Nations and Mega-Economies
- Longevity as the New Political Priority for the 21st Century

Source: www.longevity-book.com/politics

Appendix E

Longevity Industry 2.0

Book Teaser

DeepTech Engineering The Accelerated Trajectory of Human Longevity: The Blueprint and Pathway from Longevity Industry 1.0 to 2.0

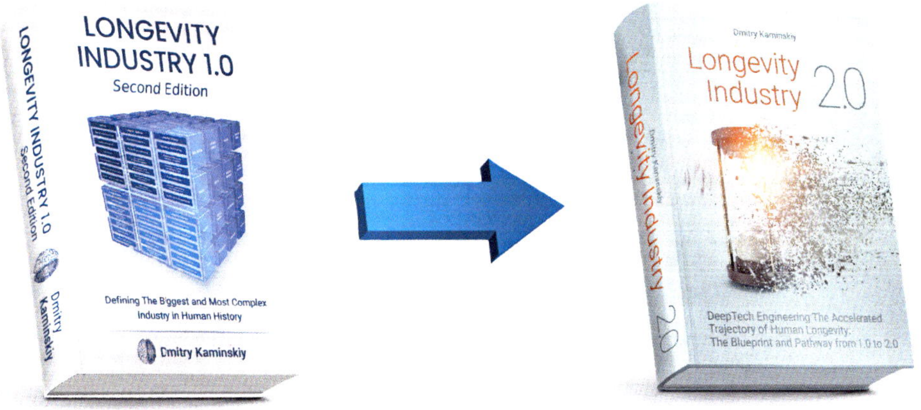

2010-2023: Evolution of the Longevity Industry from Zero to 1.0

- The Industrialization of Longevity
- The Current State of Longevity Science, Business, Finance, and Practical Applications
- Longevity Becomes National Priority Item for the Strategic Agenda of Progressive Governments
- Transforming the Challenge and Deficit of Aging into the Opportunity and Asset of Longevity
- Defining and De-Risking: Hype vs. Reality

2023-2027: DeepTech Engineering The Accelerated Trajectory of Human Longevity – The Blueprint and Pathway from 1.0 to 2.0

- Evolution from 1.0 (Defining the Industry) to 2.0 (Applying the Framework for Longevity Industrialization, Investment De-Risking and National Longevity Economy Development)
- The Evolution from Longevity Start-ups to Multi-Trillion Dollar Longevity Corporations
- Future of Longevity Finance: Financial Instruments to Bridge the Longevity Liquidity Gap
- The Rise of Longevity Technocracies and the Concept of "Longevity Valley"
- Engineering Market-Ready Pathways for Practical Human Longevity by 2030

Chapter 1. The Full-Scope Implementation of Longevity Industrialization: The Evolution from Industry 1.0 *(Defining the Industry Framework)* to Industry 2.0 *(Applying the Framework for Industry Harmonization and Investment De-Risking)*

- Forecasting the Full-Scope of Longevity Industrialization and the Achievement of Longevity Escape Velocity by 2030
- The Acceleration of Longevity Industry Development: Longevity's Explosive Exponential Growth, Cross-Sector Integration and Integral Technological Fusion
- DeepTech Engineering the Practical Pathway From 1.0 to 2.0
- Projecting Longevity Industrialization as the Natural and Necessary Pathway to the Fifth Industrial Revolution
- Technological Progress Alone, While Extreme, Will Not Be Enough To Achieve Longevity Escape Velocity by 2030
- Major Challenges, Opportunities and Recommendations for Longevity Companies, Investors, Progressive Governments and Policy-Makers

First Comprehensive Interactive Database of Global Longevity Industry Ecosystem

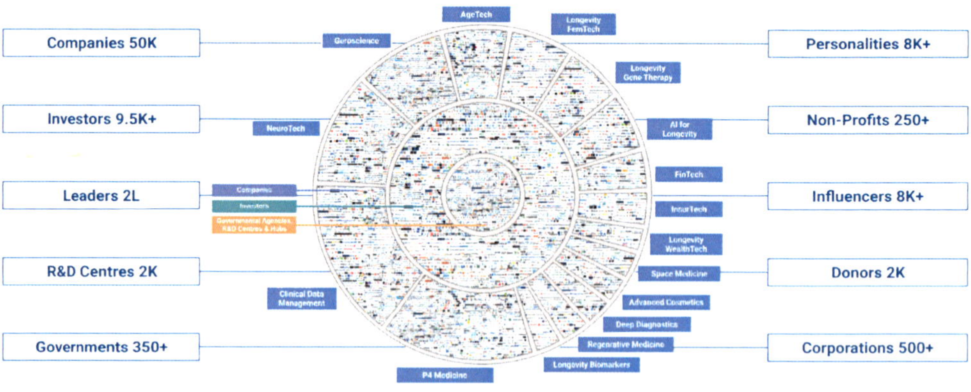

Source: www.longevity.international/global-longevity-ecosystem

Chapter 2. Reinforcement of the Longevity Industry Framework: Self-Evolving Data Science To Define and Forecast the Mega-Complex Longevity Industry in Real-Time

- The Intensification of Longevity Industry Mega-Complexity (From Natural Selection to Natural Competition): The Necessary Co-Evolution of Industry Analytics and the Exponentially Evolving Complexity of the Industry Itself
- How Increasing Sector Intersectionality and Technological Fusion Presents Even Greater Challenges to Reliable Forecasting Than Longevity Mega-Complexity
- The Evolution of Deep Knowledge Group's Longevity Analytics: The Results of a Decade of Analytical Self-Development (2013-2023)
- AI and Data Science as the Central Catalyst for Longevity Forecasting, Analytics and Benchmarking Frameworks Keeping Pace with Industry Diversification
- Overview of Deep Knowledge Group Longevity Industry Big Data Analytical AI-Driven Systems and Dashboards
- Projecting the Transition from Longevity Industrialization to the 5th Industrial Revolution
- Using Tangible Longevity Industry Analytics to De-Risk Investments, Execute Relevant Industrial Strategies and National Development Plans, and Achieve Tangible Results in Practical Human Longevity by the Year 2030

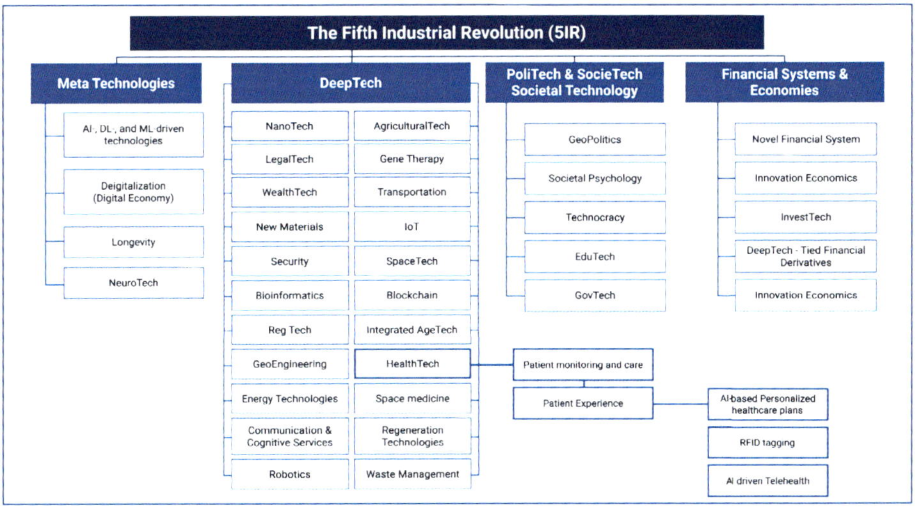

Source: www.5revolution.tech

Chapter 3. Biomarkers of Human Longevity as the Systemic Foundation for Longevity Industry 2.0: Tangible Metrics for Aging Research Acceleration, Industry Stabilization and Maturation, Practical Human Longevity, Tangible Investment Decision Making and De-Risking

- Biomarkers of Human Longevity as the Systemic Foundation of Any Frameworks Designed to Forecast, Optimize, Stabilize, and Actualize the Evolution of Longevity Industry 2.0
- Modern Tools for Investment De-Risking and Accelerating R&D
- Defining Hype vs. Reality and Human-Centered Practical Applications in the Longevity Industry
- Longevity Economy 2.0: Biomarkers of Human Longevity as Major Catalyst To Increase Investments and Ensure Sustainable Growth of Longevity Finance
- Tangible Metrics and Frameworks for Company Valuation and Due Diligence and the Paradigm Shift from Model Organism (Mouse) Validation to Human Validation
- DeepTech Engineering Toward the Concept of Digital Human Avatar 1.0 - 4.0

Humanity Has Reached the Era of the Social Genome

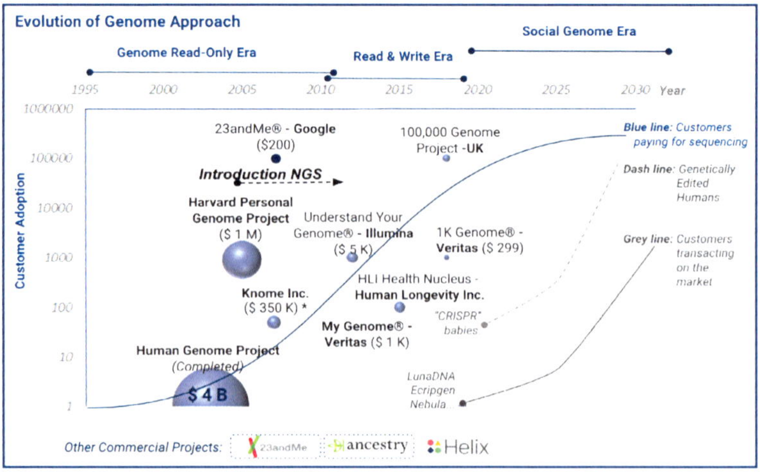

Source: www.longevity.international/biomarkers

Chapter 4. Longevity Science: AI and Data Science for Accelerating Aging Research and R&D

- Forecast on Major Developments in Longevity Science: 2025, 2027 and 2030
- Top Underleveraged Opportunities in Longevity Science: 2023-2030
- Projecting Major Roadblocks in Bottlenecks in Longevity Science: 2023-2030
- Modern approach to Longevity Clinical Trials
- Frameworks, Strategies and Approaches to Neutralize Critical Threats for the Future of Longevity Science
- DeepTech Engineering the Accelerated Trajectory of Longevity Science
- AI as the Major Accelerative Factor for Longevity Science R&D

Artificial Intelligence Framework

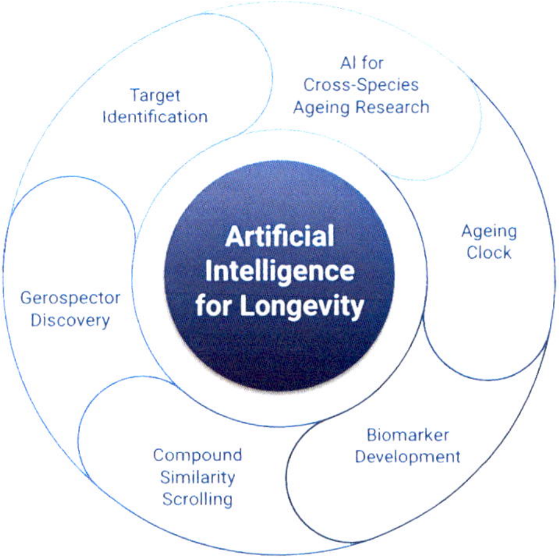

Source: www.longevity.international/global-longevity-ecosystem

Implementation of Artificial Intelligence in 2022

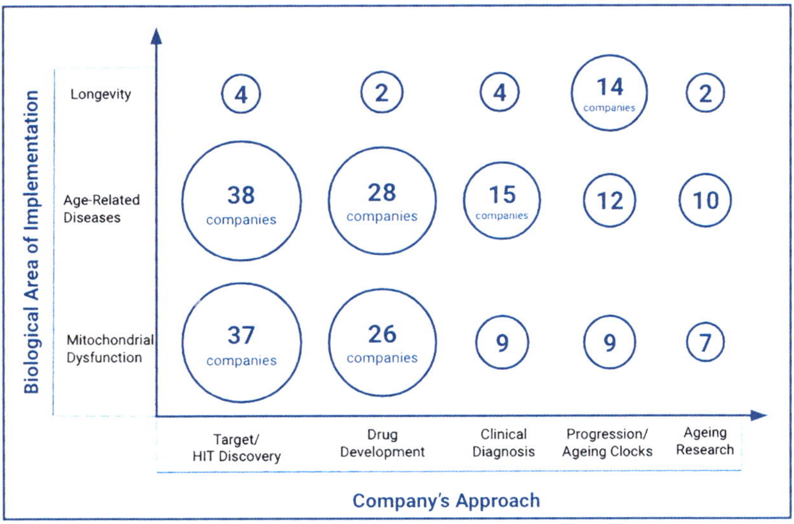

Chapter 5. Longevity Medicine: Longevity-as-a-Service, and From Precision Medicine to Precision Health

- Longevity-as-a-Service
- From Precision Medicine to Precision Health
- AI as the Major as the Major Doctor in the Longevity Clinic
- Forecast on Major Developments in Longevity Medicine: 2025, 2027 and 2030
- Top Underleveraged Opportunities in Longevity Medicine: 2023-2030
- Projecting Major Roadblocks in Bottlenecks in Longevity Medicine: 2023-2030
- Frameworks, Strategies and Approaches to Neutralize Critical Threats for the Future of Longevity Medicine
- DeepTech Engineering the Accelerated Trajectory of Longevity Medicine

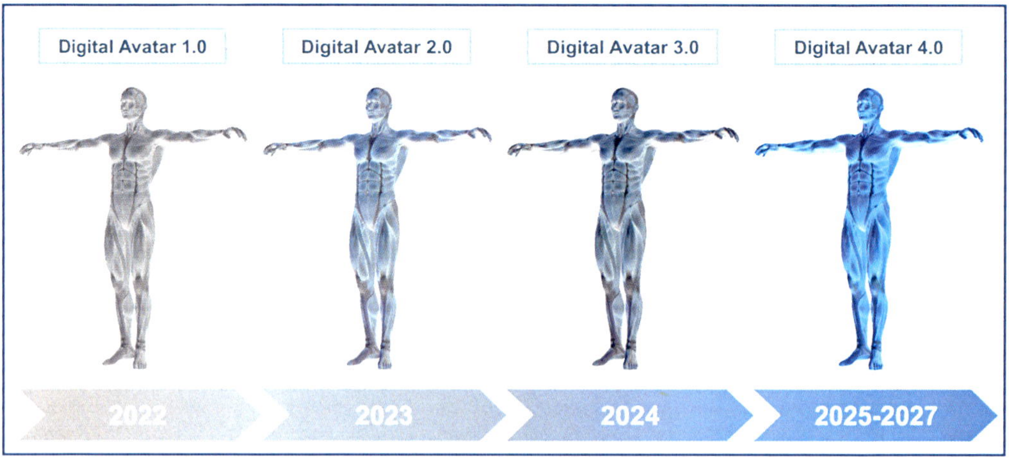

Source: www.longevity-book.com/biomarkers

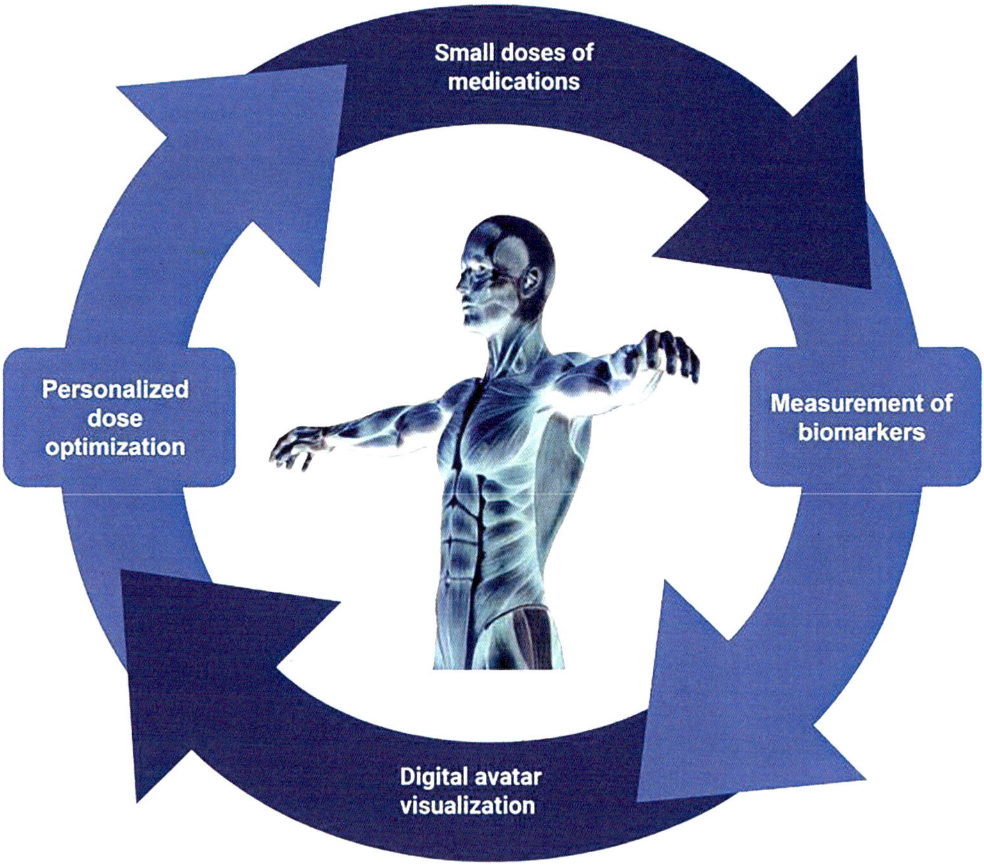

Source: www.data.longevity.international/biomarkers-landscape-overview-2021.pdf

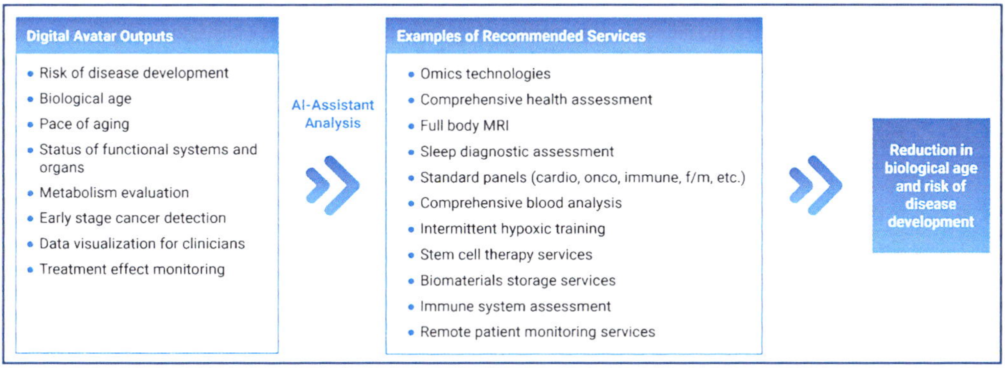

Chapter 6. Longevity AI: The Fundamental Role of AI and Data Science for the Future of Longevity Industrialization

- Why Artificial Intelligence is Required for Managing Longevity Mega-Complexity
- AI as the Ultimate Accelerative Factor in Advanced Biomedicine and Aging R&D
- The Fundamentally Necessary Role of AI in Practical Longevity: Biomarkers of Human Longevity and The Evolution of Digital Avatars from 1.0 to 4.0
- AI in Longevity Finance: Integrated AgeTech, Wealthtech and HealthTech Solutions
- AI in Longevity Governance: Data-Driven National Longevity Development Plans

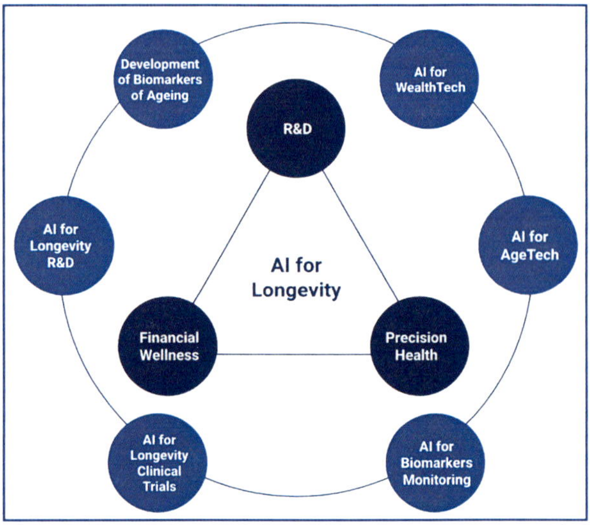

Source: www.longevity.international/global-longevity-ecosystem

Key Areas of Artificial Intelligence: From Longevity R&D to Longevity Medicine

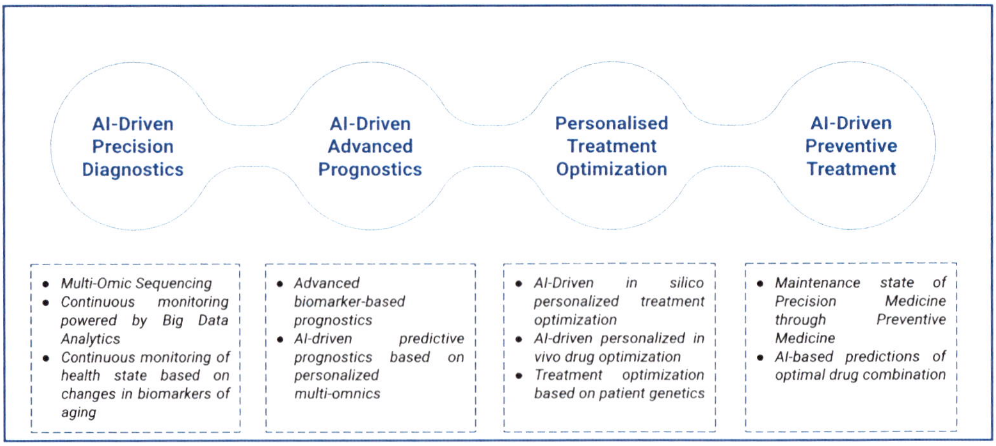

Source: AI as the Critical Catalyst for Accelerating Longevity Industrialization, 2022

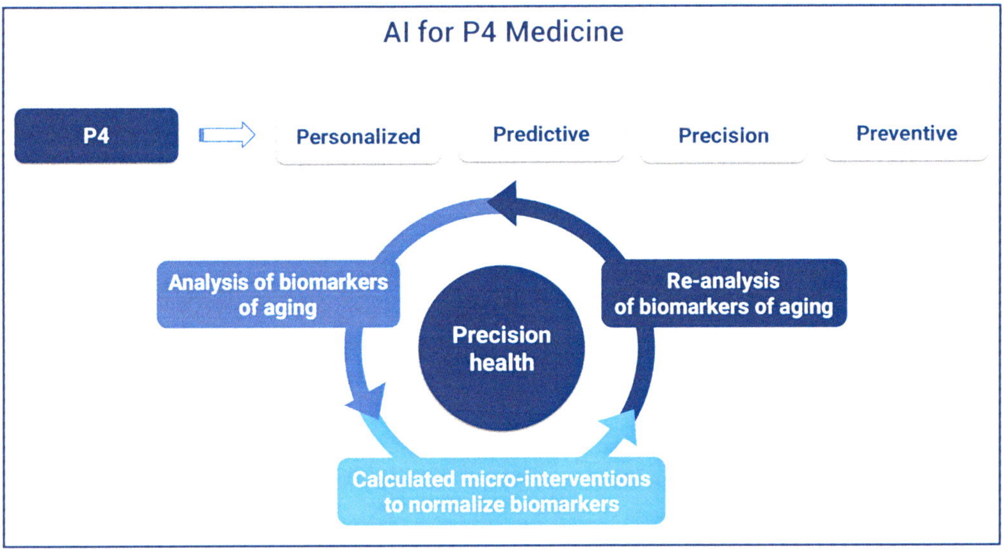

Source: www.analytics.dkv.global/data/pdf/Conferences/AI-for-Longevity-Overview-Role-of-AI-and-Data-Science-for-Healthy-Longevity.pdf

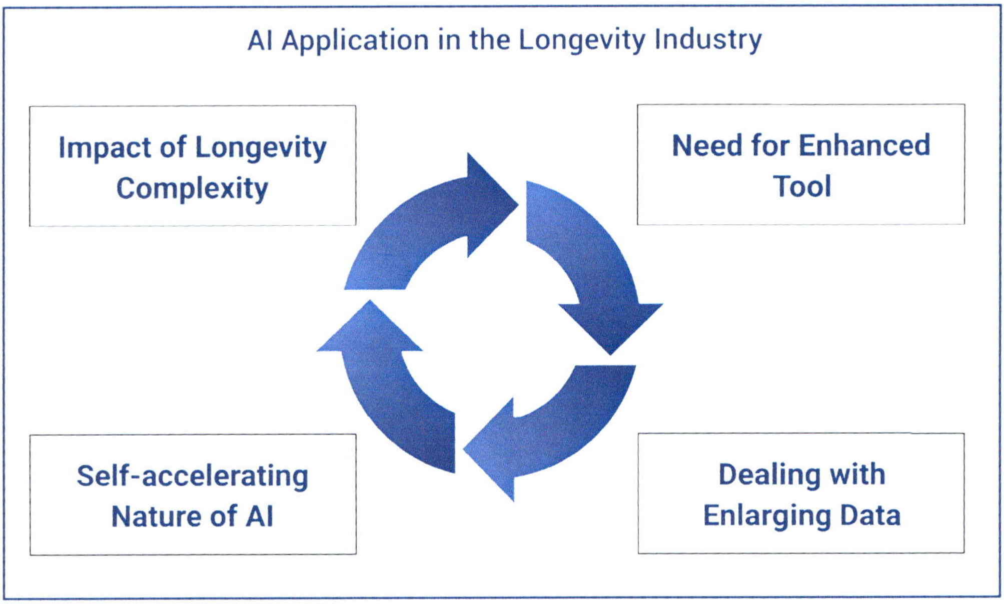

Source: AI as the Critical Catalyst for Accelerating Longevity Industrialization, 2022

Chapter 7. AgeTech and Silver Economy: The Multitrillion Longevity Industry's Sector of 1 Billion People in Retirement

- The Rise of Market-Ready Solutions for Optimizing Population Health, Performance and Productivity
- Forecast on Major Developments in AgeTech: 2025, 2027 and 2030
- Top Underleveraged Opportunities in AgeTech: 2023-2030
- Projecting Major Roadblocks in Bottlenecks in AgeTech: 2023-2030
- Frameworks, Strategies and Approaches to Neutralize Critical Threats for the Future of AgeTech
- DeepTech Engineering the Accelerated Trajectory of AgeTech
- Ageism and AgeTech: The Elderly as the Most Politically Powerful But Also Most Underrepresented, Unsupported and Marginalized Demographic of Electorate
- The Emergence of the Longevity Tech Trifecta: The Integral Fusion of AgeTech with WealthTech and HealthTech by 2030

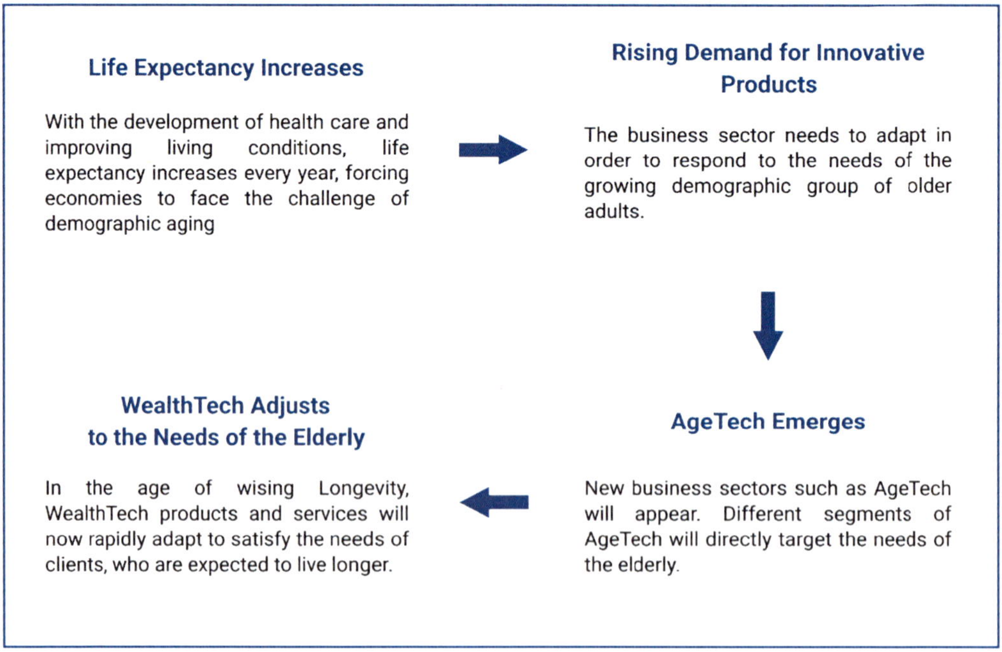

Source: www.aginganalytics.com/financial-longevity-industry

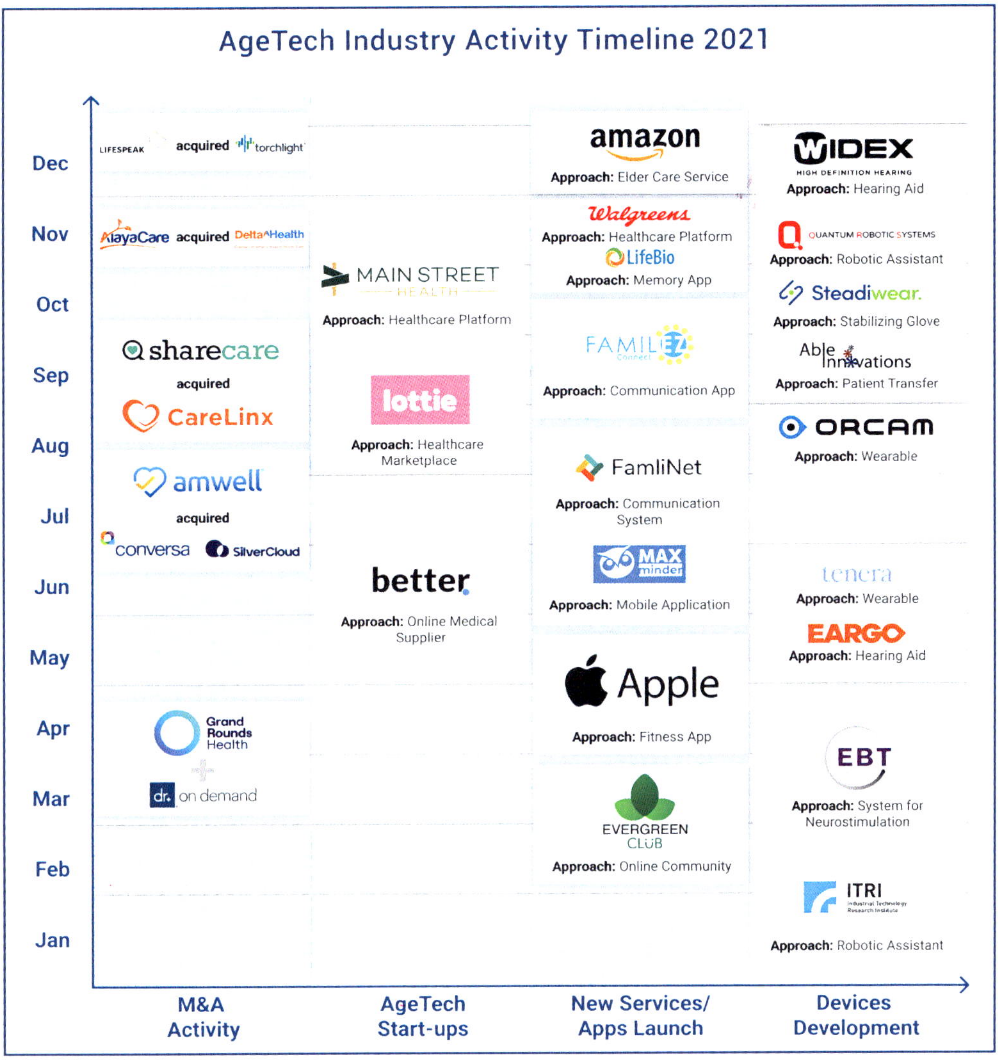

Source: www.longevity-journal.info/timelines

Chapter 8. Longevity Financial Industry: Health as the New Wealth and the Rise of Novel Financial System

- Forecast on Major Developments in Longevity Financial Industry: 2025, 2027 and 2030
- The Integral Fusion of Longevity FinTech, WealthTech, InvestTech, InsurTech, and AgeTech
- New Asset Class: Commoditization of Longevity medicine technologies and products, industry sectors and entire National Healthcare Systems
- Novel Financial Instruments, Structured Products and Derivatives 2.0: Tied to Tangible and Stable Assets in the Longevity Industry
- Longevity Futures, ETFs, Indices, Index Funds, Hedging and Shortening, Quant Strategies
- Longevity FinTech and InvestTech, Longevity Retail and Investment Banks
- Longevity InsurTech and Pension Tech
- Specialized Longevity Stock Exchange for Public Markets, Technology and OTC InvestTech Platforms for Private Markets
- Global Longevity Index as the Most Stable and Ultimate Financial Asset
- The Paradigm Shift From National Economies to National Longevity Economies
- Longevity Financial Industry as the Key Mechanism for Securing the Future of Longevity Industrialization and the Stability of National Economies
- The Concept of the Longevity Finance Valley: Global Benchmarking of Top Contenders to Create the First Municipal Longevity Finance Hub by 2025
- The Evolution of the Longevity Finance State: Global Benchmarking of Top Contenders to Create the First National Longevity Finance Hub by 2030
- Top Underleveraged Opportunities in Longevity Financial Industry 2023-2030
- Major Roadblocks in Bottlenecks in Longevity Financial Industry: 2023-2030
- Frameworks, Strategies and Approaches to Neutralize Critical Threats for the Future of the Longevity Financial Industry
- DeepTech Engineering the Accelerated Trajectory of Longevity Finance
- Forecast on the DeepTech Future of Longevity Financial Industry: 2025-2030

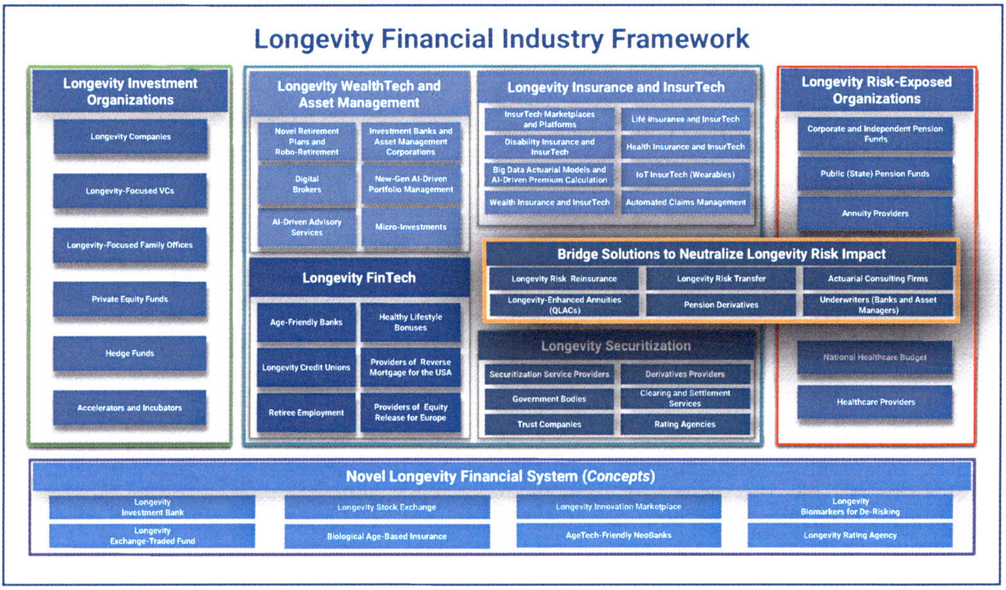

Source: www.frameworks.technology/longevity-financial-industry

Financial Commoditization of DeepTech and Longevity

The inevitable rise of DeepTech as a new asset class

Turning DeepTech into a new asset class would unlock trillions in capital, resulting in massive economic opportunities and a positive impact on humanity

Accelerated DeepTech Innovation, Practical Applications and Market Maturation Require Equally Sophisticated Approaches in InvestTech and Financial Industry Infrastructure (Markets, Exchanges, Indices and Financial Instruments)

In order to facilitate the sustainable development of the Longevity Industry, Deep Knowledge Group is pursuing an ecosystemic approach by working to establish integrative financial infrastructure to support the industry as a whole.

Chapter 9. Longevity Policy and Governance: The Rise of Longevity Technocracy and Longevity as the New Political Reality of the 21st Century

- Technocratic Nations and Municipalities Will Emerge as Global Competitors in the Race to Create the First Longevity Valley by 2025 and Longevity State by 2030
- Full-Scope Longevity Industrialization (Biomedicine + Tech + Finance) is National Governments' Most Powerful Tool for Boosting Economic Growth, Maintaining Socioeconomic Stability, Preventing Economic Decline and Maximizing Society Benefit
- 21st Century Politics Will Place Human Longevity at the Center of Economy
- The Set of Factors That Will Make Citizens Demand the Delivery of Additional Health-Adjusted Life Expectancy (HALE) and Quality-Adjusted Life Years (QALY) From Their Governments and Politicians as a Basic Resource and Human Right
- National Healthy Longevity and Longevity Industrialization is the Most Powerful Tool at National and Municipal Governments' Disposal to Achieve Economic Growth and Stability and Societal Prosperity
- The Rise of Longevity as a Fundamentally New Type of Political Asset Class at the Heart of National Economies Most Pressing Challenges and Opportunities that Will Determine the Future Socioeconomic Health and Wealth of Nations
- Longevity as the Main Driver of the Ongoing Transition toward the 5th Industrial Revolution (2030 - 2035), Characterized by the Maturation and Commoditization of DeepTech, and the Socially-Inclusive Realization of its Humanitarian Benefits
- Summary of Deep Knowledge Group Longevity Policy and Governance Activities

Semi-Automated SWOT Analysis Longevity Governance Progressiveness Benchmarking and SWOT Analysis

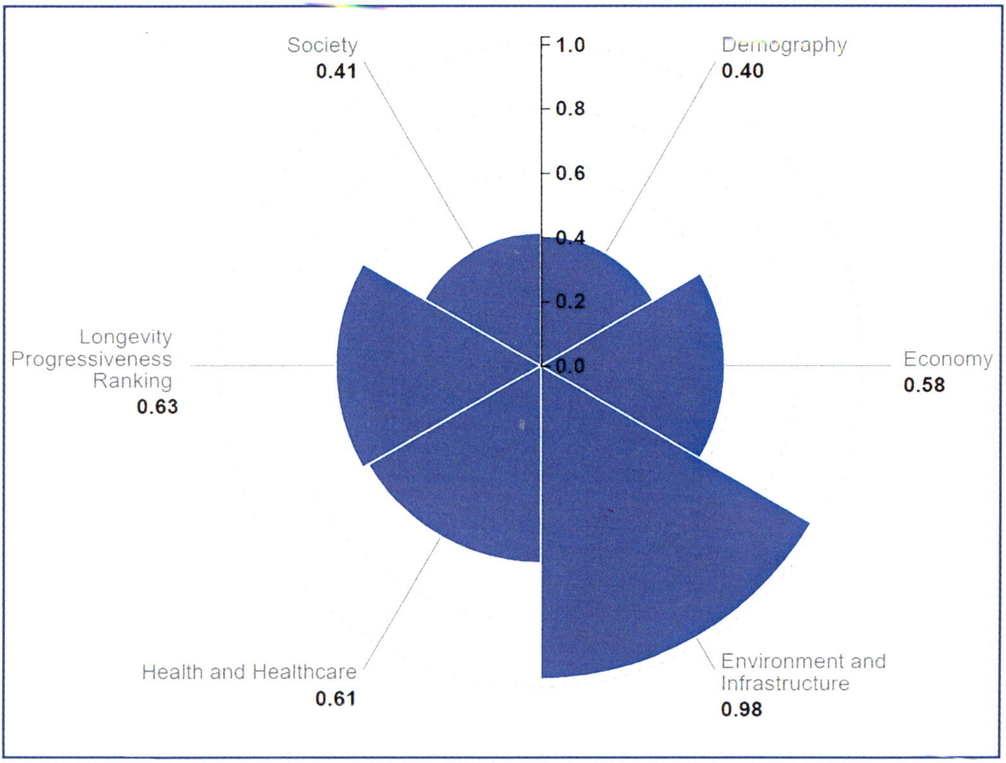

HALE is affected by various factors, with their impact varying across different regions. This is the reason why Aging Analytics Agency based its Longevity Progressiveness Ranking on the quantitative analysis of five factors: economy, health and healthcare, environment and infrastructure, society, and demography. The diagram above shows a quantitative analysis of the top-ranking countries in Longevity progressiveness in the Longevity Governance Dashboard.

Source: www.deep-innovation.tech/longevity-governance

Chapter 10. Longevity Ethics: Frameworks for Optimizing the Impact of Longevity on Global Humanity (2023 - 2030)

- Neutralizing the Potential Negative Outcomes of Longevity
  - Overpopulation (The Earth Needs More People, Not Less)
  - Famine (Technological Solutions to Unlimited Agriculture)
  - Pollution (Technological Solutions to Global Toxification)
  - Equality: The Rich vs Poor Divide
- Social Psychology as the Major Roadblock or Accelerating Factor Impacting the Societal and Ethical Outcomes of Longevity Industry
- Conservatism and Fear of Innovation as the Root Cause of Societal Risk, Stagnation and Decline
- The Framework for Optimizing Positive Outcomes and Neutralizing Potential Negative Consequences of Global Longevity Industrialization

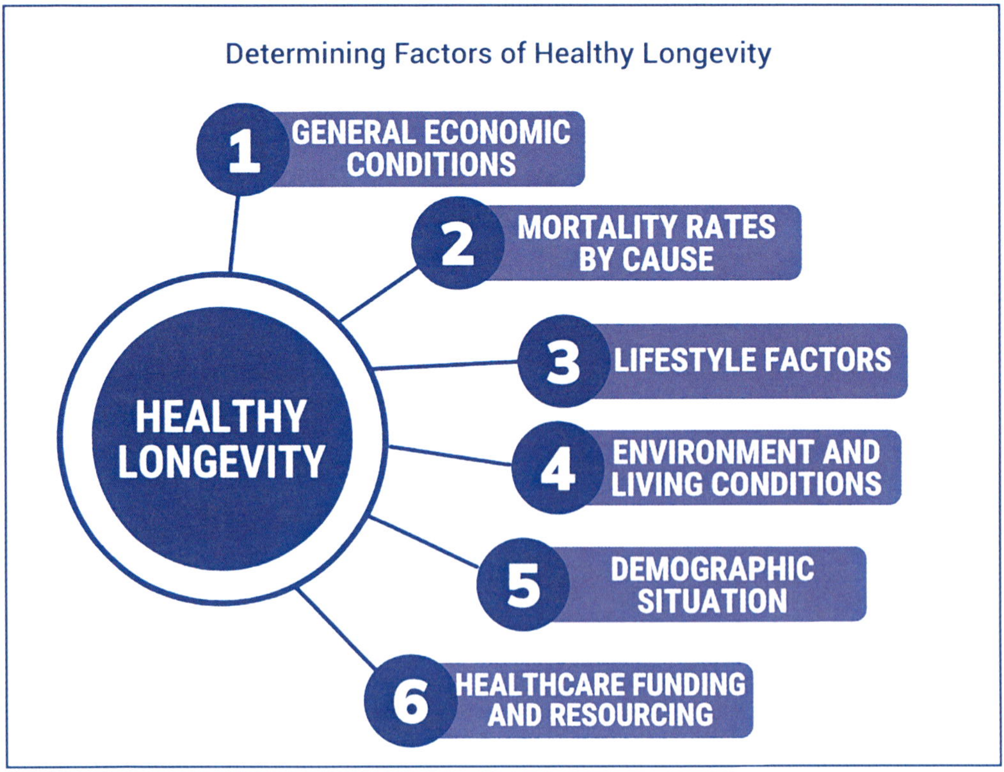

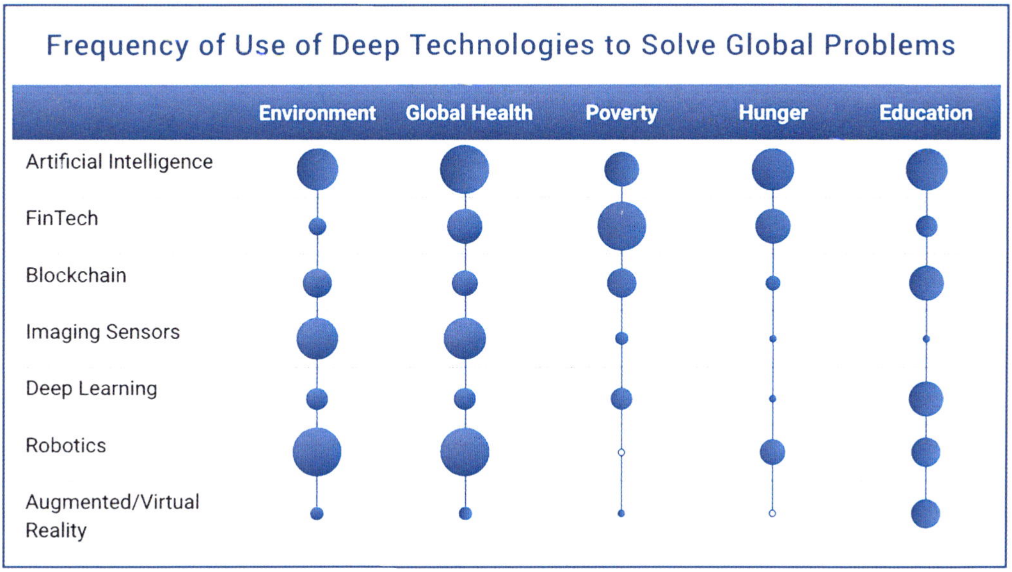

ATLAS

Assistive Technology, Longevity and Ageing Society

We are an Advocacy for the prioritisation of frontier technologies in AssistiveTech and AgeTech to impact lives on a global scale. We believe that technology is a major enabler of social inclusion in the world, and so we leverage partnerships via tech founders, venture philanthropy and impact investment to deliver a future of technology for all.

ATLAS pledges to:

- Champion the AssitiveTech industry and its potential to drive social inclusion
- Promote healthy Longevity and its importance to society and industry
- Restore the Technological Rights of our senior community

Source: *www.at-las.org*

Chapter 11. Technocratic Ethics: Embracing the Engineering and DeepTech Mindset of Human Societal, Technological and Industrial Development (2023 - 2030)

- Technological Progress Alone, While Extreme, Will Not Be Enough To Achieve Longevity Escape Velocity by 2030
- More Sophisticated and Robust Ethical Frameworks are Required To Achieve the Real World, Practical Dividends of Longevity Industrialization
- Technocratic Utilitarianism: Why Decisive, Centralized Government Efforts to Bootstrap Socially-Inclusive Longevity Industrialization Creates Net-Positive Social Good for Humanity
- If Pandemics Can Ethically Expand the Scope of Government Power to Prevent Death, Why are the 100k Deaths Per Day Globally From Ageing Any Different?
- Accelerating Progress via Massively International Manhattan Project for Longevity: Technocratic Unity Using the Best Strengths of Each Nation

Chapter 12. Technologizing Human Performance: DeepTech Engineering the Pathway from Human 1.0 to Human 2.0

- Convergences Between Longevity and Performance Enhancement
- The Role of NeuroTech in Longevity Industry 2.0: Optimizing Cognitive Performance
- The Role of SpaceTech and Space Medicine in Longevity Industry 2.0: The Pathway from Mice on Earth to Men on Mars
- AgeTech for All: Market Ready Approaches for Optimizing Non-Biomedical Aspects of Human Capacity and Performance
- The Inevitable Intersection Between Human Health and Performance

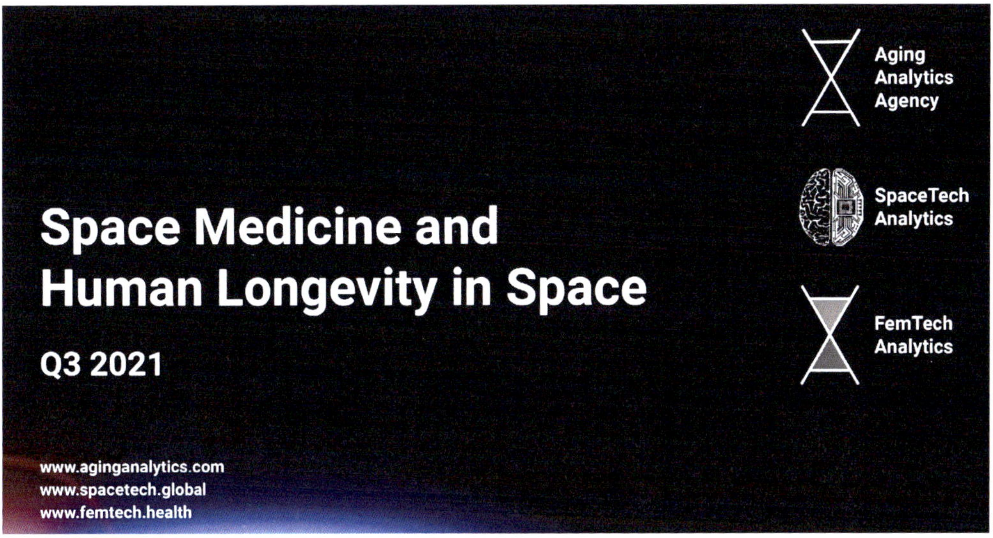

Source: *Space Medicine and Human Longevity in Space Q3 2021*[1] ***(SpaceTech Analytics)***

[1] www.analytics.dkv.global/spacetech/Space-Medicine-and-Human-Longevity-in-Space-2021.pdf

Appendix A, Chapter 1. Longevity Industrialization as the Major Catalyst for the 5th Industrial Revolution: Longevity as the Fundamental Engine for the Rise of the 5th Industrial Revolution and the Expansion of Humanity Beyond Earth

- Introduction to the 5th Industrial Revolution: The Future is Faster, bBut Much More Complex Than You Think
- 5th IR Technological Framework: DeepTech, MetaTech, SocieTech, and Finance
- Building the AI Analytical Engine for Long-Term Technological Forecasting
- Longevity Escape Velocity as the Launch Pad Booster for 5th IR and the Key to Sustainable Human Development Beyond Earth

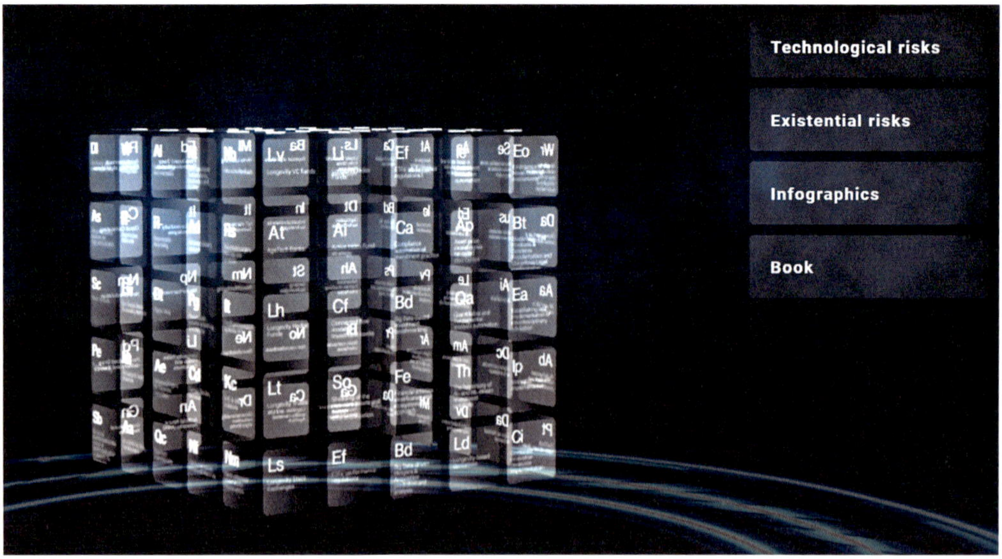

Source: www.5revolution.tech

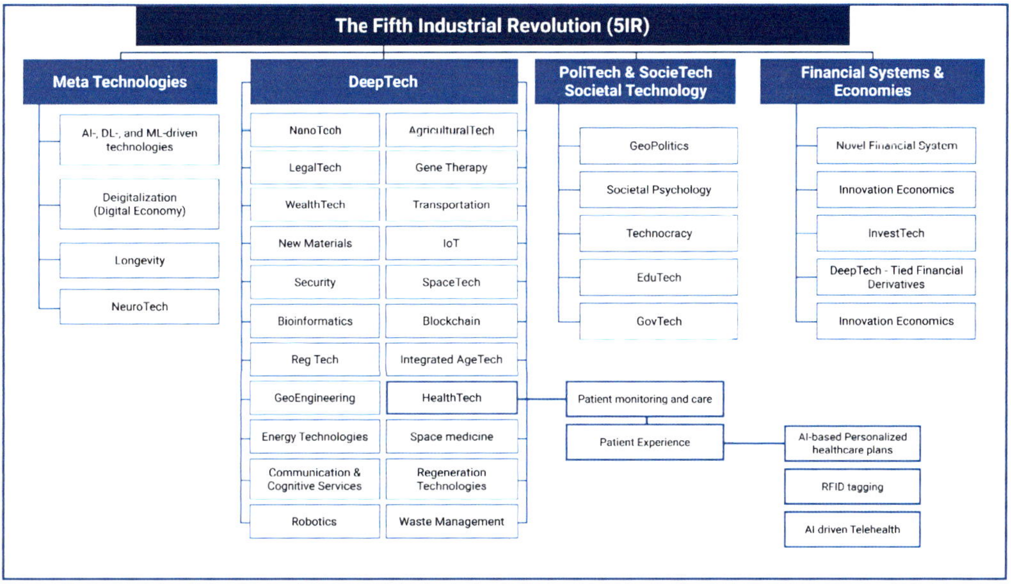
The Fifth Industrial Revolution (5IR)
Meta Technologies
AI-, DL-, and ML-driven technologies
Deigitalization (Digital Economy)
Longevity
NeuroTech
DeepTech
NanoTech
AgriculturalTech
LegalTech
Gene Therapy
WealthTech
Transportation
New Materials
IoT
Security
SpaceTech
Bioinformatics
Blockchain
Reg Tech
Integrated AgeTech
GeoEngineering
HealthTech
Energy Technologies
Space medicine
Communication & Cognitive Services
Regeneration Technologies
Robotics
Waste Management
PoliTech & SocieTech Societal Technology
GeoPolitics
Societal Psychology
Technocracy
EduTech
GovTech
Financial Systems & Economies
Novel Financial System
Innovation Economics
InvestTech
DeepTech - Tied Financial Derivatives
Innovation Economics
Patient monitoring and care
Patient Experience
AI-based Personalized healthcare plans
RFID tagging
AI driven Telehealth

The 5th Industrial Revolution
The 5th Industrial Revolution
Dmitry Kaminskiy
The Future is Faster
But Much More Complex Than You Think
Dmitry Kaminskiy

Appendix A, Chapter 2. DeepTech Engineering the Pathway to Longevity Industry 3.0: Longevity for All and Practical Humanitarian Impact on the Global Scale by the Year 2030

- Evolution from Longevity Industry 2.0 (Stabilizing the Industry) to Longevity Industry 3.0
- Optimizing the Industry and Accelerating the Realization of its Practical Impacts Ahead of Standard Timelines
- Distribution of Longevity Technologies on a Global Scale
- Guaranteed Basic Income and Guaranteed Basic Longevity
- How Longevity will Drive a Global Paradigm Shift From Capital-based Economies to Health-Adjusted Life Expectancy (HALE) and Quality-Adjusted Life Year (QALY) Economies
- Tangible Frameworks to Enable Governments to Use Longevity Industrialization as the Launch Pad Booster and Fusion Reaction to Enter the 5th Industrial Revolution
- DeepTech Engineering of Stable Acceleration of Longevity Industrialization to Achieve its Practical Humanitarian Impact on the Global Scale by the Year 2030

PoliTech, SocieTech and Societal Psychology

GeoPolitics
Data as a Political Asset
Data as Political Intelligence
Data as Political Influence
PoliTech Data-driven Risks
Power re-stratification

EduTech
AI in EdTech
AR and VR in EdTech
Big Data for Education

GovTech
AI-enhanced behavioral and predictive analytics
Public services customization and Smart cities

Societal Psychology
Executive Decentralization
Diversement as a social approach
Attempts to manufacturing individualization

Technocracy
AI-based social and technical solution implementation
Sustainability-centered economical approach
Workforce restructuring due to the ubiquitous automation

Source: www.5revolution.tech

2023-2027: DeepTech Engineering The Accelerated Trajectory of Human Longevity – The Blueprint and Pathway from 1.0 to 2.0

- Evolution from 1.0 (Defining the Industry) to 2.0 (Applying the Framework for Longevity Industrialization, Investment De-Risking and National Longevity Economy Development)
- The Evolution from Longevity Start-ups to Multi-Trillion Dollar Longevity Corporations
- Future of Longevity Finance: Financial Instruments to Bridge the Longevity Liquidity Gap
- The Rise of Longevity Technocracies and the Concept of 'Longevity Valley'
- Engineering Market-Ready Pathways for Practical Human Longevity by 2030

2027-2030: Stable Acceleration of Longevity Industrialization to Achieve its Practical Humanitarian Impact by the Year 2027

- Evolution from 2.0 (Stabilizing the Industry) to 2.0 (Optimizing the Industry and Accelerating the Realization of its Practical Impacts Ahead of Standard Timelines)
- How Longevity will Drive a Global Paradigm Shift From Capital-Based Economies to Health-Adjusted Life Expectancy (HALE) and Quality-Adjusted Life Year (QALY) Economies
- Tangible Frameworks to Enable Governments to Use Longevity Industrialization as the Launch Pad Booster and Fission Reaction to Enter the 5th Industrial Revolution

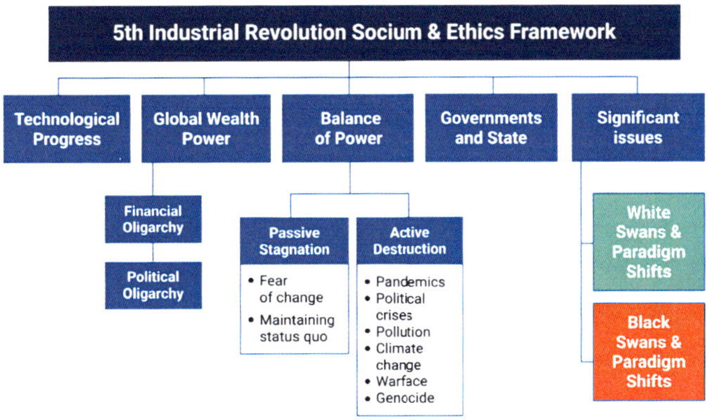

Source: www.5revolution.tech

Appendix A, Chapter 3. DeepTech Engineering of Longevity Industry 4.0: Pathway from Humanity 1.0 to Humanity 2.0 and Interplanetary Civilization

- Advanced Longevity Technologies and Space Medicine 2.0 to Create Super Humans
- Crucial Requirements for Human Immunity Establish Colonies on Mars and Moon
- Establishment Colonies on Mars as the Pathway to Protect Human Civilization from Extinction in Case of Existential Catastrophe on Earth
- Technological Progress and Increasing Global Geopolitical Instability and Existential Threats
- Next Generation of Longevity Technologies as the Pathway to Protect Humanity from Critical Global Catastrophes and Risks on Earth and Other Planets
- Beyond 5th Industrial Revolution and Humanity 2.0

DEEP KNOWLEDGE GROUP OVERVIEW

Knowledge Is Power

Deep Knowledge Is Transcending Power

About Deep Knowledge Group

Deep Knowledge Group[1] is a data-driven consortium of commercial and non-profit organizations active on many fronts in the realm of DeepTech and Frontier Technologies (AI, Longevity, BioTech, Pharma, FinTech, GovTech, SpaceTech, FemTech, Data Science, InvestTech), ranging from scientific research to investment, entrepreneurship, analytics, consulting, media, philanthropy and more.

As a mathematical corporation focused on building the bridge to the 5th Industrial Revolution, Deep Knowledge Group is decisively committed to DeepTech for Social Good, Techno-Philanthropy and DeepTech and Longevity Industry Financial Commoditization (establishing the core investment and financial industry infrastructure needed for the emergence of DeepTech and Longevity as fundamentally new asset classes).

Deep Knowledge Group has been committed since 2014 to the corporate thesis of DeepTech for social impact, techno-ethical business for social good, and the continual reinvestment of profits into the expansion of our core Data Science, AI and analytics assets and the refinement of our financial-ecosystemic development projects (which we see as the major tool driving our Group's further growth as a Data Science driven financial corporation with focus on Longevity and DeepTech). Our unique approach lies in combining ecosystemic financial innovations with sophisticated AI and Data Science technology development for DeepTech and Longevity Industry analysis, benchmarking and forecasting.

About Deep Knowledge Group

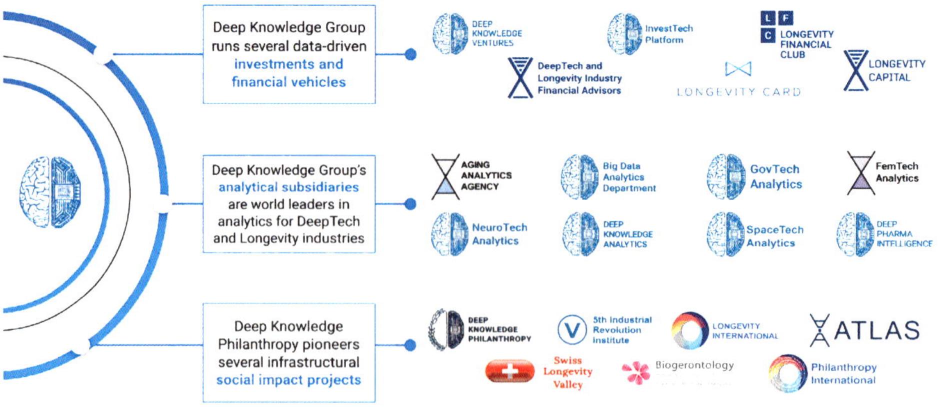

Source: *Deep Knowledge Group Overview*[1]

[1] www.dkv.global

Deep Knowledge Group: 2022 Accomplishments and Outlook for 2023

DeepTech and Longevity Financial Commoditization

Developing the ecosystemic infrastructure (platforms, indices, exchanges, financial instruments and derivatives) to support the emergence of DeepTech and Longevity Industries as fundamentally new asset classes.

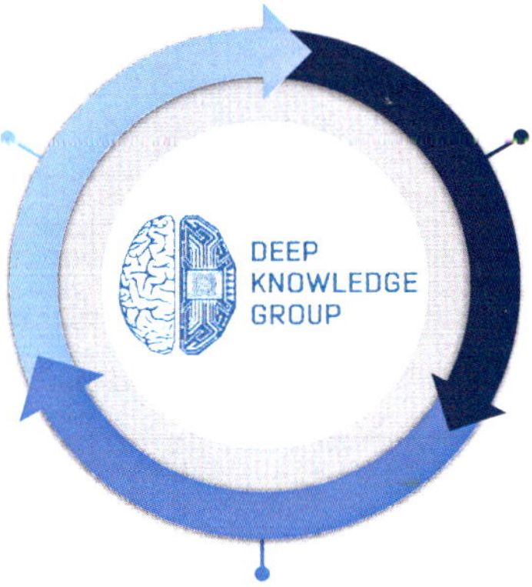

Big Data Aggregation, Analysis, and Visualization

Automatized data parsing, aggregation, optimization of machine learning and deterministic scoring algorithms for competitor matching, and due diligence result

Financial Insights and Investment Strategies

Modern InvestTech solutions and DeepTech investment approaches driven by Data Science and Big Data Analytics

Source: *www.dkv.global*

Deep Knowledge Group Organisational Structure

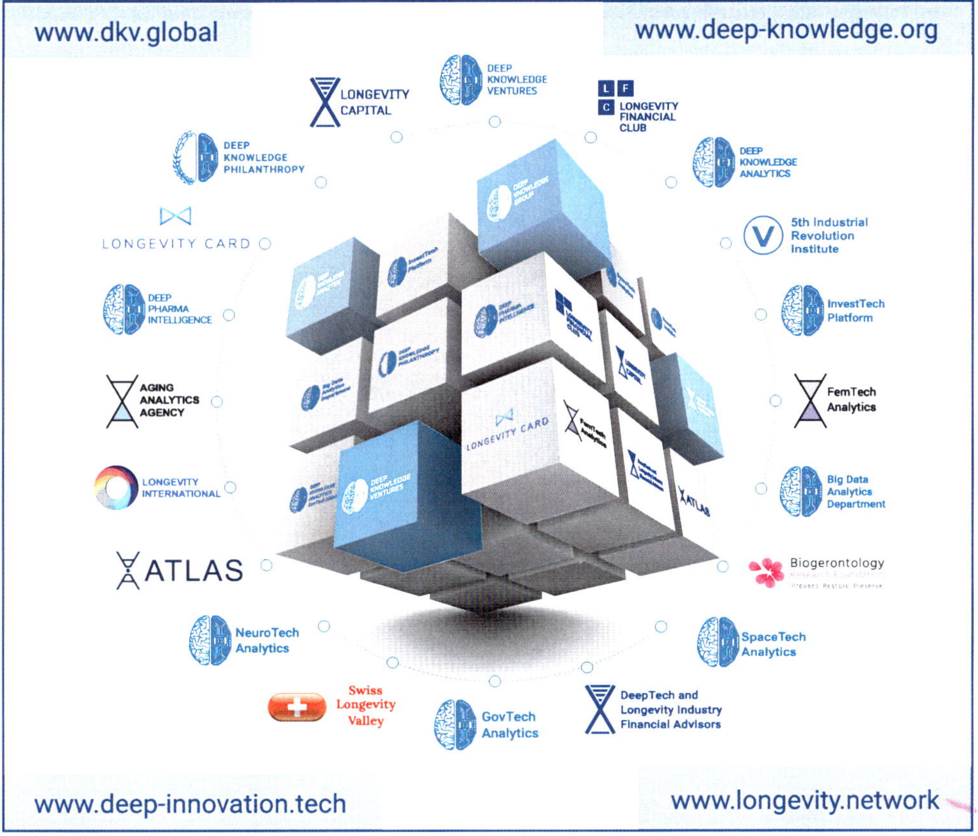

Source: *www.dkv.global*

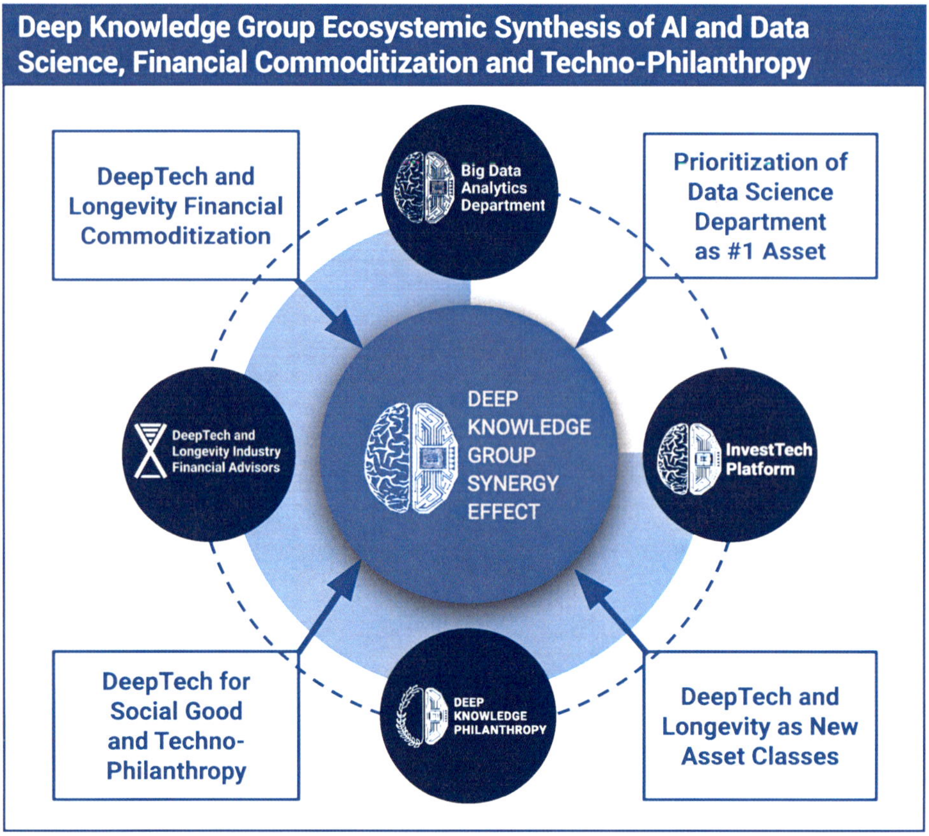

Source: *www.dkv.global*

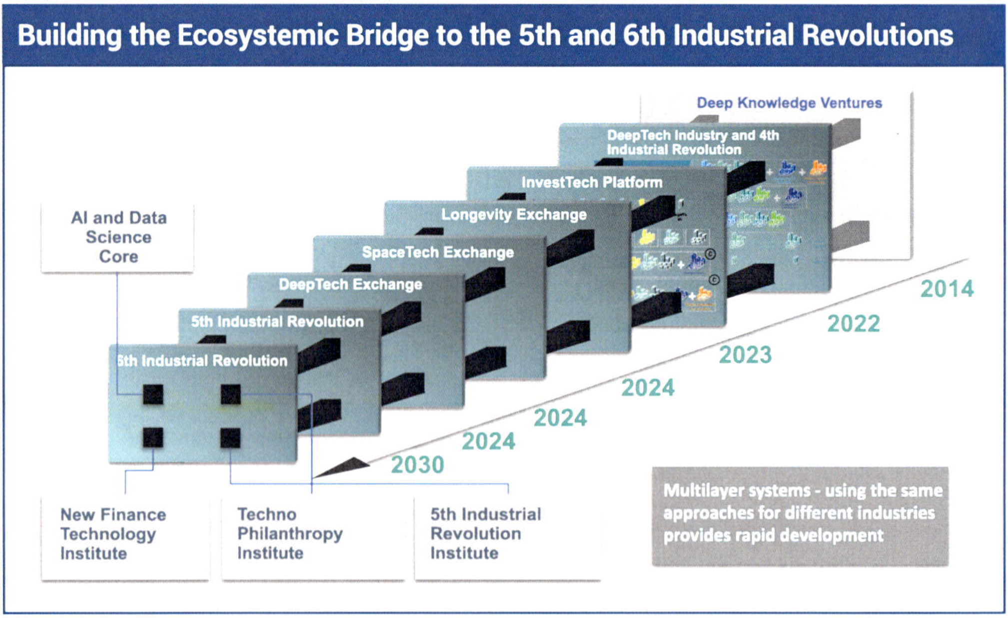

Source: *www.dkv.global*

Our continuous, systematic internal development of advanced data science, AI, and IT solutions for DeepTech analytics, predictive forecasting, and benchmarking across our investment, entrepreneurship, analytics and philanthropic activities as the main tool for coping with the notorious complexity of DeepTech domains is well known, receiving coverage even as early as 2014 in bestselling books on the future of technology including Homo Deus by Yuval Noah Harari and The Fourth Industrial Revolution by World Economic Forum Founder and Chairman Klaus Schwab.

«In May 2014 Deep Knowledge Ventures — a Hong Kong venture-capital firm specialising in regenerative medicine — broke new ground by appointing an algorithm called VITAL to its board.

VITAL makes investment recommendations by analysing huge amounts of data on the financial situation, clinical trials and intellectual property of prospective companies. Like the other five board members, the algorithm gets to vote on whether the firm makes an investment in a specific company or not.»

Yuval Noah Harari "Homo Deus: A Brief History of Tomorrow", page 322

«Deep Knowledge Ventures, a Hong Kong- based venture capital fund that invests in life sciences, cancer research, age-related diseases and regenerative medicine has appointed an artificial intelligence algorithm called VITAL (Validating Investment Tool for Advancing Life Sciences) to its board of directors.»

Klaus Schwab "The Fourth Industrial Revolution", page 149

Source: *Deep Knowledge Group Overview*[1]

The Longevity Industry has always been a heavily prioritized sector for the Group, which made its first investment in the sector in 2014 (long before it rose to its current status as a mainstream megatrend embraced by large conservative investors, financial corporations, and national governments), providing the ***seed funding***[2] for Insilico Medicine and the AI for Drug Discovery company, which has gone on to raise over ***$300 million in follow-on financing***[3].

Deep Knowledge Group was also the first entity to define the Longevity Industry in its full depth and breadth and to formulate actionable frameworks for industry analytics, forecasting, and benchmarking to manage the industry's unprecedented complexity, multidimensionality, and technological intersectionality. The Group's starting point, back in 2013, was to establish its dedicated Longevity analytical subsidiary, ***Aging Analytics Agency***[4], in order to comprehensively structure the industry's breadth and complexity by identifying, classifying, and profiling all participants in the Global Longevity Industry ecosystem, applying AI, modern data science, and Big Data analysis for indus-

[1] www.dkv.global/about
[2] www.fiercebiotech.com/it/deep-knowledge-ventures-and-insilico-medicine-enter-into-a-convertible-bridge-note-agreement-to
[3] www.techcrunch.com/2021/06/22/a-i-drug-discovery-platform-insilico-medicine-announces-255-million-in-series-c-funding
[4] www.aginganalytics.com

try analytics to make this information available through a variety of open-access reports and analytics.

The company quickly established itself as the world-leading provider of Longevity Industry analytics, forecasting, and benchmarking, producing two landmark reports in 2017 and 2018 which defined the industry for the first time and coined the term "Longevity Industry" (which, at that time, still referred primarily to life extension and anti-aging). Since then the company has produced dozens of open-access and proprietary analytical reports, case studies, IT-platforms, and Big Data Analytical Dashboards. The company serves as an Official Member Organization of the United Nations NGO Committee on Aging, a Strategic Partner of the Metabesity Conference Series in Washington, D.C. (the largest in the USA), and a Founding Partner of the original UK *All-Party Parliamentary Group for Longevity*[1] Secretariat in 2019.

Global Longevity Ecosystem by Sector

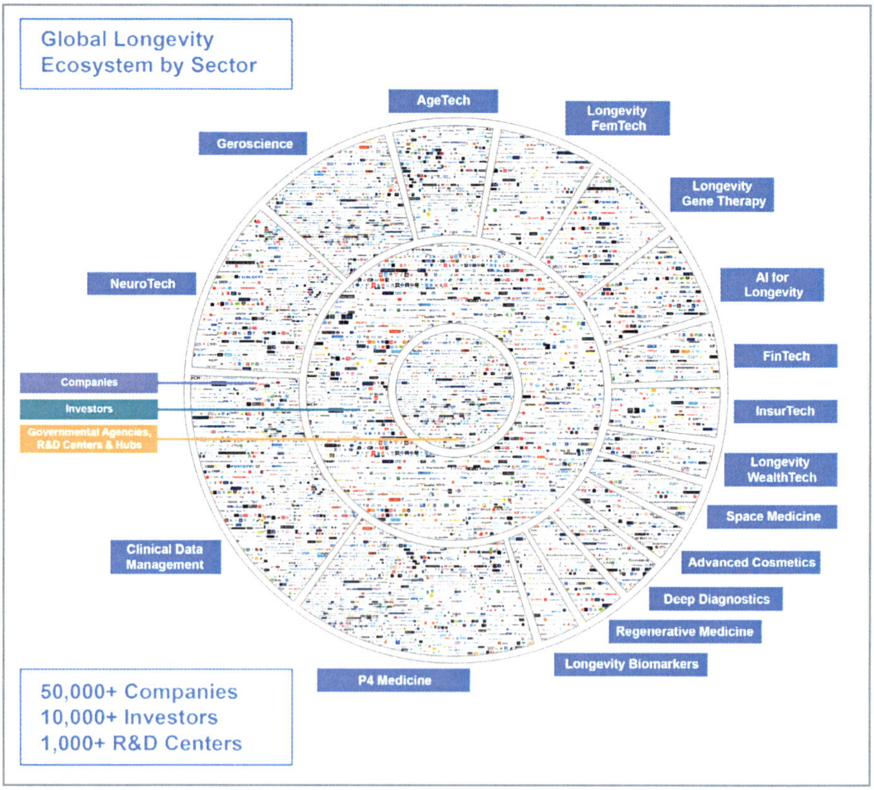

Source: *Longevity.International's Global Longevity Ecosystem IT-Platform*[2]

[1] www.appg-longevity.org
[2] www.longevity.international/global-longevity-ecosystem

As a result, Deep Knowledge Group now owns the world's largest Longevity Industry database, encompassing 50,000+ companies, 12,000+ investors, and 1,200+ R&D hubs across 20 sectors. Backed by the Big Data analytics and sophisticated data visualization capabilities of Deep Knowledge Group, this project took the form of a single coherent Global Longevity Big Data Analytics IT System.

Deep Knowledge Group also established *10 other analytical subsidiaries*[1] with specialized focus on different DeepTech sectors of strategic interest:

- *AI Analytics Agency*[2]
- *Deep Knowledge Analytics*[3]
- *Deep Pharma Intelligence*[4]
- *InvestTech Advanced Solutions*[5]
- *COVID-19 Analytics*[6]
- *FemTech Analytics*[7]
- *NeuroTech Analytics*[8]
- *GovTech E-governance Analytics*[9]
- *SpaceTech Analytics*[10]
- *Interactive Mindmaps*[11]

Deep Knowledge Group is also active in the arena of technological philanthropy and DeepTech for social good via its dedicated subsidiary *Deep Knowledge Philanthropy*[12], a data-driven nonprofit dedicated to the support, development, and advancement of DeepTech for impact philanthropy, ethical investment, and the wider social good. It is founded on the principle that DeepTech innovation is the most efficient driver of "social profit," technological humanitarianism, and societal progress

---

[1] www.dkv.global/analytics
[2] www.ai-industry.tech
[3] www.dka.global
[4] www.deep-pharma.tech
[5] www.invest-solutions.tech
[6] www.dkv.global/covid-assessments
[7] www.femtech.health
[8] www.neurotech.com
[9] www.govtech.global
[10] www.spacetech.global
[11] www.dka.global/interactive-mind-maps
[12] www.deep-knowledge.org

and that venture philanthropy is a profitable long-term investment for individuals, national economies, and humanity itself.

In January 2023 Deep Knowledge Philanthropy launched the Philanthropy.International Platform with the assistance of Deep Knowledge Group's AI and Data Science Division as the first full-scope tool for matching charities, donors, and volunteers. The unique approach is developed to eliminate the concerns of philanthropy industry players regarding the proper money allocations in the charity sector. They profiled 4,950 businesses with purpose, 6,900 charities, 14,995 non-profits, 3,665 investors, > 530k data points and 50 parameters and built the worlds-first Philanthropy Big Data Analytical System & Dashboard delivering on-demand market intelligence analytics through data coverage across the Philanthropy Industry.

In late May 2022, Dmitry Kaminskiy will be delivering the Keynote Address for the Longevity Forum component of the 2022 Davos Forum.

Source: *Dmitry Kaminskiy's website*[1]

Philanthropy.International is a next-generation platform built by Deep Knowledge Philanthropy, a subsidiary of Deep Knowledge Group, burdened with the vision of deploying frontier technologies to efficiently maximize social impact and create social good. We are driven by the core belief that technology is an effective tool to drive social change and impact, herein

[1] www.kaminskiy.info/gallery

lies the core mandate of Philanthropy International - to deploy frontier technologies (Big Data Analytics, Artificial Intelligence, Blockchain) to tackle the vastly different social challenges in global philanthropy to build a more equitable world. We are creating a secure traceability system for maximum visibility and transparency, coupled with a smart matching tool that streamlines the entire process - making it easier than ever before to support good causes.

Deep Knowledge Group's activities have been the subject of widespread top-tier media coverage across 500 articles in the press globally. The Group's Founder, Dmitry Kaminskiy, has extensive experience delivering keynote presentations at first-class DeepTech and Longevity-themed conferences, meetings, and events organized by world-renowned brands including Julius Baer; The Economist; the Financial Times; the Reform Club in London; Glaxosmithkline; the UK All-Party Parliamentary Groups for Longevity, AI, and Blockchain; the University of Oxford; the University of Cambridge; and many others.

Information about the Group's activities can be found in a variety of world-famous magazines and media brands ranging from Forbes, the Financial Times, Business Insider, the BBC, CNBC, Der Spiegel, The Telegraph, NIkkei Asian Review, Deutsche Welle, Esquire, Fox News, The Washington Post, The Straits Times, The National Post, and many others.

Top-Tier Media Coverage of Deep Knowledge Group

Source: *Deep Knowledge Group Media Overview*[1]

Meanwhile, the Group's analytics have also been the subject of numerous notable international acknowledgments from various governments and heads of state including Alain Berset (former President of Switzerland), Benjamin Netanyahu (Prime Minister of Israel), Hong Nam-ki (Deputy Prime Minister of South Korea), Karin Prien (German Minister for Education, Science, and Culture), Hemann Gröhe (Former

[1] www.dkv.global/media-news

German Minister of Health), Judith Varga (Hungarian Minister of Justice), and a number of formal government bodies and ministries including the House of Switzerland; the Swiss Embassy in Ireland; the Swiss Embassy in France; the Israeli Ministry of Foreign Affairs, Ministry of Strategic Affairs, and Prime Minister's Office; and many others.

Notable Acknowledgements of Deep Knowledge Group

Benjamin Netanyahu

Prime Minister of Israel

In a tweet, the Office of the Prime Minister of Israel acknowledged Israel's first-place ranking in the COVID-19 Health Safety Countries Ranking.

Hong Nam-ki

Deputy Prime Minister of South Korea

Hong Nam-ki posted about Deep Knowledge Group's COVID-19 Safety Ranking published on August 26, 2020.

Karin Prien

German Minister for Education, Science and Culture

"Compared to the other countries, Germany currently has the best security and stability ranking in Europe and is also one of the world's leading nations when it comes to crisis management."

Judith Varga

Hungarian Minister of Justice

"The latest IMF forecast sees Hungary registering the second smallest GDP drop in the EU, and we are the #4 State by European Covid19 Safety Countries Ranking, according to the research of the Deep Knowledge Group."

Alain Berset

Former President of Switzerland

"Deep Knowledge Group's report, based on 130 parameters, ranks Switzerland #1 due to its economy's resilience and careful ways to relax lockdown without sacrificing public health and safety."

Source: *Notable Acknowledgements of Deep Knowledge Group*[1]

Much of the Group's Longevity Industry activities in particular have been driven by its General Partner, ***Dmitry Kaminskiy***[2], who has a long-standing personal interest in the sector, believing Longevity Industrialization to be the most effective and efficient means to achieve massive humanitarian benefit and optimize the health and wealth of citizens; and the key to ensuring the stability of national economies.

Through his efforts both within Deep Knowledge Group and externally (e.g., as Co-Founder and Head of International Development for the original UK All-Party Parliamentary Group for Longevity Secretariat, and as Managing Trustee for the Biogerontology Research Foundation, the UK's oldest Longevity-focused charity founded in 2008), Kaminskiy has devoted significant portions of his personal and professional activities to advancing Longevity Industrialization.

His 2020 book, ***Longevity Industry 1.0: Defining the Biggest and Most Complex Industry in Human History***[3], distilled the complex assembly of deep market intelligence and industry knowledge that ***Deep Knowledge Group***[4] and its Longevity-focused subsidiaries (including

[1] www.dkv.global/dkg/international-acknowledgements
[2] www.kaminskiy.info
[3] www.longevity-book.com
[4] www.dkv.global

Longevity.Capital[1] and *Aging Analytics Agency*[2]) have developed from 2013 to 2020 into a full-scope understanding of the Global Longevity Industry, effectively presenting the first truly comprehensive definition and framework, and providing major industry participants with a relevant and robust introductory text for decision-makers of all types across many industries, from science to medicine, finance, investment, technology, and more, on the current state, promise, and potential of Longevity Industrialization.

Meanwhile, his 2021 book, *Biomarkers of Human Longevity: Data Science for Accelerating Aging Research and R&D and the Critical Catalyst for Practical Human Longevity, Tangible Investment Decision-Making, and De-Risking*[3], urges major Longevity Industry decision-makers to adopt Biomarkers of Human Longevity as the only fundamentally market-ready framework of metrics for tangible investment decision-making and de-risking industry stabilization, tangible decision-making, predictable forecasting of clinical trial outcomes, public market consistency, investment de-risking, and reliable validation of Longevity technologies and therapies.

Kaminskiy also has a number of other topics and sector-specific books forthcoming as part of his larger *Longevity Book Series*[4].

Name of the Book	Short Description	Availability
LONGEVITY INDUSTRY 1.0[5]	Defining the Biggest and Most Complex Industry in Human History	*AVAILABLE*[6]
BIOMARKERS OF HUMAN LONGEVITY[7]	Data Science for Accelerating Aging Research and R&D, The Critical Catalyst for Practical Human Longevity, Tangible Investment Decision-Making and De-Risking	*AVAILABLE*[8]
LONGEVITY INDUSTRY 2.0[9]	DeepTech Engineering The Accelerated Trajectory of Human Longevity: The Blueprint and Pathway from Longevity Industry 1.0 to 2.0	AVAILABLE SOON

[1] www.longevity.capital
[2] www.aginganalytics.com
[3] www.ca.movies.yahoo.com/biomarkers-human-longevity-essential-investment-082000885.html
[4] www.longevity-book.com
[5] www.longevity-book.com/one
[6] www.longevity-book.com/order
[7] www.longevity-book.com/biomarkers
[8] www.longevity-book.com/order-biomarkers-book
[9] www.longevity-book.com/two

LONGEVITY POLITICS[1]	Longevity Technocracy, Modern Approaches to Policy, Governance and National Industrial Strategies, and Longevity as the New Political Priority of the 21st Century	AVAILABLE SOON
LONGEVITY FINANCIAL INDUSTRY[2]	Health as New Wealth. Longevity FinTech, WealthTech, InvestTech, InsurTech, AgeTech and the Coming Rise of Longevity Banks, Investment Banks, Stock Exchanges, ETFs, Indices, Novel Financial Instruments and Derivatives	AVAILABLE SOON
PRACTICAL LONGEVITY[3]	The Complete Guide on How to Celebrate Your 123rd Birthday	AVAILABLE SOON
WOMEN LONGEVITY	Women Longevity, Female Health, the Rise of FemTech and Why the Future Health and Wealth of Humanity (on Earth and Beyond) Depends on Female Longevity	AVAILABLE SOON

Deep Knowledge Group and its various subsidiaries are also the owners of several dozen granted and registered/pending trademarks within the Longevity Industry, including (but not limited to) Health as the New Wealth (granted), Longevity FinTech, Swiss Longevity Valley, and others.

Source: *Longevity Library*[4]

[1] www.longevity-book.com/politics
[2] www.longevity-book.com/finance
[3] www.longevity-book.com/practical-longevity
[4] www.longevity-library.store

Big Data Analytics Systems and Dashboards

Over time the analytical assets produced via Deep Knowledge Group's 10+ analytical subsidiaries continued to grow, transforming from isolated analytical case-studies into truly comprehensive, fully-integrated interactive industry databases augmented with AI Analytics (Reinforcement Learning, Natural Language Processing) and sophisticated data visualization techniques, and culminating in a series of specialized, industry-specific professionally structured Big Data Analytical Dashboards.

These Big Data Analytical Dashboards are professionally-structured data-driven analytical platforms, and the closest analogy to them would be a combination of Bloomberg Terminal, CB Insights, and Pitchbook, but backed by the data science and AI assets of DeepMind.

Longevity Industry Big Data Analytics Dashboards

Source: *Longevity Industry Big Data Analytics Dashboards*[1]

Deep Knowledge Group implements cutting-edge technological solutions in order to create a competitive advantage in market assessment by providing an unseen level of precision and multidimensionality of the data being processed. Deep Knowledge Group developed a series of proprietary AI/ML algorithms that enable users to scan companies with unmet speed and depth. Our algorithms are capable of joining financial, scientific, IP and other types of data in order to provide a full understanding of the development stage of the companies. That, in turn, creates

[1] www.deep-innovation.tech/longevity-dashboard

wide applicability of the Systems from market business intelligence and due diligence, up to portfolio optimization and M&A. The Dashboards provide a holistic view on the Longevity and DeepTech Industries, analyzing even the governmental policies applied by different countries in the corresponding area. Deep Knowledge Group analyzes millions of scientific articles, patents, companies, financials, news, and other data in order to optimize multiple areas of the contemporary investment processes and create solutions that will allow people to form the future.

Big Data Analytical Dashboards Cover:

Longevity Industry:

- *Longevity Investment Big Data Analytics Dashboard1*
- *Public Longevity Companies Investment Big Data Analytics Dashboard1*
- *Longevity Finance Big Data Analytics Dashboard2*
- *Longevity Governance Big Data AnalyticsDashboard*[3]

Pharma Industry:

- *Deep Pharma Intelligence Big Data AnalyticsDashboard*[4]

DeepTech Ecosystem:

- *Digital UK Ecosystem Big Data Analytics Dashboard5*
- *Global GovTech Big Data Analytics Dashboard6*
- *UK GovTech Dashboard7*
- *Asia Digital Ecosystem Big Data Analytics Dashboard8*
- *Digital Europe Ecosystem Big Data Analytics Dashboard9*
- *Switzerland DeepTech Big Data Analytics Dashboard10*
- *SpaceTech Industry Big Data Analytics Dashboard*[11]

[1] www.deep-innovation.tech/public-longevity-investment-dashboard
[2] www.deep-innovation.tech/longevity-finance-dashboard
[3] www.deep-innovation.tech/longevity-governance
[4] www.deep-innovation.tech/pharma-dashboard
[5] www.deep-innovation.tech/deeptech-uk
[6] www.deep-innovation.tech/govtech-global
[7] www.deep-innovation.tech/govtech-uk
[8] www.deep-innovation.tech/deeptech-asia-ecosystem
[9] www.deep-innovation.tech/deeptech-digital-europe-dashboard
[10] www.deep-innovation.tech/deeptech-switzerland-dashboard
[11] www.deep-innovation.tech/deeptech-spacetech-dashboard

- *Gulf Region Tech Industry Big Data Analytics System and Dashboard*[1]

Philanthropy Activities:

- *Philanthropy Industry Big Data Analytics System and Dashboard*[2]

DeepTech and Longevity Financial Activities

Deep Knowledge Group has established several dedicated subsidiaries to oversee the group's financial and investments activities.

Longevity.Capital[3] is Deep Knowledge Group's flagship Longevity-focused hybrid investment platform and securitization vehicle. It structures its portfolio on sophisticated industry intelligence and comparative analytics provided by Aging Analytics Agency.

DeepTech and Longevity Industry Financial Advisors[4] is a niche advisory boutique, authorized and regulated by the Financial Conduct Authority ("FCA") in the United Kingdom, focused on the DeepTech and Longevity Industries.

Deep Knowledge Group is decisively committed to establishing the core investment and financial industry infrastructure (platforms, indices, exchanges, financial instruments and derivatives) needed to execute the concept of DeepTech and Longevity Financial Commoditization, reflected in its pursuit of a truly ecosystemic approach through its various financial service divisions to establish an integral financial infrastructure to support the commoditization of the DeepTech and Longevity Industries as fundamentally new asset classes.

In 2022 Deep Knowledge Group established DeepTech and Longevity Industry Financial Advisors[5], authorised and regulated by the Financial Conduct Authority ('FCA') in the United Kingdom, to act as the practical vehicle for purpose, to support the accelerated development of DeepTech and Longevity Industrialization, the establishment of an integral financial infrastructure for their further growth and stabilisation, in order to expedite the emergence of DeepTech and Longevity as new asset classes in and of themselves.

1 www.deep-innovation.tech/gcc
2 www.philanthropy.international/philanthropy-dashboard
3 www.longevity.capital
4 www.longevity.finance
5 www.longevity.finance

Ariane Deep Knowledge AI Pharma Index

Deep Knowledge Group is one of the pioneers in AI in the Pharma sector. Backed by its subsidiaries' analytical and computational capabilities, Deep Knowledge Group provides market insights for the AI in Pharma and Longevity sectors.

The Group's investment and analytical experience allow it to foresee and project future dynamics of related markets. In order to facilitate growth of the AI in Pharma market and contribute to the development of the Longevity Industry, Deep Knowledge Group is creating an Actively Managed Certificate (AMC) "*Ariane Deep Knowledge AI Pharma Index*[1]."

The main goal of the Ariane Deep Knowledge AI Pharma Index is to reflect the development dynamics of the AI in the Pharma market, and to indicate all important movements and sector-related features. The ultimate value proposition of the Index is rooted in the objective assessment of the underlying assets, i.e., the companies comprising the AMC.

Cancer Vaccines Actively Managed Certificate Issued by UBS and Deep Knowledge Group

The complexity of contemporary markets creates a pressing and unmet need for niche market products that can provide adequate and feasible solutions for capital allocation. Understanding that, Deep Knowledge Group has launched the first-of-its-kind Cancer Vaccines AMC.

The main goal of the AMC is to reflect the cancer vaccine development dynamics, and to indicate all important movements and sector-related features.

The Cancer Vaccines AMC was issued in cooperation with UBS, one of the biggest and most advanced global investment banks.

Special Purpose Vehicles LegalTech IT-Platform

Deep Knowledge Group aims to remove existing barriers to investing in private equity markets and to add value for all stakeholders with its InvestTech Platform[2]. This end-to-end solution allows investors to deploy capital quickly in an efficient, timely manner enabling startups to raise funds and keep growing. Innovative LegalTech tools and e-doc-

---

[1] www.investerest.vontobel.com/en-fi/product/CH1124674461/ariane-deep-knowledge-ai-pharma-index

[1] www.invest-platform.tech

umentation provide a clear, step-by-step investment process which is enhanced by additional service integrations such as investment and financial analytics, specialized CRM and corporate payment solutions.

The strategic advantages of InvestTech Platform is the following:

- Deep Industry Expertise. Platform relies on the extensive experience of seasoned professionals which are well aware about needs of investors and startups.
- Wide Range of Supportive Solutions. Platform is not limited to providing solely venue for making investments. Its users get the legal, advisory support at every step and leverage the variety of modern tech tools.
- Exclusive Investment Deals. Users get the access to vetted fundraising opportunities which are of great potential to generate positive outcomes in future in case of successful deals completion.

Rating Agency

Deep Knowledge Group intends to develop the methodology which would provide a comprehensive assessment of the companies in DeepTech actors and in the Longevity Industry in particular, providing the right means for the creation of a corresponding Rating Agency. The methodology underlying this project encompasses a comprehensive and complex range of indicators of company performance, rather than focusing exclusively on financial metrics. The competitive strengths of this approach are further strengthened by the following factors:

- Ownership of a first-of-its-kind database of more than 50,000 Longevity-focused companies, Including qualitative and quantitative parameters
- Understanding of market dynamics and technological trends based on Deep Knowledge Group analytical research at the market, country, sub sectoral, and company levels
- High human capital including experts qualified in the different aspects of the ratings and an extended network of experts in the industry
- Parallel supplementary projects related to the Longevity financial ecosystem such as Longevity Stock Exchange, Longevity Index, Longevity Bank etc.

Longevity Index

Deep Knowledge Group is also working on the development of a first-of-its-kind Longevity Index. The Index will consist of publicly traded companies, including but not limited to Big Pharma and AI in Pharma companies as well as companies participating in the Longevity Medicine industry, aging-related chemical and manufacturing companies, IT companies that provide technological solutions for aging-related sectors, etc. This approach allows for the construction of a truly comprehensive, full-scope profile of the Longevity Industry, encompassing all of its subsectors and subindustries, simultaneously enabling participants to seek exposure to the industry as a whole (as opposed to a few specific subsectors) while providing significant potentials for de-risking and diversification.

Longevity Derivatives

The primary objective of Deep Knowledge Group's ongoing Longevity Derivatives 2.0 project is to establish an evidence-based approach for pricing Longevity-derived financial products, substituting the traditional one-dimensional data sources that are being used in insurance, pension, and other related sectors with multidimensional data that is truly reflective of the complexity and sophistication of the Longevity Industry, which will consist of, among others, human Longevity biomarkers-derived information. These innovations allow for significant degrees of customization compared to more standard approaches, and provide significant optimization potentials for existing Longevity-focused financial products, ranging from options to Longevity-enhanced annuities.

Longevity Card

Longevity Card[1], a Longevity FinTech and Banking portfolio company of Deep Knowledge Group, is the new way to lead a healthy lifestyle and manage your money in one secure app in partnership with Mastercard.

They strive to make mobile banking easier and safer for people of all ages by developing new features in combination with cutting edge HealthTech, AgeTech and FinTech designed for users who are planning to live healthy, extra long lives and remain financially stable.

[1] www.longevity.cards

The app has its very own mHealth fitness, wellbeing and nutrition console as well as the longevity marketplace of the best health and wellness brands in the world. Longevity Card also offers a retail B2C app, and a corporate B2B app enabling access to Longevity Card B2B marketplaces (including LegalTech, MarTech and Investment options) for the corporate clients.

Personal Health Dashboard

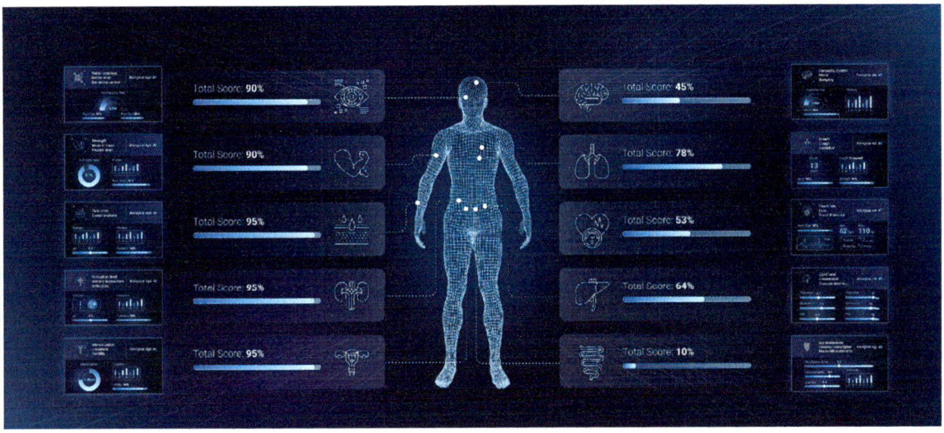

The app is currently in beta testing and will be ready to launch in Q2, 2022. Round A is planned by the end of 2023. The app shall be launched in the UK with the plans for expansion in the European market, the USA and other countries in 2023-2024.

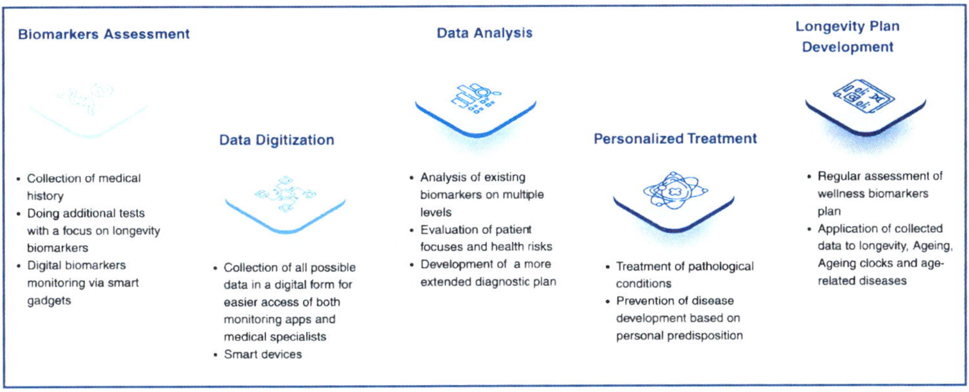

Longevity Club Card is a pass to access a unique ecosystem of numerous Longevity services. The Longevity Club is building the daily habit in longevity commerce, offering a vast mobile and online marketplace where people discover and save on health and wealth related products, maximising their Healthspan (healthy period of life), using the optimal configurations of cutting edge personalised, preventive therapeutics, diagnostics and prognostics, as well as their Wealthspan (period of financial stability) to achieve financial wellness over the course of extended healthy lifespans.

The membership of the Longevity Club allows discovering cutting-edge longevity products promoting an extended healthspan available to a very limited number of people. Our Members will have the unique opportunity to choose between a Membership Card, Banking Card with the full suite of features, products, and exclusive discounts. The Longevity Club will include the Virtual Longevity Clinic and Virtual Longevity Assistant.

The 5th Industrial Revolution

Deep Knowledge Group aims to integrate all of its developments in order to a bridge to the 5th Industrial Revolution that is going to take place in upcoming decades. We aim to create a financial ecosystem that will provide feasible solutions for companies and products that aim to change our understanding of the world as it is now in a better way. In order to achieve this goal, Deep Knowledge Group leverages its own analytical resources, covering thousands of technologies that might create a breakthrough in the near future.

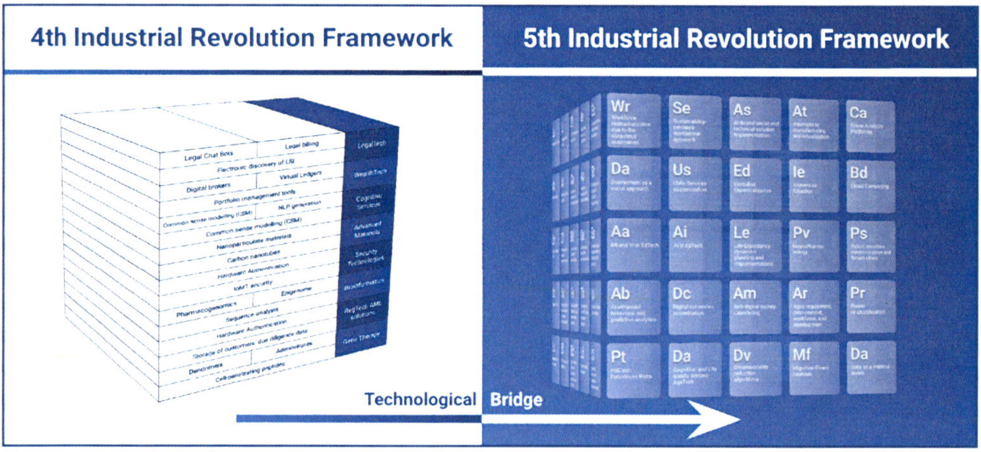

Source: www.5revolution.tech

The 5th Industrial Revolution is grounded at its very foundation in the recognition of the development-to-scale and socially-inclusive adoption of advanced technologies (DeepTech) as the key driver of quality of life, national and global economic stability, and geosocial stability. It recognizes health, wealth and knowledge as the most precious asset classes, and their preservation and optimization across citizens and economies as the highest good.

Deep Knowledge Group remains committed not just to develop and deploy reliably sophisticated and multidimensional Big Data analytics, AI, Machine Learning and Reinforcement-Learning driven technological forecasting and benchmarking of DeepTech industries, pioneering the fundamental tools required for strategic decision making in these unprecedentedly complex and innovation-driven sectors, laying the foundation and basis for defining, thinking about and understanding the 5th Industrial Revolution. More than this, the Group is equally committed to the practical execution of many of these forecasted developments by establishing the core investment and financial industry infrastructure (platforms, indices, exchanges and financial instruments and derivatives) needed to execute the concept of DeepTech Commoditization[1] and facilitate the transition into the 5th Industrial Revolution. Indeed, fostering and accelerating the emergence of the 5th Industrial Revolution is the definitive and decisive long-term goal of Deep Knowledge Group, and the only ROI it will settle for in its mission to support, develop, deploy and accelerate the advancement and emergence of DeepTech for the good of global society.

Long-Term Technological Forecasting System

Deep Knowledge Group and its analytical associates, have constructed intricate analytical frameworks competent enough to analyze, define and foretell DeepTech industries of exceptional breadth, depth, and sophistication which have in 2021 been transformed into an incorporated whole: a manner of Mega-Analytical Framework worthy of characterizing the 5th Industrial Revolution for the first time, and of predicting the most practical means of advancing, optimizing, and coordinating the trajectory of its constant advancement and the careful, de-risked and socially responsible delivery of its advantages for global humanity.

For this purpose, Deep Knowledge Group is building the Long-Term Technological Forecasting System aiming to systemize the technological changes of the 5th Industrial Revolution and conduct sophisticated long-term technology forecasting to help ensure

[1] www.dkv.global/post/deeptech-commoditization

stable 5-IR technology development via AI-driven SWOT analyses to identify and neutralise risk factors.

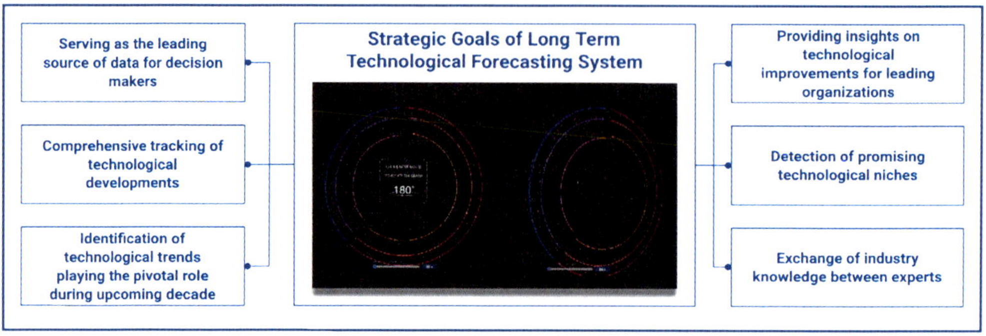

Developed by the 5th Industrial Revolution Institute and DKG's AI and Data Science Department, the Long-Term Technological Forecasting System employs cutting-edge AI and data science and mathematical technologies for monitoring, assessing, benchmarking and realistically forecasting the development of core DeepTech technologies and their impact on the global economy and society in order to de-risk, stabilize and accelerate the emergence of the 5th Industrial Revolution. The first of its kind in the world, the system provides a framework and suite of AI-driven data science tools for informed decision-making by governments, companies, investors and individuals. Detailed specifications and documentation are available on request to select counterparties.

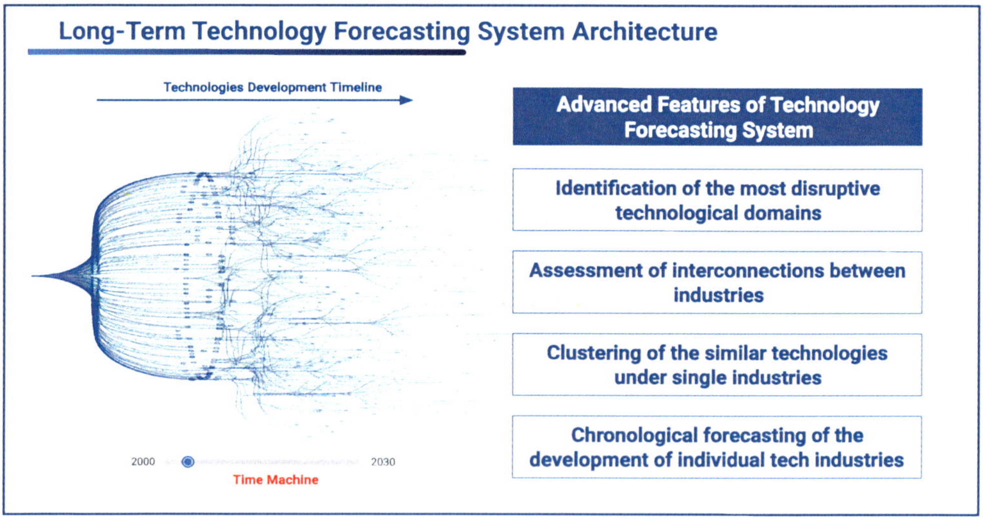

About the Author

Dmitry Kaminskiy[1] is an innovative entrepreneur and investor dedicated to the acceleration of Longevity and DeepTech Industrialization and Financial Commoditization. He is also the author of several books including The 5th Industrial Revolution and a Techno-Philanthropist devoted to hastening the pace of technological progress for the benefit of global society.

Mr. Kaminskiy is Co-Founder and General Partner of Deep Knowledge Group, a data-driven consortium of commercial and non-profit organizations active on many fronts in the realm of DeepTech and Frontier Technologies (AI, Longevity, BioTech, Pharma, FinTech, GovTech, SpaceTech, FemTech, Data Science, InvestTech), ranging from scientific research to investment, entrepreneurship, analytics, consulting, media, philanthropy and more.

Dmitry leads the activities of the consortium's venture arms - Deep Knowledge Ventures, an investment fund focused on DeepTech and advanced science projects, Longevity.Capital, which prioritizes the convergence of Longevity and Artificial Intelligence, areas in which it has unparalleled investment and exit strategies, and DeepTech and Longevity Industry Financial Advisors, the consortium's dedicated DeepTech and Longevity Financial Commoditization subsidiary launched with the mandate of establishing the integral financial infrastructure (platforms, indices, exchanges, financial instruments and derivatives) needed for accelerated and de-risked DeepTech and Longevity Industrialization, and their emergence as fundamentally new asset classes.

Mr. Kaminskiy is a frequent speaker on the topics of AI and Longevity at conferences organized in London by The Economist "Aging Societies and The

[1] www.kaminskiy.info

About the Author

Business of Longevity", Financial Times "Smart Machines vs Smart People", at the Future Finance Forum in Seoul "AI in Finance", "Precision Medicine World Conference" in Silicon Valley, as well as several others at Oxford and Cambridge Universities.

In addition to his business activities, Dmitry is involved in several scientific endeavors. He strongly believes that humans should live longer healthier lives. He has a major interest in Healthy Longevity which is reflected in his business, research and public activities. As Managing Trustee of The Biogerontology Research Foundation (BGRF), the UK's oldest charity devoted to accelerating Longevity science, technology and policy, he was actively involved in instigating the BGRF's successful initiative of adding a new extension code for "ageing-related diseases" (XT9T) accepted in 2018 by the World Health Organization during the last revisions of its International Classification of Diseases framework.

His $1 million prize for the first person to reach their 123rd birthday went viral, and was widely covered by the world's top media. The current lifespan record is 122.5 years, achieved by Jeanne Calment in 1997, which has remained unchanged despite the tremendous advances in biomedicine and in our understanding of the aging process that have occurred since that time.

Made in United States
North Haven, CT
05 November 2026

21758666R60215